Dingyü Xue
Calculus Problem Solutions with MATLAB®

Also of Interest

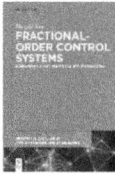

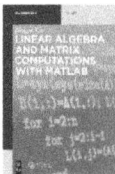

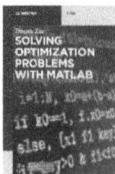

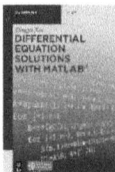

Dingyü Xue

Calculus Problem Solutions with MATLAB®

—

DE GRUYTER

TSINGHUA UNIVERSITY PRESS

Author
Prof. Dingyü Xue
School of Information Science and Engineering
Northeastern University
Wenhua Road 3rd Street
110819 Shenyang
China
xuedingyu@mail.neu.edu.cn

MATLAB and Simulink are registered trademarks of The MathWorks, Inc. See www.mathworks.com/ trademarks for a list of additional trademarks. The MathWorks Publisher Logo identifies books that contain MATLAB and Simulink content. Used with permission. The MathWorks does not warrant the accuracy of the text or exercises in this book. This book's use or discussion of MATLAB and Simulink software or related products does not constitute endorsement or sponsorship by The MathWorks of a particular use of the MATLAB and Simulink software or related products. For MATLAB® and Simulink® product information, or information on other related products, please contact:

The MathWorks, Inc.
3 Apple Hill Drive
Natick, MA, 01760-2098 USA
Tel: 508-647-700
Fax: 508-647-7001
E-mail: info@mathworks.com
Web: www.mathworks.com

ISBN 978-3-11-066362-4
e-ISBN (PDF) 978-3-11-066697-7
e-ISBN (EPUB) 978-3-11-066375-4

Library of Congress Control Number: 2019955233

Bibliographic information published by the Deutsche Nationalbibliothek
The Deutsche Nationalbibliothek lists this publication in the Deutsche Nationalbibliografie; detailed bibliographic data are available on the Internet at http://dnb.dnb.de.

Preface

Scientific computing is commonly and inevitably encountered in course learning, scientific research, and engineering practice of each scientific and engineering student and researcher. For the students and researchers in the disciplines which are not pure mathematics, it is usually not a wise thing to learn thoroughly low-level details of related mathematical problems, and also it is not a simple thing to find solutions of complicated problems by hand. It is an effective way to tackle scientific problems, with high efficiency and in accurate and creative manner, with the most advanced computer tools. This method is especially useful in satisfying the needs for those in the area of science and engineering.

The author had made some effort towards this goal by addressing directly the solution methods for various branches in mathematics in a single book. Such a book, entitled "MATLAB based solutions to advanced applied mathematics", was published first in 2004 by Tsinghua University Press. Several new editions were published afterwards: in 2015, the second edition in English by CRC Press, and in 2018, the fourth edition in Chinese were published. Based on the latest Chinese edition, a brand new MOOC project was released in 2018,[1] and received significant attention. The number of registered students was about 14000 in the first round of the MOOC course, and reached tens of thousands in later rounds. The textbook has been cited tens of thousands times by journal papers, books, and degree theses.

The author has over 30 years of extensive experience using MATLAB in scientific research and education. Significant amount of material and first-hand knowledge has been accumulated, which cannot be covered in a single book. A series entitled "Professor Xue Dingyü's Lecture Hall" of such works is scheduled with Tsinghua University Press, and the English editions are included in the DG STEM series with De Gruyter. These books are intended to provide systematic, extensive, and deep explorations in scientific computing skills with the use of MATLAB and related tools. The author wants to express his sincere thanks to his supervisor, Professor Derek Atherton of Sussex University, who first brought him into the paradise of MATLAB.

The MATLAB series is not a simple revision of the existing books. With decades of experience and material accumulation, the idea of "revisiting" is adopted in authoring these books, in contrast to other mathematics and other MATLAB-rich books. The viewpoint of an engineering professor is established and the focus is on solving various applied mathematical problems with tools. Many innovative skills and general-purpose solvers are provided to solve problems with MATLAB, which is not possible by other existing solvers, so as to better illustrate the applications of computer tools in solving mathematical problems in every mathematics branch. It also helps the readers

---

[1] MOOC address: https://www.icourse163.org/learn/NEU-1002660001

https://doi.org/10.1515/9783110666977-201

broaden their viewpoints in scientific computing, and even in finding innovative solutions by themselves to scientific computing which cannot be solved by other existing methods.

The first title in the MATLAB series, "MATLAB Programming", can be used as an entry-level textbook or reference to MATLAB programming, so as to establish a solid foundation and deep understanding for the application of MATLAB in scientific computing. Each subsequent volume tries to cover a branch or topic in mathematical courses. Bearing in mind "computational thinking" in authoring the series, deep understanding and explorations are made for each mathematics branch involved. These MATLAB books are suitable for the readers who have already learnt the related mathematical courses, and want to revisit the courses to learn how to solve the problems by using computer tools. It can also be used as a companion in synchronizing the learning of related mathematics courses, and viewing the course from a different angle, so that the readers may expand their knowledge in learning the related courses, so as to better learn, understand, and practice the material in the courses.

This book is the second in the MATLAB series. A brand new viewpoint is established for presenting topics in calculus using traditional order, where the focus is on directly solving calculus computation problems with the well-established MATLAB. In this book, the description and graphical representation of functions and sequences are proposed, followed by the direct solutions of problems involving limits, derivatives, and integrals of univariate and multivariate functions. Then series and function approximation techniques are addressed. Numerical differentiation and integrals are also thoroughly discussed. Finally, the concepts and solutions in special branches such as integral transforms and fractional calculus are also introduced, equipped with essential MATLAB solution methods.

At the time the books are published, the author wishes to express his sincere gratitude to his wife, Professor Yang Jun. Her love and selfless care over the decades provided the author immense power, which supports his academic research, teaching, and writing.

September 2019   Xue Dingyü

Contents

1 Introduction to calculus problems

1.1 A brief history of calculus

It is widely recognized that modern calculus appeared in the seventeenth century. The British scientist, Sir Isaac Newton (1643–1727) and German mathematician Gottfried Wilhelm Leibniz (1646–1716) established the subject independently. In the earlier stages of the establishment, different methods and objects were used, while the independent variable t of a function was regarded as time by Newton, and the variable x was regarded as a coordinate by Leibniz. The notations Newton used for fluxions were $\dot{x}$, $\ddot{x}$, while the derivatives Leibniz used was $d^n f(x)/dx^n$, which is still used today. Also, the integral notation $\int$ was introduced by Leibniz.

The study of calculus can be traced to the work of the era of the Greek philosopher, mathematician, and physicist Archimedes (287BCE–212BCE)[1, 19]. In 250BCE, the rudiments of integral calculus was established by Archimedes, and infinitesimals were studied in the principle of lever balance and the method of exhaustion. The area and volumes of objects were computed with infinitely thin slices. Many theorems were also proposed and proved. In the third century CE, Chinese mathematician Liu Hui (225–295) published a book, *The Notes to the Nine Chapters on the Mathematical Art*, in 265, and the area of a circle was computed. In the fifth century, Zu Gen, the son of the greatest Chinese mathematician Zu Chongzhi, formulated the concept of the volume of a sphere, and a theorem was proposed, which can be translated in today's language as follows: for two objects of the same height, if the areas at all horizontal slices are identical, the volumes of the two objects are the same.

Some milestones in modern calculus were summarized in [19], and can be extended as follows:

In 1666, fluxions were introduced by Newton to describe instant velocities. In his work, *Method of Fluxions*, completed in 1671 and published in 1736, continuous variables were referred to as fluents, and the derivatives were referred to as fluxions. The differentiation method to compute the velocity from given displacement and integral method to compute the displacement from velocities were formulated. Infinite series were also studied in Newton's work. A brand new mathematics branch – mathematical analysis – was founded.

In 1676, the notation $\int y(x)dx$ was introduced by Leibniz, which was a stretched letter S, for the Latin word *summa*, meaning sum. The area under the curve $y(x)$ can be obtained.

In 1684, the first paper on calculus was published by Leibniz. Unlike the physical background of Newton, the topic studied by Leibniz was geometry. The tangent lines of a curve were considered. The definition of first-order derivative was explicitly proposed, with the notations dx and dy.

https://doi.org/10.1515/9783110666977-001

In 1687, the great scientific work of Newton, *Philosophiae naturalis principia mathematica* (Mathematical principles of natural philosophy), was published in Latin[10].

In 1691, the well-known L'Hôpital's rule was proposed by French mathematician Guillaume François Antoine L'Hôpital (1661–1704), however, it was regarded that the rule was found and taught to L'Hôpital by Swiss mathematician Johann Bernoulli (1667–1748)[15]. L'Hôpital's work, *Analyse des Infiniment Petits pour l'Intelligence des Lignes Courbes* (Infinitesimal calculus with applications to curved lines), had profound influence at that time.

In 1693, the fundamental theorem in calculus was proposed by Leibniz.

In 1715, British mathematician Brook Taylor (1685–1731) proposed an approach to expand functions as infinite series, later known as Taylor series. The original idea of the series was proposed by Scotish mathematician and astronomer James Gregory (1638–1675).

In 1754, limits were used to replace infinitesimals by French mathematician Jean-Baptiste le Rond d'Alembert (1717–1783).

In 1772, French mathematician Joseph-Louis Lagrange (1736–1813) was the first to use the words "derived function", which were later unified as "derivative". The mean-value theorem was proposed by Lagrange in 1797.

In 1817, the mathematician Bernard Bolzano (1781–1848) in the Kingdom of Bohemia (now the Czech Republic) proposed the concept of continuity of functions, and the $\varepsilon-\delta$ definition. Also the intermediate-value theorem was proposed.

In 1853, German mathematician Georg Friedrich Bernhard Riemann (1826–1866) proposed Riemann integral of real functions, which is an extension to the Cauchy integral, proposed in 1823 by the French mathematician Baron Augustin-Louis Cauchy (1789–1857).

In 1861, German mathematician Karl Theodor Wilhelm Weierstrass (1815–1897) proposed the extreme-value theorem, and improved the limit definition and function continuity concepts by Bolzano in 1874. A more rigorous mathematical theoretical system was established.

The differential and integral calculus established by Newton and Leibniz is the foundation of many scientific and engineering branches. The main topics in calculus are limits, derivatives, and integrals of univariate and multivariate functions, series, Taylor and Fourier series expansions, ordinary differential equations.

1.2 Main topics in the book

The fundamentals and basic knowledge in MATLAB programming are fully covered in the first volume of the series. In this volume, the focus is on how to find solutions to calculus problems with MATLAB. The viewpoint from a professor of engineering discipline is adopted to explore calculus-related problems.

Since the studied objects in calculus are functions and sequences, the systematic descriptions of them in MATLAB are presented first in Chapter 2. The graphics facilities of various functions and sequences are also presented, so as to establish a solid foundation for calculus problems.

In Chapter 3, the computation of univariate, single-sided, and multivariate limits, as well as numerical and functional sequence limits, are proposed, with the help of MATLAB Symbolic Toolbox. Better explorations of the limit problems can also be made with graphical facilities.

In Chapter 4, computation methods for derivatives and differentiations are introduced. Starting from the definition of derivatives, MATLAB-based functions are introduced, for finding various derivatives and partial derivatives. New universal MATLAB functions are written to find derivatives of parametric equations and implicit functions. The concepts and computations of fields are also introduced.

The analytical computations of various indefinite, definite, multivariate and improper integrals are introduced in Chapter 5. Path and surface integrals are also discussed, with the universal functions written by the author.

In Chapter 6, the computations of the sums and products of numeric and functional sequences are introduced, followed by the systematic convergence tests for infinite sequences. Taylor series approximation of univariate and multivariate functions, Fourier series approximations, continued fraction expansions and Padé approximants are also introduced. For complex functions, Laurent series expansions are also explored.

Most of the material summarized above is mainly tackled with analytical approaches. Computer algebra is employed, which is usually not possible with the use of numerical approaches. For the readers who are not familiar with the tools in computer algebra, it may not be possible to solve these problems with computer languages such as C. Computer mathematics languages must be employed for such problems.

In practical scientific and engineering research, analytical solutions of calculus problems may also face difficulties. For instance, if the functions to be studied are not known, while only a set of experimental samples are available, the above-mentioned analytical approaches cannot be used, numerical approaches must be used to find the derivatives and integrals. In Chapter 7, numerical differentials for univariate and multivariate functions are studied, while in Chapter 8, where analytical expressions to some functions may not exist, numerical integration must be carried out. Also for integral problems, when the integrands are known, the analytical solutions may still not exist. In this case, numerical integral algorithms and tools should be employed for finding the approximate solutions. Numerical solutions for multiple integrals are also considered. Apart from that, interpolation techniques are adopted.

In Chapter 9, fundamentals on integral transforms are introduced, and the solutions of problems involving Laplace, Fourier, Mellin, Hankel, and z transform are presented. If analytical solutions do not exist, numerical approaches are employed instead.

If the orders in calculus are not integral, a new branch – fractional calculus – is introduced. In Chapter 10, MATLAB-based numerical solutions and implementations are introduced, aiming at finding high-precision numerical solutions.

This book is by no means a strict book in pure mathematics. The eventual objective of the book is to use computer commands acceptable by MATLAB, to find the solutions of the problems of mathematics, with handy computer tools. The powerful mathematics facilities in computer languages are employed extensively, so as to directly find the solutions of complicated mathematical problems. For problems whose analytical solutions are not immediately available, numerical approaches can be adopted instead to find meaningful solutions.

Symbolic Toolbox in MATLAB is recommended for tackling the problems in this book. For certain problems with no existing MATLAB solving functions, corresponding universal MATLAB functions should be written. For other problems, third-party toolboxes by other scholars can be downloaded for finding solutions of mathematical problems directly. The ultimate goal of the book is to enable the users acquire solutions in an efficient manner, such that rapid, concise, accurate, and reliable solutions can be found.

Equipped with the approaches presented in the book, the readers may find that the problems in the terrific "Problems in mathematical analysis" by Demidovich[5] or other workbooks in the field of mathematical analysis become unbelievably easy and straightforward. Equipped with the knowledge and new viewpoints presented in this book, the problems studied in calculus courses can be reviewed, and computers can be used to creatively solve related mathematical problems, with powerful facilities provided in MATLAB.

2 Functions and sequences

Functions and sequences are essential objects in calculus studies. Mastering the methodology in representing and handling them may establish a good foundation in the study of calculus.

In Section 2.1, the definition of functions is presented first, and an introduction on MATLAB representations of univariate and multivariate functions is proposed. In Section 2.2, introductions to inverse, composite, and implicit functions and parametric equations are presented. The definition and classification of odd and even functions are presented in Section 2.3. In Section 2.4, definitions and representations on complex-valued functions are proposed, and complex plane mappings are addressed. In Section 2.5, MATLAB representations of sequences are given, and MATLAB graphic tools for sequences are demonstrated.

2.1 Functions and mappings

2.1.1 Definitions of functions

In this section, the mathematical definition of functions is proposed, and then MATLAB presentations and graphical manipulations of various functions are presented.

Definition 2.1. Assume that in a real process, there are two variables, x and y. If a value of x in a certain range is chosen, a unique value of y can be generated. Then, y is referred to as a function of x, denoted as $y = f(x)$. The variable x is referred to as an independent variable, and y is the dependent variable. Function $y = f(x)$ can also regarded as the mapping from variable x to variable y, denoted in mathematics as $f : x \mapsto y$.

This type of function is more strictly referred to as a univariate function.

Definition 2.2. If the independent variables are $x_1, x_2, \ldots, x_n$, and they are mutually independent, the function is referred to as a multivariate function, denoted as $y = f(x_1, x_2, \ldots, x_n)$.

Definition 2.3. The collection of values of an independent variable x is referred to as the domain of the function, the corresponding collection of values of y is referred to as the range of the function.

Functions are important objects in calculus. In this section, the concept and computation of transcendental functions are given, followed by the methodology in MATLAB representation of functions, inverse functions, implicit functions, and so on.

https://doi.org/10.1515/9783110666977-002

2.1.2 MATLAB computation of commonly used transcendental functions

Simple algebraic manipulations can be performed in MATLAB by the operators, such as "+", "−", "*", "/", and "^". For instance, the MATLAB command $a+b*c/(d+e)$ can be used to express an algebraic function. Powers of a variable can be realized with "^" as well, such as $a\hat{\ }(1/3)$.

Definition 2.4. An analytic function $f(z)$ of a real- or complex-valued z is transcendental if it is algebraically independent of that variable.

Several commonly used transcendental functions are supported in MATLAB with the syntax y=fun(x), where x can be a scalar, vector, matrix, symbolic variable or even a multidimensional array. It can be real or complex as well. The date type and size of returned argument y is the same as those of x.

(1) Trigonometric functions. Sinusoidal, cosine, tangent, and cotangent functions can be evaluated with MATLAB functions such as sin(), cos(), tan(), and cot(); secant (reciprocal of cosine), cosecant (reciprocal of sine) can be evaluated with functions sec() and csec(); hyperbolic sine, $\sinh x = (e^x - e^{-x})/2$ and hyperbolic cosine, $\cosh x = (e^x + e^{-x})/2$, can be evaluated with sinh() and cosh(). The default unit of all these functions is radian. If the degree unit is expected, one may do the conversion with $x_1 = 180x/\pi$, or with another set of functions such as sind().

(2) Arc trigonometric functions. An a can be placed in front of a trigonometric function name, such as asin();

(3) Exponential functions. Exponential function e^x can be evaluated with exp() function;

(4) Logarithmic functions. Natural logarithmic function $\ln x$ can be evaluated with log(), common logarithmic function $\lg x$ with log10() function. For function $\log_a x$, the command log(x)/log(a) can be used.

2.1.3 MATLAB representation of functions

For an ordinary function $y = f(x)$, the following direct methods in MATLAB can be used. One is to declare x as a symbolic variable, then use symbolic expression to input the function. An alternative way is to define function expressions directly under symbolic data types.

Example 2.1. Input the following functions in MATLAB in symbolic form

$$f(x) = ax^2 + bx + c, \quad g(x,y) = (x^2 - 2x)\,e^{-x^2-y^2-xy}.$$

Solutions. The following two formats can be used to input the two functions

```
>> syms x y a b c
   f1=a*x^2+b*x+c; f2(x)=a*x^2+b*x+c;
```

```
g1=(x^2-2*x)*exp(-x^2-y^2-x*y);
g2(x,y)=(x^2-2*x)*exp(-x^2-y^2-x*y);
```

If the values of $f(5)$ and $g(a^2, a + b)$ are expected, two ways can be used to call the functions, as follows. It is obvious that the second way is more convenient, and closer to mathematical representations.

```
>> subs(f1,x,5), f2(5)
   subs(g1,x,y,a^2,a+b), g2(a^2,a+b)
```

Essentially, there is no difference in the two ways in describing functions in MATLAB. In practical applications, one can chose either way, whichever is more suitable. Since independent variables can be declared with the function-like description method, it is closer to the mathematical counterpart and recommended for use whenever possible.

It should also be noted that if $f(x)$ is defined as a symbolic function, the use of $f(x, y)$ may lead to errors. Therefore, clear f command should be issued first before $f(x, y)$ is redefined.

2.1.4 Curves and surfaces of functions

Generally speaking, univariate functions can be represented graphically by curves, functions with two independent variables can be represented by surfaces. Here curve and surface representations of functions in MATLAB are described.

(1) Univariate functions. If a univariate function is defined as $y = f(x)$, function fplot() can be used to draw directly the curves, with the syntax

fplot(f) %with default interval of $[-5, 5]$, or fplot$(f, [x_m, x_M])$,

where f can be an anonymous function, a symbolic expression or function. If anonymous functions are used, dot operations should be adopted.

In the old versions of MATLAB, the same plot can be drawn with ezplot() function, with the following syntax:

ezplot(f) %default interval $[-2\pi, 2\pi]$, or ezplot$(f, [x_m, x_M])$,

where f can be a symbolic expression and function, as well as a string. Similarly, axes can be automatically assigned with ezplot() function. Unfortunately, piecewise functions cannot be drawn directly with ezplot(), while fplot() function can be used.

(2) Two-dimensional functions. For a 2D function $z = f(x, y)$, function ezsurf() can be used to draw the surface, with the syntax:

fsurf(f) %default interval $[-5, 5]$, or

$$\texttt{fsurf}(f, [x_\mathrm{m}, x_\mathrm{M}]), \qquad \texttt{fsurf}(f, [x_\mathrm{m}, x_\mathrm{M}, y_\mathrm{m}, y_\mathrm{M}])$$

In the old versions, function $\texttt{ezsurf}()$ can also be used, with the syntax

$$\texttt{ezsurf}(f) \qquad \%\text{default interval } [-2\pi, 2\pi], \text{ or}$$
$$\texttt{ezsurf}(f, [x_\mathrm{m}, x_\mathrm{M}]), \qquad \texttt{ezsurf}(f, [x_\mathrm{m}, x_\mathrm{M}, y_\mathrm{m}, y_\mathrm{M}]),$$

where f can be a symbolic expression, function, or a string.

The curves and surfaces of other types of function will be addressed separately in the next section.

2.2 Descriptions of various functions

Apart from the regular functions studied earlier, some special functions are also useful in real applications. Inverse, composite, and implicit functions and parametric equations will be discussed in this section, with MATLAB implementations and graphical representations.

2.2.1 Inverse functions

Definition 2.5. If a function $y = f(x)$ is known, and for each value of y, there is a function $x = g(y)$ to find x, in a one-to-one correspondence manner, function $x = g(y)$ is referred to as the inverse function of $y = f(x)$, denoted as f^{-1}.

MATLAB function $\texttt{finverse}()$ can be used to find the inverse functions of some given functions. It should be noted that, although a certain function is defined in a one-to-one correspondence manner, it is not necessary to say that there is an analytical expression for its inverse function.

Example 2.2. Find the inverse function of $f(x) = 1 + \ln(x + 1)$.

Solutions. The original function should be entered first, and then the inverse function can be tried, and the independent variable can be substituted into y. The inverse function of the original function is $g(y) = f^{-1}(y) = e^{y-1} - 1$.

```
>> syms x y; f(x)=1+log(x+1);
   g(x)=finverse(f,x); g(y)
```

2.2.2 Composite functions

Definition 2.6. If $y = f(u)$ is a function of u, while $u = g(x)$ is a function of x, then y is a function of x, denoted as $y = f(g(x))$. This type of function is referred to as a composite function.

In practical applications, a function can be embedded into another to construct a composite function. For instance, for two given functions, $f(x)$ and $g(x)$, two composite functions, $f(g(x))$ and $g(f(x))$, can be obtained. The composite function $f(g(x))$ can also be denoted as $(f \circ g)(x)$.

Example 2.3. For the known functions

$$f(x) = \frac{x \sin x}{\sqrt{x^2 + 2}\,(x + 5)}, \quad g(x) = \tan x,$$

represent the composite functions $f(g(x))$ and $g(f(x))$.

Solutions. Three methods can be used to represent composite functions in MATLAB: the first is with variable substitution, the second is with the function compose(), provided in MATLAB Symbolic Math Toolbox, and the third is to compute directly with symbolic function expression. The three ways are demonstrated in the following MAT-LAB commands

```
>> syms x; f=x*sin(x)/sqrt(x^2+2)/(x+5); g=tan(x);
   F1=subs(f,x,g), F2=subs(g,x,f)          % variable substitution
   F3=compose(f,g,x), F4=compose(g,f,x)    % direct composite
   f(x)=x*sin(x)/sqrt(x^2+2)/(x+5); g(x)=tan(x);
   F5=f(g), F6=g(f)                        % function expression
```

The results of the three methods are exactly the same. Considering the ease of use, the third is recommended.

$$F_1 = F_3 = F_5 = \frac{\sin \tan x \, \tan x}{(\tan x + 5)\sqrt{\tan^2 x + 2}}, \quad F_2 = F_4 = F_6 = \tan\left(\frac{x \sin x}{\sqrt{x^2 + 2}\,(x + 5)}\right).$$

Example 2.4. Dawson function is a special function, defined as

$$\text{daw}(z) = e^{-z^2} \int_0^z e^{\tau^2} d\tau. \qquad (2.2.1)$$

It is known that the first-order derivative of Dawson function can be evaluated with the following formula:

$$\frac{d}{dz} \text{daw}(z) = 1 - 2z \, \text{daw}(z). \qquad (2.2.2)$$

Find the second-, third-, and fourth-order derivative functions with composite functions.

Solutions. The function $f(x)$ can be input into MATLAB first, then variable substitution method can be used in finding the high-order derivatives of Dawson function:

```
>> syms x daw(x); f=1-2*x*daw; f1=f;
   for i=1:3,
      f1=diff(f1,x);
      f1=collect(expand(subs(f1,diff(daw,x),f)),daw)
   end
```

A series of Dawson function derivatives can be obtained as follows:

$$\mathrm{daw}''(x) = (4x^2 - 2)\,\mathrm{daw}(x) - 2x,$$
$$\mathrm{daw}'''(x) = (-8x^3 + 12x)\,\mathrm{daw}(x) + 4x^2 - 4,$$
$$\mathrm{daw}^{(4)}(x) = (16x^4 - 48x^2 + 12)\,\mathrm{daw}(x) - 8x^3 + 20x.$$

2.2.3 Describing piecewise functions

Definition 2.7. If for different values of the independent variables, the expressions in function evaluations are different, the function is referred to as a piecewise function.

Example 2.5. An example of a piecewise function is given by[27]

$$p(x_1, x_2) = \begin{cases} 0.5457\exp(-0.75x_2^2 - 3.75x_1^2 - 1.5x_1), & x_1 + x_2 > 1, \\ 0.7575\exp(-x_2^2 - 6x_1^2), & -1 < x_1 + x_2 \leqslant 1, \\ 0.5457\exp(-0.75x_2^2 - 3.75x_1^2 + 1.5x_1), & x_1 + x_2 \leqslant -1. \end{cases}$$

With symbolic data types, it is not possible to use statements such as if to describe piecewise functions. In the new versions of MATLAB, function piecewise() is provided to directly describe piecewise functions. The syntax of the function is f=piecewise(var$_1$,var$_2$,...), where input arguments var$_i$ must be supplied in symbolic expressions in pairs, with the former describing the conditions, and the latter providing the function expressions. Logical conditions can be joined with the symbols &, |, and ~.

Example 2.6. Consider the saturation nonlinearity defined as follows and draw the curve of the function:

$$y = \begin{cases} 1.1\,\mathrm{sign}(x), & |x| > 1.1, \\ x, & |x| \leqslant 1.1. \end{cases}$$

Solutions. The piecewise function can be entered first, then the curve of the function can be drawn. It should be noted that, due to the limitations in symbolic computation, the functions represented with piecewise functions cannot be handled directly with function ezplot().

```
>> syms x;
   f(x)=piecewise(abs(x)>1.1,1.1*sign(x), abs(x)<=1.1,x);
   x0=-3:0.01:3; f1=double(f(x0)); plot(x0,f1) % draw function
```

Therefore, the same curve can also be obtained with

```
>> fplot(f,[-3,3])
```

If $|x| \leqslant 1.1$ is denoted mathematically as $-1.1 \leqslant x \leqslant 1.1$, it can be explained also as $x \geqslant -1.1$ and $x \leqslant 1.1$, and the string should be "$x>=-1.1$ and $x<=1.1$".

Example 2.7. Represent the piecewise function with the symbolic expression in Example 2.5, and draw the 3D surface.

Solutions. Symbolic variables are declared first, then the function $p(x_1, x_2)$ and evaluations can be input to the MATLAB workspace, next the variables can be converted to have double presentation, and finally, the surface is given in Figure 2.1.

```
>> syms x1 x2; [x10,x20]=meshgrid(-2:0.1:2,-2:0.1:2);
   p(x1,x2)=piecewise(x1+x2>1,...
            0.5457*exp(-0.75*x2^2-3.75*x1^2-1.5*x1),...
            -1<x1+x2 & x1+x2<=1,0.7575*exp(-x2^2-6*x1^2),...
            x1+x2<=-1,0.5457*exp(-0.75*x2^2-3.75*x1^2+1.5*x1))
   surf(x10,x20,double(p(x10,x20)))
```

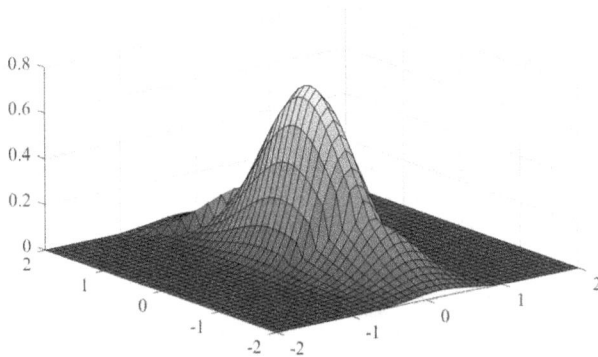

Figure 2.1: Surface of a bivariate function.

Example 2.8. The benefit of using piecewise function is that the domain of definition can be specified. Express the function $f(x) = 2 \sin 3x$ $(-\pi/6 \leqslant x \leqslant \pi/6)$ in MATLAB.

Solutions. The domain of definition can be expressed as conditions for a piecewise function. If the independent variable is not in the domain of definition, NaN (not a number) is retained.

```
>> syms x; f(x)=piecewise(-pi/6<=x & x<=pi/6,2*sin(3*x))
   f(1/7), f(sym(1/7)), f(5)
```

2.2.4 Implicit functions

For some particular functions, the equation $f(x, y) = 0$ is satisfied, however, the explicit form $y = g(x)$ cannot be obtained. This type of function is known as an implicit function. Here implicit functions with two and three independent variables will be studied.

Definition 2.8. The general form of a multivariate implicit function is defined as $f(x_1, x_2, \ldots, x_n) = 0$.

(1) Two-dimensional functions. Consider a 2D implicit function $f(x, y) = 0$. Two methods can be used to express it in MATLAB. The first way is to denote it with a string, while the other is to use symbolic expression. Function ezplot() can be used to draw directly the curve for the implicit function.

Alternatively, anonymous function f can be used to describe an implicit function, and can be drawn with fimplicit(f) function. Dot operations should be used in describing anonymous functions. The default interval is $[-5, 5]$, and the following syntax can be used to specify the intervals with

fimplicit(f, [x_m, x_M]), or fimplicit(f, [x_m, x_M, y_m, y_M]).

Example 2.9. Draw the curve for the implicit function

$$x^2 \sin(x + y^2) + y^2 e^x + y + 5 \cos(x^2 + y) = 0.$$

Solutions. Two ways can be used to express the implicit function, and ezplot() function can be called to draw the curve, as shown in Figure 2.2. It can be seen that identical curves can be drawn with the two sets of commands.

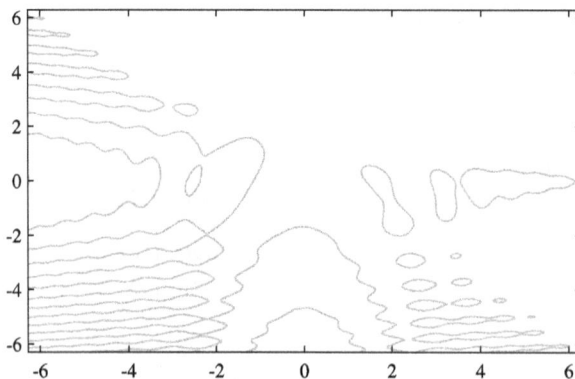

Figure 2.2: Phase-plane trajectory of an implicit function.

```
>> f1='x^2*sin(x+y^2)+y^2*exp(x+y)+5*cos(x^2+y)';
   syms x y; f2(x,y)=x^2*sin(x+y^2)+y^2*exp(x+y)+5*cos(x^2+y);
   ezplot(f1), figure, ezplot(f2)
```

If the new function `fimplicit()` is used, the following statements can be issued, and consistent results can be obtained:

```
>> fimplicit(f2,[-2*pi,2*pi])
```

(2) **Three-dimensional functions.** For the implicit function $f(x,y,z) = 0$, a new function `fimplicit3()` is available to draw the surface of the function. Anonymous functions with dot operations can be used to describe 3D implicit functions in f, then `fimplicit3(f)` can be used to draw the surface. The ranges of the variables can also be assigned with

```
fimplicit3(f,[xm,xM]),  % range of x
fimplicit3(f,[xm,xM,ym,yM]),  % ranges of x and y
fimplicit3(f,[xm,xM,ym,yM,zm,zM]),  % ranges for x, y, and z
```

Example 2.10. Draw the surface of the following 3D implicit function:

$$(x^2 + y^2 + z^2 + 2y - 1)((x^2 + y^2 + z^2 - 2y - 1)^2 - 8z^2) + 16xz(x^2 + y^2 + z^2 - 2y - 1) = 0.$$

Solutions. An anonymous function can be used to describe the 3D implicit function. Then function `fimplicit3()` can be used to draw the surface shown in Figure 2.3.

```
>> f=@(x,y,z)(x.^2+y.^2+z.^2+2*y-1).*((x.^2+y.^2+z.^2-2*y-1).^2 ...
      -8*z.^2)+16*x.*z.*(x.^2+y.^2+z.^2-2*y-1);
   fimplicit3(f,[-4 4,-4,4,-4,4])
```

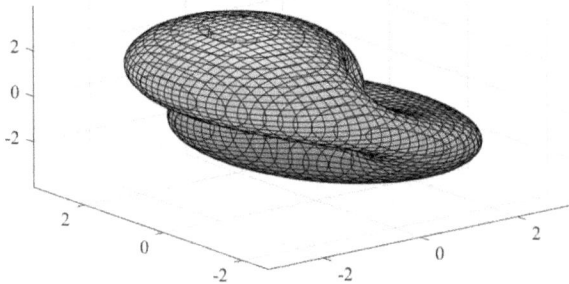

Figure 2.3: Surface of a 3D implicit function.

2.2.5 Parametric equations

Definition 2.9. If a set of variables can be defined as functions of one and more independent variables, the variables are referred to as parametric equations.

Here parametric equations with two and three independent variables are considered, together with their MATLAB representations and graphics display.

Definition 2.10. For a particle in a 2D plane, if its position is governed by the functions of time, $x = f(t)$ and $y = g(t)$, the function is referred to as a parametric equation.

There are two ways to draw the trajectories of parametric equations. The first to generate a time vector t, and compute the vectors x and y, such that `plot(x,y)` can be used to draw the phase-plane trajectory. Similar methods can also be used to handle 3D phase-space trajectory of particles. An alternative way is to express the symbolic expressions for $f(t)$ and $g(t)$, then use `fplot(f,g,[t_m,t_M])` function to draw the trajectory, for $t \in [t_m, t_M]$. Function `ezplot()` can also be used.

Example 2.11. For $a = 8$, $b = 5$, and $t \in (0, 10\pi)$, draw the parametric equation

$$x = (a + b)\cos t - b\cos((a/b + 1)t), \quad y = (a + b)\sin t - b\sin((a/b + 1)t).$$

Solutions. Symbolic expressions can be used to describe the parametric equations, and the following statements can be used to draw the function, as shown in Figure 2.4.

```
>> a=8; b=5; syms t;
   x=(a+b)*cos(t)-b*cos((a/b+1)*t);
   y=(a+b)*sin(t)-b*sin((a/b+1)*t); ezplot(x,y,[0,10*pi])
```

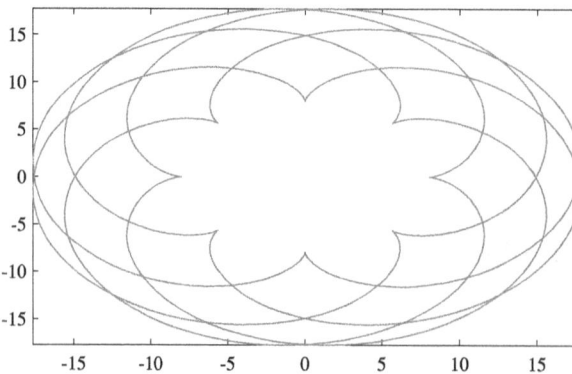

Figure 2.4: Graphical display of a parametric equation.

Example 2.12. For the parametric equations

$$x(t) = (0.5\sqrt{t} + 2)e^{-0.1t}\cos t^2, \quad y(t) = 2\sqrt[3]{|\sin t|}\sin 0.1\sqrt[3]{t^2}, \quad t \in (0, 10)$$

draw the relationship between $x(t)$ and $y(t)$.

Solutions. The first method is recommended for the phase-plane trajectory in the example, as shown in Figure 2.5. It can be seen that the trajectory is disorganized, and it can only be drawn with computers.

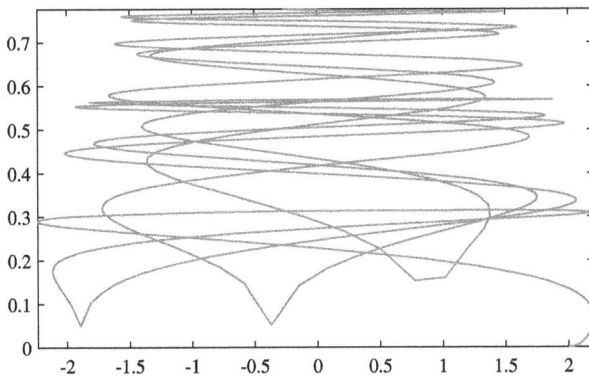

Figure 2.5: Phase plane trajectory of parametric equations.

```
>> t=0:0.001:10; x=(0.5*sqrt(t)+2).*exp(-0.1*t).*cos(t.^2);
   y=2*abs(sin(t)).^(1/3).*sin(0.1*t.^(2/3)); plot(x,y)
```

Of course, another method can be used, however, for this particular example, the trajectory is not satisfactory with the automatically selected scales and step-sizes with function fplot(). It is better to use the first method.

```
>> syms t; x=(0.5*sqrt(t)+2)*exp(-0.1*t)*cos(t^2);
   y=2*abs(sin(t))^(1/3)*sin(0.1*t^(2/3)); fplot(x,y,[0,10])
```

Definition 2.11. Parametric equations with two variables are given by

$$x = x(u, v), \quad y = y(u, v), \quad z = z(u, v) \tag{2.2.3}$$

If $u_m \leqslant u \leqslant u_M$, $v_m \leqslant v \leqslant v_M$, function fsurf($x,y,z,[u_m,u_M,v_m,v_M]$) can be used to draw 3D surfaces, with the default intervals for u and v in $(-5, 5)$. The function ezplot() can be used, with default interval of $(-2\pi, 2\pi)$.

Example 2.13. The well-known Möbius strip can be modeled with the parametric equations $x = \cos u + v\cos u\cos u/2$, $y = \sin u + v\sin u\cos u/2$, and $z = v\sin u/2$. If $0 \leqslant u \leqslant 2\pi$, $-0.5 \leqslant v \leqslant 0.5$, draw the 3D surface of the Möbius strip.

Solutions. Two symbolic variables u and v are declared first, and the parametric equations can be entered into MATLAB environment. The following MATLAB commands can be used to draw the surface of the Möbius strip, as shown in Figure 2.6.

```
>> syms u v; x=cos(u)+v*cos(u)*cos(u/2); y=sin(u)+v*sin(u)*cos(u/2);
   z=v*sin(u/2); fsurf(x,y,z,[0,2*pi,-0.5,0.5]) % Möbius strip
```

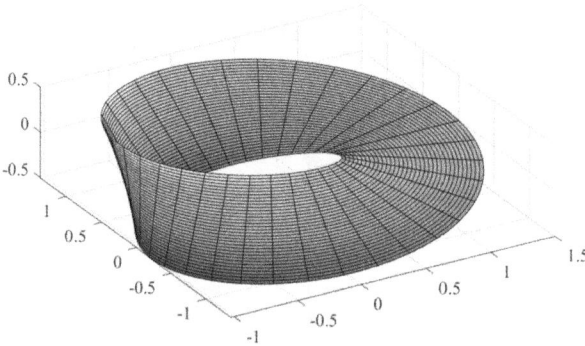

Figure 2.6: Surface plot of Möbius strip, after rotations.

2.2.6 Polar functions

The commonly used coordinate systems discussed so far are Cartesian coordinate systems. In real applications, sometimes polar coordinates are used.

Definition 2.12. From the origin in a plane, a ray can be defined by its angle θ and length ρ. With the two independent variables, a coordinate system can be set up. Such a coordinate system is referred to as a polar coordinate system. The explicit form of a polar expression is $\rho = \psi(\theta)$.

Theorem 2.1. *The coordinate conversion from polar to Cartesian is* $x = \rho\cos\theta$, $y = \rho\sin\theta$.

Theorem 2.2. *The coordinate conversion from Cartesian to polar is* $\rho = \sqrt{x^2 + y^2}$, $\theta = \arctan y/x$.

In MATLAB, function `polarplot()` can be used to draw explicit polar functions $\rho = \psi(\theta)$, with the syntax `polarplot(`**θ**`,`**ρ**`)`, where **θ** and **ρ** are vectors of the samples, and the unit of **θ** is radian. In the earlier versions, the facility of function `polar()` is the same, however, it is not recommended.

Example 2.14. Draw the curve of the polar function $\rho = e^{-0.1\theta}\sin 4\theta$.

Solutions. Letting the independent variable θ vary in the interval $\theta \in (0, 10\pi)$, the vectors can be generated, and the polar plot can be obtained, as shown in Figure 2.7.

```
>> theta=0:0.001:10*pi;
   rho=exp(-0.1*theta).*sin(4*theta); polarplot(theta,rho)
```

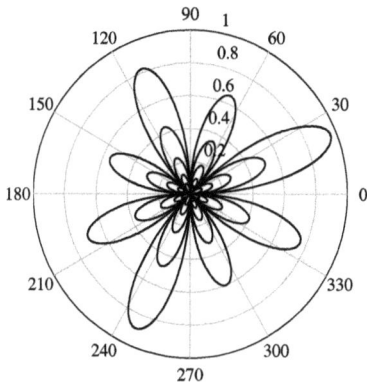

Figure 2.7: Polar plot.

Many polar functions are periodic, i. e., within a cycle, a vector θ can be created, and the polar plot can be drawn. This function is not periodic, no matter how large the interval is chosen, the complete polar plot cannot be drawn.

2.3 Odd and even functions

Definition 2.13. Assume that a function $f(x)$ is defined in the symmetric interval $-L \leqslant x \leqslant L$. If $f(x) = f(-x)$, $f(x)$ is referred to as an even function, while if $f(x) = -f(-x)$ then $f(x)$ is an odd function.

In fact, it is quite easy and straightforward to judge whether a function is even or odd in MATLAB. If $f(x) + f(-x)$ and $f(x) - f(-x)$ are simplified, one can check which of them are zero. If the former is zero, then $f(x)$ is odd, while if the latter is zero, $f(x)$ is even; if neither is zero, function $f(x)$ is neither odd nor even. An example is given below to show the classification of a given function into odd or even.

Example 2.15. Judge the parity of function $f(x) = \sqrt{1 + x + x^2} - \sqrt{1 - x + x^2}$.

Solutions. The function can be entered in MATLAB, and the two expressions are simplified. It can be seen that $f(x)$ is odd. Function fplot() can also be used to draw the curve of the function, as shown in Figure 2.8. It is also seen from the figure that $f(x)$ is an odd function.

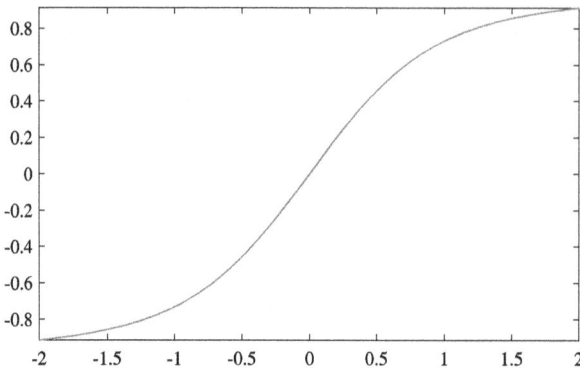

Figure 2.8: Curve of an odd function.

```
>> syms x; f(x)=sqrt(1+x+x^2)-sqrt(1-x+x^2);
   simplify(f(x)+f(-x)), simplify(f(x)-f(-x)), fplot(f,[-2,2])
```

2.4 Complex-valued functions and mapping

If the independent variables are real, the function is known as a real function, however, if the independent variables are complex, the function is referred to as a complex-valued function. In this section, the definitions and operations of complex-valued functions are given, with mapping and Riemann surface plotting presented.

2.4.1 Complex matrices and manipulations

MATLAB can be used in directly handling complex matrices. Assume that for a given complex matrix Z, the following simple functions can be used for certain purposes:
(1) Complex conjugates, Z_1=conj(Z).
(2) Real and imaginary parts, R=real(Z) and I=imag(Z).
(3) Magnitude and phase, A=abs(Z) and P=angle(Z), in radians.

2.4.2 Mapping of complex-valued functions

Definition 2.14. If the independent variable z in $f(z)$ is complex, the function is referred to as a complex-valued function.

Since complex matrix is the basic data type in MATLAB, there is no need to distinguish whether an argument is real or not, since most of the existing functions support complex matrices.

Example 2.16. For a given complex-valued function $f(z) = (z^2 + 3z + 4)/(z-1)^{-5}$, where z is complex, compute $f(-j\sqrt{5})$.

Solutions. It can be seen directly with the following commands that the result obtained is $f_1 = -7/243 + 3\sqrt{5}j/972$.

```
>> syms z; f(z)=(z^2+3*z+4)/(z-1)^5; f1=simplify(f(-sqrt(-5)))
```

A very important transformation in complex-valued functions is mapping, i. e., functions can be mapped from independent variable z into a function of variable w, where $z = g(w)$ is a known function. Commonly used mappings include translation $z = w + y$, inverse mapping $z = 1/w$, and bilinear mapping $z = (aw + b)/(cw + d)$, where y is a given complex number, while a, b, c, d are positive real numbers. The translation mapping may translate the origin to the point y, while the inverse mapping may map a point inside a unit circle to another point outside. The bilinear mapping can be used to map lines into circles.

Example 2.17. Consider the function $f(z)$ in Example 2.16. Find the mapped function $F(s)$ through bilinear mapping $z = (s-1)/(s+1)$.

Solutions. The mapping problem can be obtained directly with

```
>> syms z s; f(z)=(z^2+3*z+4)/(z-1)^5; F(s)=simplify(f((s-1)/(s+1)))
```

It can be seen that the mapped function is $F(s) = -(s+1)^3(4s^2 + 3s + 1)/16$.

Example 2.18. Map the points in the left-half of the s-plane into points in the z-domain through bilinear mapping $z = (s+1)/(s-1)$, and observe its distributions.

Solutions. Assume that the range of real axis is chosen as $(-a, 0)$, while the imaginary axis is set to $(-b, b)$. Letting $a = 1$ and $b = 3$, the mesh grid points on the left-half s-plane can be generated, as shown in Figure 2.9.

```
>> [x,y]=meshgrid(-1:0.1:0,-3:0.4:3); % generate samples
   s=x+sqrt(-1)*y; plot(s,'+');        % original points
   axis([-1.2 0.2 -3.5 3.5])
```

Through bilinear mapping, the points in the z-plane can be obtained directly, as shown in Figure 2.10.

```
>> z=(s+1)./(s-1); plot(z,'x'), hold on;
   syms x y; fimplicit(x^2+y^2==1) % mapped points
```

It can be seen that, through bilinear mapping, the points in the left-half of the s-plane can be mapped to the those in the unit circle in the z-plane. It can be seen that the

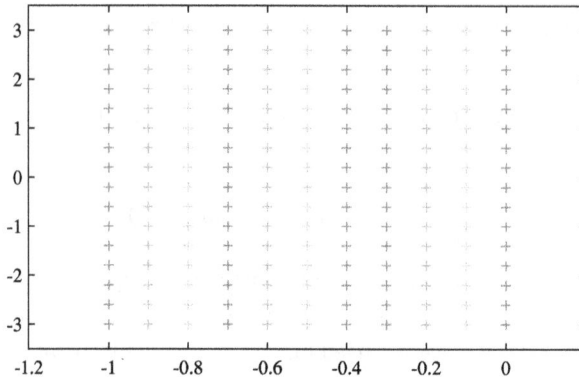

Figure 2.9: Samples in *s* domain distribution.

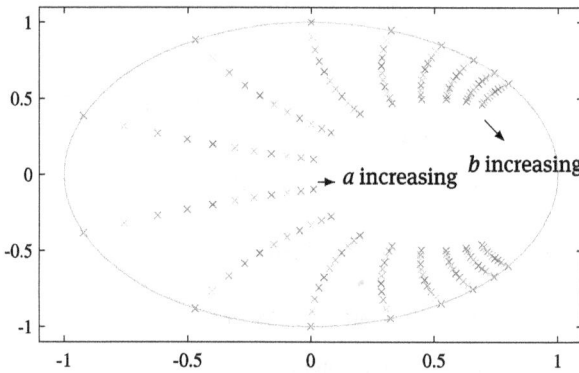

Figure 2.10: Mapping results in the *z*-domain.

mapping shown in the figure may not be complete, since the scales of *a*, *b* are not large enough. If *a* and *b* are increased, the unit circle can fully be covered with the mapped points.

Normally, it is known that through bilinear mapping, the points in left-half of the *s*-plane can be mapped into a unit circle. However, without the use of computer tools, it is usually not known how the points are mapped. With the use of MATLAB, the ways of mapping can be better observed.

2.4.3 Riemann surfaces

The 3D mappings of complex-valued functions are different from those of real functions. The function cplxgrid() should be called first to generate polar mesh grids, and based on the given single-valued formula, a matrix *f* can be computed. Function `cplxmap()` can then be used to draw the mapping surfaces. These surfaces are also known as Riemann surfaces.

The syntaxes for these functions are

z=cplxgrid(n); compute f; cplxmap(z,f).

Example 2.19. Draw the mapping surface of the complex function $f(z) = z^3 \sin z^2$.

Solutions. With the following MATLAB commands, the Riemann surface can be obtained as shown in Figure 2.11.

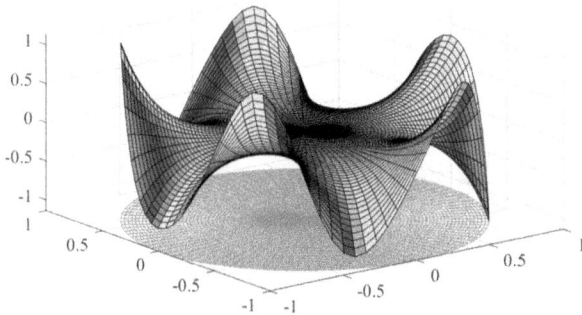

Figure 2.11: Riemann surface.

```
>> z=cplxgrid(50); f=z.^3.*sin(z.^2); cplxmap(z,f)
```

For a complex variable z, some multivalued Riemann surfaces may have several branches, known as Riemann sheets. For instance, $f(z) = \sqrt[n]{z}$ has n sheets. Root functions are provided in MATLAB to draw all the Riemann sheets with `cplxroot(n)`, for the function $\sqrt[n]{z}$.

Example 2.20. Draw the Riemann surfaces for $\sqrt[3]{z}$ and $\sqrt[4]{z}$.

Solutions. There is no need to generate polar mesh grids for these root functions. The function `cplxroot()` can be used directly to draw the surfaces for $\sqrt[3]{z}$ and $\sqrt[4]{z}$, as shown respectively in Figures 2.12 and 2.13.

```
>> cplxroot(3), figure, cplxroot(4) % draw Riemann sheets for ∛z and ∜z
```

It can be seen that the function `cplxroot()` can only be used to handle root-mapping Riemann surfaces, and cannot be used in other multivalued functions. The function `cplxmap()` can be saved as a new function `cplxmap1()`, and delete the statements `mesh()` and `hold`. The newly-modified function can then be used to draw multivalued Riemann surfaces.

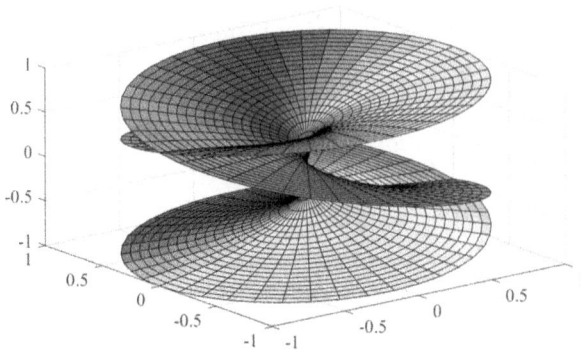

Figure 2.12: Cubic root Riemann surface.

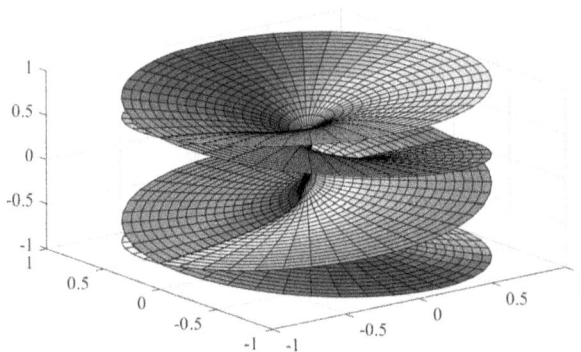

Figure 2.13: The quartic root Riemann surface.

Example 2.21. The new function can be used to draw the Riemann sheet for $\sqrt[3]{z}$.

Solutions. Consider now the function $\sqrt[3]{z}$. If function $f_1(z)$ is a branch of $f(z) = \sqrt[3]{z}$, the other two branches can be computed by $f_1(z)e^{-2j\pi/3}$ and $f_1(z)e^{-4j\pi/3}$. Therefore, the following commands can be used to draw the Riemann sheets of $\sqrt[3]{z}$, and the same results as in Figure 2.12 can be obtained.

```
>> z=cplxgrid(30); f1=z.^(1/3); a=exp(-2i*pi/3); cplxmap1(z,f1)
    hold on; cplxmap1(z,a*f1); cplxmap1(z,a^2*f1); zlim([-1 1])
```

Example 2.22. Draw the Riemann sheets for complex function $f(z) = \sqrt{z^3 \sin z^2}$.

Solutions. Complex mesh grids can be generated first, and the surface data for the first Riemann sheet can be obtained. The data for the second Riemann sheet can be obtained by multiplying scalar $e^{2\pi j/2}$ to the first Riemann sheet data. The complete Riemann surface can then be obtained as shown in Figure 2.14.

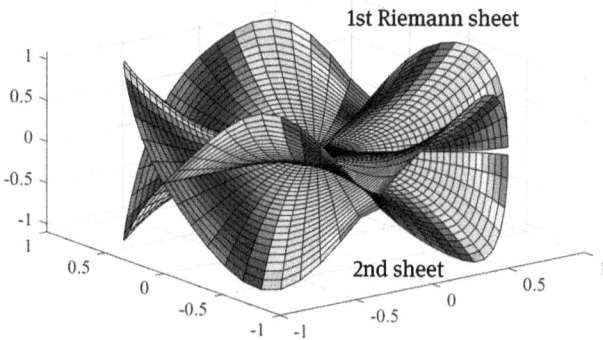

1st Riemann sheet

2nd sheet

Figure 2.14: Surfaces with two Riemann sheets.

```
>> z=cplxgrid(30); f=sqrt(z.^3.*sin(z.^2));
   C=exp(pi*1i); cplxmap1(z,f), hold on
   cplxmap1(z,f*C), zlim([-1.1 1.1])
```

2.5 Numeric and functional sequences

Definition 2.15. A set of numbers arranged in a certain way is referred to as a (numeric) sequence.

In everyday life or in scientific research, one may encounter the sequences

$$1, 2, 3, \ldots, n, \ldots,$$
$$2^0, 2^1, 2^2, \ldots, 2^{n-1}, \ldots$$

In the two sequences, n and 2^{n-1} are referred to as the general terms of the sequences. If the general terms can be expressed in MATLAB, the sequences can also be generated.

Definition 2.16. If the general term of a sequence is a function of x, as in

$$\frac{4\sin x}{2\pi}, \frac{4\sin 3x}{5\pi}, \frac{4\sin 5x}{8\pi}, \ldots, \frac{4\sin(2k-1)x}{(3k-1)\pi}, \ldots, \tag{2.5.1}$$

the sequence is referred to as a functional sequence.

Of course, a sequence can be modeled as a function of n, and it may be more flexible and easy in applications.

Example 2.23. Represent in MATLAB the functional sequence in (2.5.1), and find its 1128th term.

Solutions. In fact, it is sufficient to describe the general term of the functional sequence in MATLAB, and the 1128th term is $4\sin(2255x)/(3383\pi)$.

```
>> syms k x; f(x,k)=4*sin((2*k-1)*x)/(3*k-1)/pi, f(x,1128)
```

Example 2.24. Observe the following sequence and see how the function changes with the increase of n:

$$x_n = 1 + \frac{1}{2} + \frac{1}{3} + \frac{1}{4} + \cdots + \frac{1}{n} - \ln n.$$

Solutions. Loop structure can be used to compute the first 40 terms in the sequence. Stem plots for the terms are shown in Figure 2.15. It can be seen that the value of the sequence is decreasing in n.

```
>> s0=1; n=40; for k=2:n, s0(k)=s0(k-1)+1/k; end
   stem(1:n,s0-log(1:n))
```

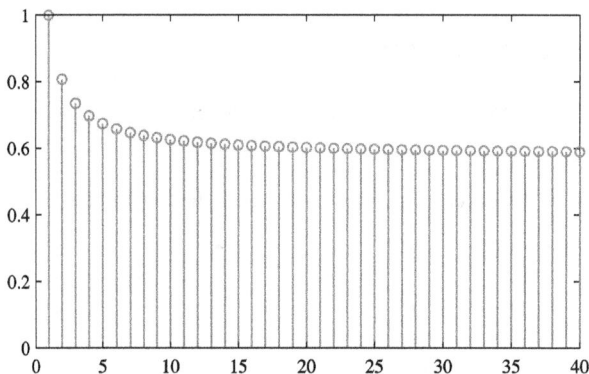

Figure 2.15: The trend for the evolution of the sequence.

It can be seen that, if n is sufficiently large, the sequence may converge to a certain value. It is known that such a value is Euler constant γ. If the first 10000 terms are computed, the last value of the sequence is 0.577265664068165, close enough to γ = 0.5772156649015328606065120900824. The solution is the so-called variable precision algorithm solution from vpa() function.

```
>> s0=1; n=10000; for k=2:n, s0(k)=s0(k-1)+1/k; end, s0(end)-log(n)
   vpa(eulergamma)
```

2.6 Exercises

2.1 Show the identities

(1) $e^{j\pi} + 1 = 0$, (2) $\dfrac{1 - 2\sin\alpha\cos\alpha}{\cos^2\alpha - \sin^2\alpha} = \dfrac{1 - \tan\alpha}{1 + \tan\alpha}$.

2.2 If $\phi(x) = (a^x + a^{-x})/2$, $\psi(x) = (a^x - a^{-x})/2$, show that

(1) $\phi(x + y) = \phi(x)\phi(y) + \psi(x)\psi(y)$, (2) $\psi(x + y) = \phi(x)\psi(y) + \psi(x)\phi(y)$.

2.3 If $f(x) = \ln\left(\dfrac{1 + x}{1 - x}\right)$, show $f(x) + f(y) = f\left(\dfrac{x + y}{1 + xy}\right)$.

2.4 If $f(x) = x^2 - x - 1$, compute $f(f(f(f(f(f(f(f(f(f(x))))))))))$. Find the highest degree of the polynomial.

2.5 The mathematical form of the Chebyshev polynomial is given by

$$T_1(x) = 1, \quad T_2(x) = x, \quad T_n(x) = 2xT_{n-1}(x) - T_{n-2}(x), \quad n = 3, 4, 5, \ldots$$

Compute $T_{10}(x)$.

2.6 Assess the parities of the following functions:[5]

(1) $\sqrt[3]{(x + 1)^2} + \sqrt[3]{(x - 1)^2}$, (2) $\ln(x + \sqrt{1 + x^2})$.

2.7 Find the inverse functions

(1) $y = \sqrt[3]{1 - x^3}$, (2) $y = \ln(x/2)$, (3) $y = 2x + 3$.

2.8 Draw the curves of the following implicit function:

$$(r - 3)\sqrt{r} + 0.75 + \sin 8\sqrt{r}\cos 6\theta - 0.75\sin 5\theta = 0,$$

where $r = x^2 + y^2$, $\theta = \arctan(y/|x|)$.

2.9 Draw the curves for the following parametric equations:

(1) $x(t) = t^2 - t$, $y(t) = t^2 - t^3$,

(2) $x(t) = a(\cos t + t\sin t)$, $y(t) = a(\sin t - t\cos t)$.

2.10 Given the parametric equations $x = \sin t$, $y = \sin at$, $z = \sin bt$. For rational and irrational values a and b, draw respectively 2D and 3D Lissajous figures. For instance, one may select (1) $a = 1/2$, $b = 1/3$, and (2) $a = \sqrt[8]{2}$, $b = \sqrt{3}$.

2.11 Express the symbolic function $f(x) = x^5 + 3x^4 + 4x^3 + 2x^2 + 3x + 6$. Let $x = (s-1)/(s+1)$, map $f(x)$ into a function of s. Generate a set of samples in the left-side of the x-plane, observe the mapping in the s-plane.

2.12 Draw the surface of the following parametric equation:

$$\begin{cases} x = (3 + \cos u/2\sin v - \sin u/2\sin 2v)\cos u \\ y = (3 + \cos u/2\sin v - \sin u/2\sin 2v)\sin u \\ z = \sin u/2\sin u + \cos u/2\sin 2v, \end{cases}$$

where $0 \leqslant u \leqslant 2\pi$, $0 \leqslant v \leqslant 2\pi$.

2.13 For the given implicit function, if the interested region is $x, y, z \in (-1, 1)$, draw the surface of the function

$$\psi(x, y, z) = x \sin(y + z^2) + y^2 \cos(x + z) + zx \cos(z + y^2) = 0.$$

2.14 Select suitable ranges of θ and draw the polar plots of
(1) $\rho = 1.0013\theta^2$, (2) $\rho = \cos 7\theta/2$, (3) $\rho = \sin \theta/\theta$, (4) $\rho = 1 - \cos^3 7\theta$.

2.15 Express in MATLAB the functional sequence with the general term

$$12 \sin x, \ \frac{3}{2} \sin 2x, \ \frac{4}{9} \sin 3x, \ \ldots, \ \frac{12}{n^3} \sin nx, \ \ldots$$

2.16 Predict with graphical method the limit value of the sequence

$$\sqrt{2}, \ \sqrt{2 + \sqrt{2}}, \ \sqrt{2 + \sqrt{2 + \sqrt{2}}}, \ \sqrt{2 + \sqrt{2 + \sqrt{2 + \sqrt{2}}}}, \ \ldots.$$

3 Limits

Limit problems are the mathematical foundation of calculus. It is even noted in [24] that "Limit is the study of calculus".

The research in calculus originally started from the limit problems. For instance, the Greek astronomer and mathematician Eudoxus of Cnidus (c408BCE–c355BCE) proposed the method of exhaustion, while in ancient China, the philosopher Zhuang Zhou (also known as Chuang Tzu, c369BCE–286BCE) stated that "For a foot-long stick, if you take half from the remaining part each day, you will never exhaust it in a million years". The mathematical problem of the above can be expressed as follows:

$$R = 1 - \frac{1}{2} - \frac{1}{2^2} - \frac{1}{2^3} - \frac{1}{2^4} - \cdots.$$

Mathematicians and philosophers, of course, can imagine "never exhaust in million years", however, scientists and engineers are more interested in the realizability of how to "take a half each day". For a foot-long stick, after one month, the size of stick left is well under the scale of atoms. How to "take a half each day" in the remaining time? Also to fold a piece of A4 paper, mathematicians may think the paper can be folded an infinite number of times, however, it was pointed out in an experiment that it is almost impossible to fold the paper more than seven times.

In "college mathematics" or "mathematical analysis" textbooks, many pages are devoted to the limit problems. Several important limits need to be memorized, followed by a huge quantity of exercises. Various skills and tactics need to be mastered by the students to solve limit problems. The solvability of the limit problems largely depends upon the knowledge, skills, or even luck of the solvers. Years after graduation, when limit problems are encountered again in real world, are you still able to solve them?

In this chapter, an alternative way for solving and exploring limit problems is proposed. The main idea is to send the limit problems to computers in the languages understandable by them, and let computers work for you. The method is quite well-organized, and the capabilities of solving limit problems do not depend upon the experience of the solvers. Apart from the solutions in mathematics, the powerful graphical facilities in MATLAB can be fully used, and better visible results rather than pure mathematical formulas can also be obtained, so as to help the readers better understand the solutions in limit problems.

In Section 3.1, limit problems of univariate functions are explored. The abstract ε–δ definition in mathematics is demonstrated with MATLAB graphical facilities. Then MATLAB-based solution strategies for limit problems to univariate, composite, and piecewise functions, as well as sequences, are presented. The concepts of infinitesimals and infinities are given. In Section 3.2, one-sided limit problem is presented, and lower- and upper-limits for certain functions are discussed. Continuity of functions is

https://doi.org/10.1515/9783110666977-003

studied. In Section 3.3, poles and singularities of complex-valued functions are studied, and the computations of residues are presented. In Section 3.4, limit problems of multivariate functions are explored, and based on bivariate functions, the problems of sequential and multiple limits are studied.

The most important MATLAB function in this chapter is `limit()` function. It is not exaggerating to say that, once you have mastered the use of such a function, you can solve with ease almost all the limit-related problems in [5].

3.1 Limits of univariate functions

3.1.1 The ε–δ definition

For a given function $f(x)$, the limit problem can mathematically be denoted as

$$L = \lim_{x \to x_0} f(x), \tag{3.1.1}$$

and the physical meaning is that, when the independent variable x is sufficiently close to the target x_0, the function value $f(x)$ is close to the limit. In calculus textbooks, the following well-explained ε–δ definition is established.

Definition 3.1 (The ε–δ definition). For any preselected $\varepsilon > 0$, if there exists a positive δ satisfying $0 < |x - x_0| < \delta$ such that $|f(x) - L| < \varepsilon$, the limit $\lim_{x \to x_0} f(x) = L$.

Example 3.1. Constant e is an irrational number, whose first 32 digits are

$$e \approx 2.7182818284590452353602874713527.$$

```
>> vpa(exp(sym(1)),32) % more digits can also be displayed
```

Example 3.2. Verify through ε–δ definition an important limit $\lim_{x \to 0}\left(1 + \dfrac{1}{x}\right)^x = e$.

Solutions. Consider the inequality $|(1 + 1/x)^x - e| < \varepsilon$, which is nonlinear in x. It is not possible to solve the inequality without the use of computers. In fact, the inequality problem can better be solved by converting it to nonlinear equations. For a given ε, solve the equation $(1 + 1/x)^x - e - \varepsilon = 0$. The boundary of solutions can be found, and the absolute value of the boundary is the expected δ.

A very handy univariate equation solver `fzero()` is provided in MATLAB. If one chooses $\varepsilon = 0.01$, then from the solver $D = 0.0073$, indicating that an x in $-0.0073 \leqslant x \leqslant 0.0073$ ensures that the distance between $(1 + 1/x)^x$ and e is smaller than $\varepsilon = 0.01$.

```
>> eps0=0.01; f=@(x)(1+x)^(1/x)-exp(1)-eps0; D=abs(fzero(f,0.1))
```

The following statements can be used to draw the curve in the interval $x \in (-0.01, 0.01)$, as shown in Figure 3.1. It can be seen that, if the error tolerance $\varepsilon = 0.01$ is chosen,

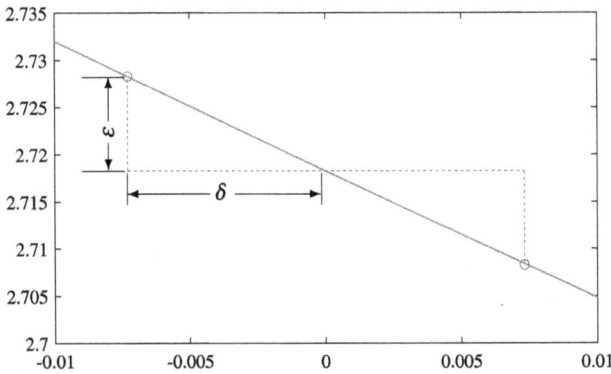

Figure 3.1: The curve in the neighborhood of $x = 0$.

a δ can be found. With $-\delta \leqslant x \leqslant \delta$, the inequalities $-\varepsilon \leqslant L - e \leqslant \varepsilon$ are ensured. If the error tolerance ε is reduced, nonlinear equation solver can be used to find the value of δ such that $-\delta \leqslant x \leqslant \delta$ is ensured.

```
>> x=-0.01:0.0001:0.01; x=x(x~=0); e=exp(1);
   f=(1+x).^(1./x); plot(x,f), hold on
   a=0.01; u1=exp(1)+a; u2=exp(1)-a; plot(-D,u1,'o',D,u2,'o')
   plot(-D*[1 1],[e u1],'--',D*[1 1],[e u2],'--',[-D D],[e e],'--')
```

Even smaller error tolerance ε can be tried, for instance, $10^{-3},\ldots,10^{-8}$. With the following loop structure, the corresponding boundaries δ can also be found, as shown in Table 3.1. It can be seen that, no matter how small the value of ε is selected, a corresponding δ can be found. It is almost noticed that δ and ε are almost linear for this example, this phenomenon cannot be witnessed without using computers.

Table 3.1: The values of δ for differently selected ε's.

ε	0.01	0.001	0.0001	10^{-5}	10^{-6}	10^{-7}	10^{-8}
δ	0.00731	7.35×10^{-4}	7.36×10^{-5}	7.36×10^{-6}	7.36×10^{-7}	7.42×10^{-8}	1.09×10^{-8}

```
>> ee=10.^(-[2:8]); N=[];
   for eps0=ee;
       f=@(x)(1+x)^(1/x)-exp(1)-eps0; D=abs(fzero(f,0.1)); N=[N D];
   end, [ee; N]
```

If one wants to further decrease ε, double precision function such as fzero() may fail to get appropriate δ. Symbolic data type should be employed instead, with δ =

7.358×10^{-11}, again it is linear with the change in ε. Even if the symbolic framework is used, appropriate δ cannot be obtained with even smaller ε's.

```
>> syms x; f=(1+x)^(1/x)-exp(sym(1))-1e-10; x1=vpasolve(f)
```

3.1.2 Limit computing with MATLAB

It is a very complicated task to compute limits with the ε–δ definition. In mathematics textbooks, different approaches, formulas, and skills are presented for computing limits. For instance, two important limits need to be memorized: $\lim\limits_{x \to 0} \sin x / x = 1$ and $\lim\limits_{x \to \infty} (1 + x)^{1/x} = $ e. Different approaches, such as algebraic methods or the middle-value theorem may be tried. One can also use the methods such as L'Hôpital's rule but only after learning the concept of a derivative. Even more advanced approaches such as Taylor series expansion, definite integral, or Stolz–Cesàro theorem may be applied later. A great amount of time and practice are needed to learn how to compute limits. In solving a particular limit problem, its solvability is largely dependent upon the skills of the solver.

In this book, such low-level trivial details are avoided, and a universal way is introduced to feed the problems to computers, and let computers work them out for you. The task then left for you is just waiting for the final results.

Now let us consider the limit problem

$$L = \lim_{x \to x_0} f(x), \tag{3.1.2}$$

where x_0 can be a finite or infinite value, e. g., $x \to \infty$, or another known function or constant. Limit problems can be solved directly with the `limit()` function, provided in the Symbolic Math Toolbox in MATLAB, with the syntaxes

L=limit(f,x_0), %default syntax
L=limit(f,x,x_0), % normal syntax

To solve a problem, symbolic variables such as x must be declared, and symbolic expressions are needed to represent the original function f. If x_0 is infinity ∞, the constant `inf` can be used directly.

If there is only one symbolic variable in the expression, or $f(x)$ is expressed in symbolic function format, the independent variable x in the syntax can be omitted. Also, the `symvar()` function can be used to extract the symbolic variables from expression f with `list=symvar(f)`.

The following examples are used to demonstrate the solutions of limit problems with MATLAB.

Example 3.3. Now, solve directly another important limit problem $\lim\limits_{x \to 0} \dfrac{\sin x}{x}$.

Solutions. The following three procedures are needed to solve limit problems:
(1) Declare symbolic variables. In this case, x should be declared;
(2) Express $f(x) = \sin x / x$ using MATLAB;
(3) Call the MATLAB function `limit()` to compute the limit directly.

The following commands for the three procedures can be issued in MATLAB working window, and the final result of $L = 1$ can be obtained.

```
>> syms x; f(x)=sin(x)/x; L=limit(f,0) % solve limit problems directly
```

Since $f(x)$ is expressed in the function format f already, it is not necessary to use x again in the calling syntax. Of course, the following syntax is also valid, and the same result can be obtained.

```
>> L=limit(f,x,0)
```

With the powerful facilities in MATLAB, the curve of the function in the interval $x \in (-0.1, 0.1)$ can be drawn, as shown in Figure 3.2. It can be seen that curves can be used to visualize the limit process.

```
>> x=-0.1:0.001:0.1; x=x(x~=0);                    % exclude x = 0 in vector x
   plot(x,f(x),0,1,'o'), ylim([0.99,1.001]) % set the vertical axis
```

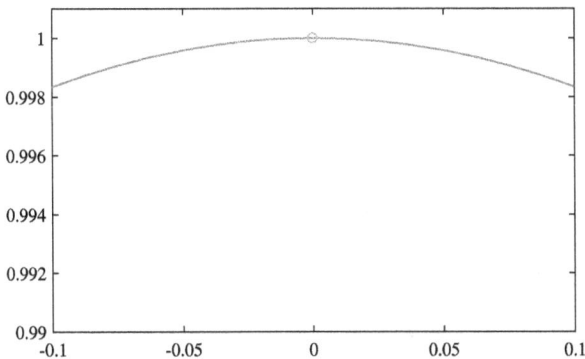

Figure 3.2: Limit process in the neighborhood of $x = 0$.

Example 3.4. Solve the limit problem $\lim\limits_{x \to \infty} x \left(1 + \dfrac{a}{x}\right)^x \sin\left(\dfrac{b}{x}\right)$.

Solutions. With MATLAB language, declare a, b, and x as symbolic variables, then input the function, and finally call the `limit()` function to directly find the needed limit. The final result is $L = e^a b$. It can be seen that the limit problem solution process for the user is as simple as in Example 3.3.

```
>> syms x a b; f(x)=x*(1+a/x)^x*sin(b/x);
   L=limit(f,inf) % direct computation
```

Although three symbolic variables are used in the function, obtainable with v=symvar (f) as v = $[a, b, x]$, and since the original function is expressed in the function format, x can be omitted in the function call.

Example 3.5. Find the limit

$$\lim_{x \to 1} \frac{1}{2(1 - \sqrt{x})} - \frac{1}{3(1 - \sqrt[3]{x})}.$$

Solutions. To solve the problem, no skills or first-hand knowledge are required. All what is needed for the user is to feed the problem to MATLAB, and use limit() function to find the final result. For the example, $L = 1/12$.

```
>> syms x; f(x)=1/2/(1-sqrt(x))-1/3/(1-x^(1/3)); L=limit(f,1)
```

Example 3.6. Calculate the limit

$$\lim_{x \to \infty} \ln(1 + 5^x) \frac{(x + a)^{x+a}(x + b)^{x+b}}{(x + a + b)^{2x+a+b}} \ln(1 + 7/x).$$

Solutions. Although the original function seems a bit complicated, MATLAB can be used to describe it, and limit() function can be called to find the final result $L = 7 \ln 5 e^{-a-b}$.

```
>> syms x a b
   f(x)=log(1+5^x)*(x+a)^(x+a)*(x+b)^(x+b)/...
        (x+a+b)^(2*x+a+b)*log(1+7/x);
   L=limit(f,inf)
```

3.1.3 Limits of composite functions

If a composite function is given, or a function is a functional of several other functions, and the limits of the components are given, the limit of the composite function can be obtained.

Example 3.7. For $\lim_{x \to a} f(x) = 3$, $\lim_{x \to a} g(x) = -1$, compute $\lim_{x \to a} \sqrt[3]{g(x)} [f(x) + 3]$.

Solutions. The following can be used to find the result $L = 6\sqrt[3]{-1}$. In the area of real functions, the result can be simplified to -6.

```
>> syms a f(x) g(x); F=g^(1/3)*(f+3);    % input composite function
   L=limit(F,a); L=subs(L,{f(a),g(a)},{3,-1}) % find the limit
```

3.1.4 Limits of sequences

If a sequence can be expressed in MATLAB, `limit()` function can also be used, and normally it is even not necessary to declare n as an integer-valued symbolic variable.

Example 3.8. Find the limit of the sequence

$$\sqrt{2}, \quad \sqrt{2\sqrt{2}}, \quad \sqrt{2\sqrt{2\sqrt{2}}}, \quad \sqrt{2\sqrt{2\sqrt{2\sqrt{2}}}}, \quad \dots$$

Solutions. It can be seen that the first term of the sequence is $2^{1/2}$, the second term is $2^{1/2+1/2^2}$, the third and fourth terms are respectively $2^{1/2+1/2^2+1/2^3}$ and $2^{1/2+1/2^2+1/2^3+1/2^4}$. It can be concluded easily that the nth term is

$$a_n = 2^{1/2+1/2^2+1/2^3+1/2^4+\cdots+1/2^n}.$$

It is obvious that the power is a geometric sequence, with the first term $a_0 = 1/2$, ratio $q = 1/2$, and number of terms n. Hence the nth term can be expressed as $a_n = 2^{a_0(1-q^n)/(1-q)}$. The following statements can be used to find the limit of the sequence, $L = 2$.

```
>> syms n; a0=sym(1/2); q=a0; F(n)=2^(a0*(1-q^n)/(1-q));
   L=limit(F,n,inf) % find the general term and then compute the limit
```

The geometric sequence formula was used to determine the power in the sequence in this example. In fact, even if you do not know the formula, you can still solve the problem directly with the method shown in Chapter 6.

Selecting a series of values for n, the evolution process of the sequence can be observed as shown in Figure 3.3. It can be seen that the errors are within 1 % boundaries from the seventh term onward.

```
>> n0=1:20; f=F(n0); stem(n0,f)
```

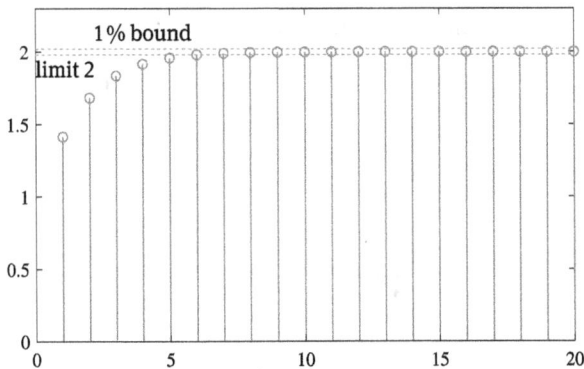

Figure 3.3: Evolution of the sequence.

Example 3.9. Find the limit

$$\lim_{n\to\infty} \frac{\sqrt[3]{n^2}\sin n!}{n+1}.$$

Solutions. The solutions of limit problems for sequences are exactly the same as those when finding the limit of a function in MATLAB. All what is needed is to declare symbolic variables, then express the sequence in MATLAB, and finally call the limit() function. It can be found that $F = 0$.

```
>> syms n; f=n^(2/3)*sin(factorial(n))/(n+1);
   F=limit(f,n,inf) % compute directly the limit
```

Example 3.10. Find the limit

$$\lim_{n\to\infty} n\arctan\left(\frac{1}{n(x^2+1)+x}\right)\tan^n\left(\frac{\pi}{4}+\frac{x}{2n}\right).$$

Solutions. It can be seen that the object here is a sequence, and the general term is a function of x. However, this does not add any difficulties for the user. All what is needed is to express the object in a similar way, and the solution of the problem is as simple as the solution of the easy $\sin x/x$ problem. The solution of $e^x/(x^2+1)$ can be found with the following statements.

```
>> syms x n; f(x)=n*atan(1/(n*(x^2+1)+x))*tan(pi/4+x/2/n)^n;
   L=limit(f,n,inf)
```

For sequence limit problems, it is even not necessary to declare n as an integer-type symbolic variable.

Many sequence limit problems are related to sequence products of some problems, which can be solved with MATLAB directly. In Sections 6.1 and 6.3, these problems will be solved. Here, only an example with code is given, without further explanations.

Example 3.11. Find the limit

$$\lim_{x\to 0} \frac{1-\cos x\sqrt{\cos 2x}\sqrt[3]{\cos 3x}\cdots\sqrt[n]{\cos nx}}{x^2}.$$

Solutions. With the following MATLAB statements, the solution is obtained as $L = n(n+1)/4$.

```
>> syms n x k; L=limit((1-symprod(cos(k*x)^(1/k),k,1,n))/x^2,x,0)
```

3.1.5 Limits of piecewise functions

In learning calculus, sometimes one has to consider several different cases for certain problems. With a powerful tool such as MATLAB, these low-level problems are considered by the computer, and what you need to do is to write commands as usual. The limit can be found, in this case, as a piecewise function.

Example 3.12. Compute $\lim\limits_{x\to\infty} x^n$ and $\lim\limits_{n\to\infty} x^n$.

Solutions. If one wants to solve the problems by hand, several cases need to be considered. In MATLAB-based solutions, the following statements can be used directly:

```
>> syms x n real; f=x^n; L1=limit(f,n,inf), L2=limit(f,x,inf)
```

and the results are all piecewise functions, where L_2 is expressed as

```
piecewise([n == 0,1],[0 < n,Inf],[n < 0,0])
```

The results of the two limits can further be explained as (where, in the last condition in L_1, the point $x = 0$ should be included such that the condition can be changed to $-1 < x < 1$)

$$
L_1 = \begin{cases} 1, & x = 1, \\ \infty, & x > 1, \\ \text{no limit}, & x < -1, \\ 0, & 0 < x < 1 \text{ or } -1 < x < 0, \end{cases}
\qquad
L_2 = \begin{cases} 1, & n = 0, \\ \infty, & n > 0, \\ 0, & n < 0. \end{cases}
$$

3.1.6 Infinitesimals and infinity

Some related definitions will be presented first, then examples will be given to show the solutions of the problems.

Definition 3.2. If $\lim\limits_{x\to x_0} f(x) = 0$, $f(x)$ is referred to as infinitesimal when $x \to x_0$.

Definition 3.3. For infinitesimal functions, if $\lim\limits_{x\to x_0} f(x)/(x - x_0)^p = c$, where c is a constant, $p > 0$, $f(x)$ is referred to as a pth order infinitesimal.

Definition 3.4. If $f(x)$ and $g(x)$ are both infinitesimals, and $\lim\limits_{x\to x_0} f(x)/g(x) = c$, with $c = 0$, $f(x)$ is referred to as a higher-order infinitesimal of $g(x)$, denoted as $f(x) = o(g(x))$; If c is a nonzero constant, then $f(x)$ and $g(x)$ are regarded as infinitesimals of the same order, or $f(x)$ is equivalent to $cg(x)$, denoted as $f(x) \sim cg(x)$.

Definition 3.5. If $\lim\limits_{x\to x_0} f(x) = \infty$, then $f(x)$ is referred to as infinity when $x \to x_0$.

Definition 3.6. If $\lim\limits_{x\to\infty} f(x)/x^p = c$, where c is a constant, $p > 0$, $f(x)$ is referred to as a pth order infinity.

Example 3.13. Show that when $x \to 0$, $\arcsin\dfrac{x}{\sqrt{1-x^2}} \sim \ln(1-x)$.

Solutions. In order to show that the two infinitesimals are equivalent, the easiest way is to show that the ratio of them tends to a nonzero constant. The following statements can be used directly to show that the two functions are equivalent:

```
>> syms x; f(x)=asin(x)/sqrt(1-x^2)/log(1-x); limit(f,0)
```

3.2 Single-sided limits and continuity of functions

3.2.1 Left and right limits

In the problems discussed earlier, $x \to x_0$ normally allows x to approach x_0 from any direction. In real applications, sometimes x is only allowed to approach x_0 from the left- or from the right-hand side. Such limit problems are known as single-sided limit problems.

Definition 3.7. The left- and right-hand limits of a function are usually denoted as

$$L_1 = \lim_{x \to x_0^-} f(x) \quad \text{and} \quad L_2 = \lim_{x \to x_0^+} f(x). \tag{3.2.1}$$

The physical explanation to the former single-sided limit problem is that the independent variable x approaches x_0 from the left side only, while the latter allows x to approach x_0 only from the right side. Single-sided limits can be solved also with `limit()` function, with the syntaxes

L=limit $(f,x,x_0,\text{'left'})$, % left limit
L=limit $(f,x,x_0,\text{'right'})$, % right limit

Example 3.14. Solve the following single-sided limit problem:

$$\lim_{x \to 0^+} \sqrt{\frac{1}{x} + \sqrt{\frac{1}{x} + \sqrt{\frac{1}{x} + \sqrt{\frac{1}{x} + \sqrt{\frac{1}{x}}}}}} - \sqrt{\frac{1}{x} - \sqrt{\frac{1}{x} + \sqrt{\frac{1}{x} - \sqrt{\frac{1}{x} + \sqrt{\frac{1}{x}}}}}}.$$

Solutions. Finding the single-sided limit of a function is as simple as finding the normal limit discussed earlier. The procedures are also exactly the same: declare symbolic variables, input the function, and then call the `limit()` function. For this particular example, the right-sided limit is 1. In fact, the users can try to solve problems with more folds, and the results should be the same.

```
>> syms x positive
   f(x)=sqrt(1/x+sqrt(1/x+sqrt(1/x+sqrt(1/x+sqrt(1/x)))))-...
```

```
       sqrt(1/x-sqrt(1/x+sqrt(1/x-sqrt(1/x+sqrt(1/x)))));
   l=limit(f,x,0,'right')
```

For this example, if $x < 0$, the original function is not defined under the real-number framework. The variable x must be real, and cannot be 0 to avoid "0 in denominator" problem. The case of $x > 0$ is the only possible one. Therefore x should be defined as a positive symbolic variable. The curve of the function with in a neighborhood of $x = 0$ can be obtained as shown in Figure 3.4. It can be seen that when $x \to 0^+$, the function indeed approaches 1.

```
>> ezplot(f,[-0.001,0.01]), hold on, plot(0,1,'o')
```

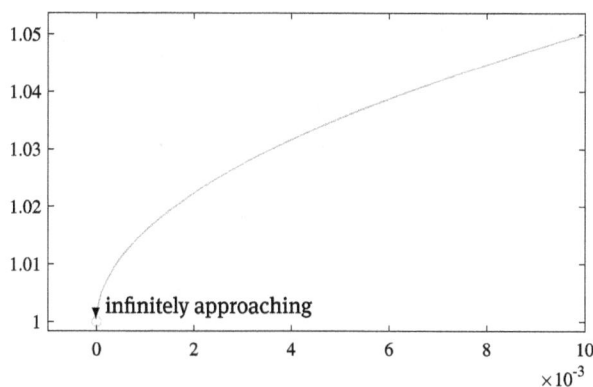

Figure 3.4: The curve in the neighborhood of $x = 0$.

Example 3.15. Solve the single-sided limit problem

$$\lim_{x \to 0^+} \frac{e^{x^3} - 1}{1 - \cos \sqrt{x} - \sin x}.$$

Solutions. With the `limit()` function, it is found that the right-sided limit is 12.

```
>> syms x; f(x)=(exp(x^3)-1)/(1-cos(sqrt(x-sin(x))));
   c=limit(f,x,0,'right')
```

The curve of the function in the interval $(-0.1, 0.1)$ can be obtained as shown in Figure 3.5.

```
>> x0=-0.1:0.001:0.1; x0=x0(x0~=0); y0=f(x0); plot(x0,y0,0,c,'o')
```

It can be seen from the example that, even though `limit(f,x,0)` command is used, the same limit can be found.

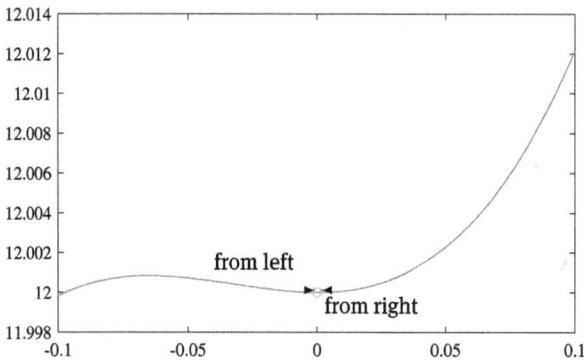

Figure 3.5: The curve around $x = 0$.

Let us go back to the original problem. The single-sided limit as $x \to 0^+$ was used simply to ensure that the quantities under the square root sign are non-negative. What will happen if the quantity is negative? Considering $\cos j\alpha = \cosh \alpha = (e^\alpha + e^{-\alpha})/2$, it can be seen that cosine functions with imaginary argument also exist, and the value can be real. Therefore, there is no problem even if the function comes with an imaginary argument. For some other piecewise functions, the single-sided limits may be different, when approached from different directions.

Example 3.16. Find the right- and left-sided limits of $\tan t$ as $t \to \pi/2$.

Solutions. The following commands can be used to find the left and right limits for the tangent function, and it is found that $L_1 = \infty$ and $L_2 = -\infty$.

```
>> syms t; f=tan(t);
   L1=limit(f,t,pi/2,'left'), L2=limit(f,t,pi/2,'right')
```

3.2.2 Continuity of functions

Definition 3.8. If a function $f(x)$ is defined at $x = x_0$ and in a certain neighborhood, and $\lim\limits_{x \to x_0} f(x) = f(x_0)$, $f(x)$ is continuous at x_0 point.

Definition 3.9. If one of the following cases appears, then $x = x_0$ is a discontinuity in $f(x)$:
(1) Function $f(x)$ is undefined at $x = x_0$, or $\lim\limits_{x \to x_0} f(x)$ does not exist;
(2) Function $f(x_0)$ is defined, while $\lim\limits_{x \to x_0} f(x) \neq f(x_0)$.

Definition 3.10. At a certain point x_0, if the left and right limits of $f(x)$ are the same, it is referred to as a discontinuity of the first type, otherwise, the discontinuity is of the second type.

Example 3.17. Assess the continuity for the following piecewise function

$$f(x) = \begin{cases} 1/(x+1), & -\infty < x < -7, \\ x, & -7 \leqslant x \leqslant 1, \\ (x-1)\sin[1/(x-1)], & 1 < x < \infty. \end{cases}$$

Solutions. For the given function, it can be seen that $f(x)$ within the three intervals is continuous. The next step is to analyze whether the function is continuous at the bounders of the intervals, i. e., at $x = -7$ and $x = 1$. One needs to check if, in the first interval, the left limit as $x \to -7^-$ equals -7^+ or not, and, in the third interval, whether the right limit as $x \to 1^+$ equals 1 or not. With the commands

```
>> syms x;
   f(x)=piecewise(x<-7,1/(x+1), x>=-7 & x<=1,x,...
        x>1,(x-1)*sin(1/(x-1)));
   L1=limit(f,x,-7,'left'), L2=limit(f,x,1,'right')
   fplot(f,[-8 2])
```

it is found that $L_1 = -1/6$, and $L_2 = 1/2$, which are different from those expected. In other words, the function is not continuous at those two points. The above statements can also be used to draw the curve of the function, as shown in Figure 3.6. It can be seen that there are jumps at the two points, therefore, the function is not continuous.

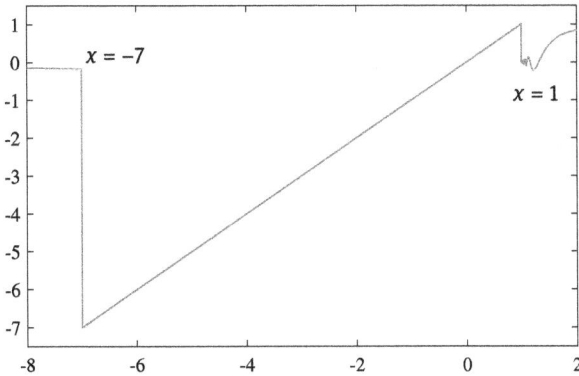

Figure 3.6: Continuity assessment of the function.

Example 3.18. Select a proper a such that the following function is continuous:

$$f(x) = \begin{cases} e^x, & x < 0, \\ a+x, & x \geqslant 0. \end{cases}$$

Solutions. The key point in the problem is at point $x = 0$. It can be seen from $f(x)$ that $f(0) = a$. If the limit $\lim_{x \to 0^-} f(x)$ happens to equal a, the function is continuous. It is found by the following statements that $a = 1$.

```
>> syms x; f(x)=exp(x); a=limit(f,x,0,'left')
```

3.2.3 Interval limits

For some particular functions, such as $\sin x$, the limit when $x \to \infty$ does not exist, however, through a low-level MuPAD function `limit()`, finding interval limits is also possible, with the syntax given below

$$L=\text{feval}(\text{symengine}, '\text{limit}', f, 'x=\text{infinity}', '\text{Intervals}')$$

where `feval()` is the function to evaluate strings. The symbolic engine `symengine` in MuPAD can be accessed, the low-level `limit()` function in MuPAD is called, and the input arguments are in the format of MuPAD as well.

Example 3.19. Assume that $a, b > 0$ and find the limit of $f(t) = a \sin 8x^2 + b \cos(2x - 2)$ as $x \to \infty$.

Solutions. With the following low-level MuPAD commands, the interval limit is obtained as $(-a - b, a + b)$.

```
>> syms a b positive, syms x; f=a*sin(8*x^2)+b*cos(-2*x+2);
   L=feval(symengine,'limit',f,'x=infinity','Intervals') % interval limit
```

Definition 3.11. The lower and upper bounds in the interval limits are also respectively denoted as $\underline{\lim}_{x \to x_0}$ and $\overline{\lim}_{x \to x_0}$, known also as lower and upper limits.

3.2.4 Applications of continuity – assessment of equation solutions

Theorem 3.1. *If $f(x)$ is continuous in the interval (a, b), and $f(a)f(b) < 0$, there exists at least one $\xi \in (a, b)$ such that $f(\xi) = 0$.*

You may still remember that your math teacher was very proud when talking about this result. In this way, the existence of the solution to equation $f(x) = 0$ is guaranteed. However, the existence issue may not be very important to scientists and engineers, since they are more interested in where those ξ's are, and how many of them exist.

In fact, with a powerful tool such as MATLAB, equation solving tasks become very simple, even though you are expecting all the possible solutions. Details on this topic will be presented in Volume IV of the series. For a simple univariate equation, the `fzero()` function discussed in Section 3.1 is adequate to handle the problems.

Example 3.20. Judge whether or not equation $f(x) = \sin(x^3 + 1/x) = 0$ has solutions in $(\epsilon, 0.1)$. If they exist, how many solutions are there, and where?

Solutions. Substituting points ϵ and 0.1 into the function, it is found that $f(\epsilon) = 0.8742$, $f(0.1) = -0.5449$. They are of different signs and the function is continuous. Mathematicians may tell you that there is at least one ξ satisfying $f(\xi) = 0$.

```
>> a=eps; b=0.1; sin(a^3+1/a), sin(b^3+1/b)
```

Unfortunately, this is all the information mathematicians know, if they do not use computer tools. They curve of the function can be obtained as shown in Figure 3.7, from which it can be seen that there are infinitely many solutions.

```
>> x0=-0.1:0.00001:0.1; plot(x0,sin(x0.^3+1./x0))
```

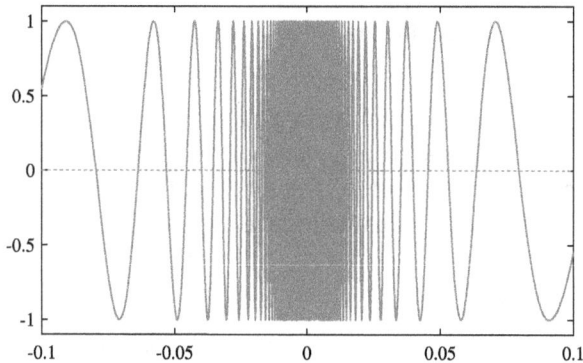

Figure 3.7: Illustration of the equation $\sin(x^3 + 1/x) = 0$.

Of course, scientists and engineers may not be satisfied with the information provided by mathematicians. They are more interested in how many solutions an equation has and how to find all of them. To find one solution is easy – simply use the searching function `fzero()`, and the solution can be found. This function can be used to find equation solutions, one at each time, and it might be very complicated to find more solutions. With the MATLAB solvers `more_sols()` to be introduced in [28], one command is adequate to find many solutions immediately, as shown in Figure 3.8.

```
>> f=@(x)sin(x^3+1/x); more_sols(f,zeros(1,1,0),0.2)
   x=X(1,1,:); hold on, plot(x(:),0,'o'); xlim([-0.1,0.1])
```

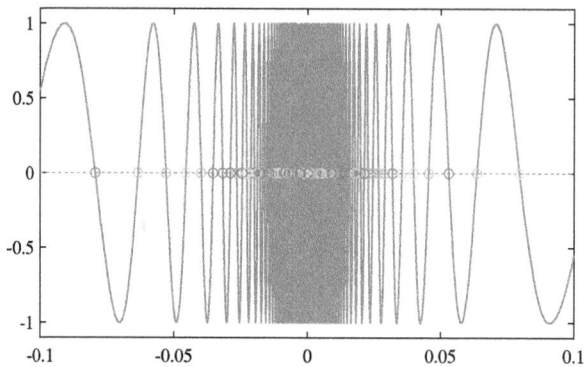

Figure 3.8: Signs of the solutions (each circle indicates one solution).

3.3 Singularities, poles and residues of complex functions

3.3.1 Computation of singularities and poles

The material in this section traditionally belongs to a branch of mathematics called "Complex-valued Functions". Before introducing the concepts of residues, some of the concepts of complex functions are presented.

Definition 3.12. If a function $f(z)$ is single-valued at any point in a complex plane, and the derivatives are finite, $f(z)$ is referred to as an analytic function. The points which make $f(z)$ not analytic are called singularities. In particular, the singularities which make denominator polynomials of $f(z)$ equal zero are referred to as poles of the function.

Definition 3.13. Assuming that $z = a$ is a singularity in $f(z)$, and there exists a smallest integer m such that $(z - a)^m f(z)$ is analytic at $z = a$, the singularity $z = a$ is referred to as a pole of order m (or multiple singularity of order m).

In particular, if the denominator of function $f(z)$ is a polynomial, the poles and multiplicities can be evaluated directly from $[p,m]$=poles(f). If the number of poles is larger than 1, the poles and multiplicities will be returned in vectors p and m. If the poles within the interval (a, b) are expected, the function can also be called with $[p,m]$=poles(f,a,b).

Example 3.21. Find the poles and their multiplicity for the complex function

$$f(z) = \frac{1}{z^3(z-1)} \sin\left(z + \frac{\pi}{3}\right)e^{-2z}.$$

Solutions. It can be seen from $f(z)$ that the denominator is a polynomial of z, function poles() can be called directly to find the poles and multiplicities, with $p = [1,0]^T$ and

$m = [1, 3]^T$, indicating that there are two poles, located at $z = 1$ and $z = 0$, with multiplicities of 1 and 3, respectively. The results are exactly the same as those observed from the given function.

```
>> syms z; f(z)=sin(z+pi/3)*exp(-2*z)/(z^3*(z-1));
   [p,m]=poles(f)
```

Example 3.22. Find the singularities of function $f(z) = 1/(z \sin z)$.

Solutions. The function poles() can be tried, while a warning message "Warning: Unable to determine poles" will be returned, indicating that the original function does no have denominator polynomials. Therefore, the poles and singularities cannot be evaluated in this way.

```
>> syms z; f=1/z/sin(z); [p m]=poles(f)
```

3.3.2 Residues of complex functions

If there are singularities in the function, finite substitutes to the function values at those points are expected. In this section, the concepts and solutions of residues will be presented.

Definition 3.14. If $z = a$ is a single singularity of $f(z)$, the residue is defined as

$$\text{Res}[f(z), a] = \lim_{z \to a}(z - a)f(z). \tag{3.3.1}$$

Definition 3.15. If $z = a$ is a multiple singularity of $f(z)$, with multiplicity of m, the residue at this point is defined as

$$\text{Res}[f(z), a] = \lim_{z \to a} \frac{1}{(m-1)!} \frac{d^{m-1}}{dz^{m-1}}[f(z)(z - a)^m]. \tag{3.3.2}$$

It can be seen from the definitions that the computation of residues is very easy when MATLAB is used. Assume that when the singularity and its multiplicity, a and m, are known, the following MATLAB statements can be used to find the residues

```
c=limit(F*(z - a),z,a), %single singularity
c=limit(diff(F*(z - a)^m,z,m - 1)...
        /prod(1:m - 1),z,a), %m-multiple singularity
```

Example 3.23. Compute the residues for the function $f(z)$ in Example 3.21.

Solutions. It can be seen through simple analysis that $z = 0$ is a triple-singularity, while $z = 1$ is a single singularity. The following statements can be used to compute the residues by definition. It can be found that the residue at $z = 0$ is $F_1 = 1/2 - \sqrt{3}/4$, while that at $z = 1$ is $F_2 = e^{-2}\sin(\pi/3 + 1)$.

```
>> syms z; f(z)=sin(z+pi/3)*exp(-2*z)/(z^3*(z-1));
   F1=limit(diff(f*z^3,z,2)/factorial(2),z,0)
   F2=limit(f*(z-1),z,1)
```

Based on the above, a new MATLAB function, residuesym(), can be written to find residues, poles, and multiplicities of a given function, with the syntax of $[r,p,m]$=residuesym(f,a,b), where the boundaries a and b can also be omitted.

```
function [r,p,m]=residuesym(f,a,b)
z=symvar(f); % extract the independent variables
if nargin==1, [p,m]=poles(f); else, [p,m]=poles(f,a,b); end % find poles
for k=1:length(p)                  % process each pole individually
    F=diff(f*(z-p(k))^m(k),z,m(k)-1)/factorial(m(k)-1);
    try, r(k)=limit(F,z,p(k)); % find the residues
    catch, r(k)=subs(F,z,p(k)); % variable substitution is necessary
end, end
```

Example 3.24. Find the residues for the function $f(z) = (\sin z - z)/z^6$.

Solutions. It seems that $z = 0$ is a pole with multiplicity of 6. The following MATLAB commands can be used, and the residue found is 1/120, with the multiplicity of $m = 3$.

```
>> syms z; f=(sin(z)-z)/z^6; [r,p,m]=residuesym(f) % pole and residue
```

One may try from $k = 1$ onward and find the smallest k such that

$$\lim_{z \to a} \frac{d^{k-1}[(z-a)^k f(z)]}{dz^{k-1}} < \infty$$

from which the multiplicity can be obtained. For this example, considering $k = 2$, it is found that F_1 is infinite, and so larger k should be tried. If k is 3, the value of F_2 is 1/120, the same as the residue found. Therefore, $k = 3$ is the multiplicity of the pole. For larger values of k, for instance, $k = 20$, the same results $F_3 = 1/120$ can be found.

```
>> syms z; f(z)=(sin(z)-z)/z^6;
   F1=limit(diff(f*z^2,z,1)/prod(1:1),z,0)
   F2=limit(diff(f*z^3,z,2)/prod(1:2),z,0)        % increase the order
   F3=limit(diff(f*z^20,z,19)/prod(1:19),z,0) % further increase
```

It can also be seen that if the selected value of n is larger than the actual multiplicity, the residue found is still the same. Normally, a larger n can be tried, without introducing too many difficulties in computation.

Example 3.25. Find the residue in Example 3.22.

Solutions. Since the denominator of $f(z)$ is not a polynomial, as indicated in Example 3.22, function poles() cannot be used to find the singularities. It can be seen from the function that $\sin z = 0$ and $z = 0$ are the singularities, where $z = 0$ is a double singularity, and the residue can be found as $c_0 = 0$.

```
>> syms z; f=1/(z*sin(z));
   c0=limit(diff(f*z^2,z,1),z,0)   % residue at z = 0
```

Further analysis of the function $f(z)$ shows that $z = k\pi$ are singularities, where k is an integer, and they are all single singularities. The following statements can be tried:

```
>> k=[-4 4 -3 3 -2 2 -1 1]; c=[];              % select sample in k
   for kk=k; c=[c,limit(f*(z-kk*pi),z,kk*pi)]; end; c %z = kπ residue
```

For vector $k = [-4, 4, -3, 3, -2, 2, -1, 1]$, the residues can be found as

$$c = \left[-1/(4\pi), 1/(4\pi), 1/(3\pi), -1/(3\pi), -1/(2\pi), 1/(2\pi), 1/\pi, -1/\pi \right]$$

This can be summarized with $\mathrm{Res}[f(z), k\pi] = (-1)^k/(k\pi)$.

In fact, if integer symbolic data type in MATLAB is used, the same results can be found directly.

```
>> syms k; assume(k,'integer'), assumeAlso(k~=0); % k nonzero integers
   R=simplify(limit(f*(z-k*pi),z,k*pi))           % find the residues
```

It should be noted that the function residuesym() cannot be used or such functions. Specific methods involving the definition should be used instead.

3.4 Limits of multivariate functions

There are two categories of limits for a multivariate function: one is the sequential limit, and the other is the multiple limit. The concepts of the two limits will be given here first, and the remaining part discusses how to find them.

3.4.1 Sequential limits

Definition 3.16. Given a multivariate function $f(x_1, x_2, \ldots, x_n)$, when the independent variables approach their targets in a particular order, the limit is referred to as a sequential limit.

Considering the 2D function $f(x, y)$, there are two sequential limits

$$L_1 = \lim_{x \to x_0}\left[\lim_{y \to y_0} f(x, y)\right] \quad \text{and} \quad L_2 = \lim_{y \to y_0}\left[\lim_{x \to x_0} f(x, y)\right], \tag{3.4.1}$$

where x_0 and y_0 can either be specific values or functions.

Consider the limit L_1. The physical interpretation of the sequential limit is as follows: the inner limit $f(x, y)$ as $x \to x_0$ is found first, and then for the result, the outer limit as $y \to y_0$ is taken. Here, x_0 may be a value, infinity, or even a function of y.

The sequential limits can easily be evaluated in MATLAB by the nested `limit()` function with the syntaxes

L_1=limit(limit(f,x,x_0),y,y_0), %first x then y
or L_1=limit(limit(f,y,y_0),x,x_0), %first y then x

Example 3.26. Find the sequential limit for the 2D function

$$\lim_{y \to \infty}\left[\lim_{x \to 1/\sqrt{y}} e^{-1/(y^2+x^2)} \frac{\sin^2 x}{x^2}\left(1 + \frac{1}{y^2}\right)^{x+a^2 y^2}\right].$$

Solutions. Since $\sqrt{y}$ is involved, y must be declared as a positive symbolic variable. The following statements can be used, and the result is e^{a^2}.

```
>> syms x a; syms y positive;  % declare symbolic variables and make y positive
   f(x,y)=exp(-1/(y^2+x^2))*sin(x)^2/x^2*(1+1/y^2)^(x+a^2*y^2);
   L1=limit(f,x,1/sqrt(y))        % find the inner limit
   L=limit(L1,y,inf)              % find the final sequential limit
```

By the way, the inner limit can also be found as

$$L_1(y) = y \sin^2 \frac{1}{\sqrt{y}} \exp\left[\frac{(\ln(y^2 + 1) - 2\ln y)(a^2 y^{5/2} + 1)}{\sqrt{y}} - \frac{y}{y^3 + 1}\right].$$

3.4.2 Multiple limits and computations

Definition 3.17. If all the independent variables approach their targets simultaneously from any possible direction, the limit is referred to as a multiple limit.

The multiple limit of the 2D function $f(x, y)$ can be represented as

$$L = \lim_{(x,y) \to (x_0, y_0)} f(x, y). \tag{3.4.2}$$

The physical meaning of the multiple limit is that the independent variables (x, y) approach the target (x_0, y_0) from any possible direction. It is almost impossible to implement "any direction" on the current computers, therefore, some specific directions can be probed, to see whether they yield the same results. Normally, imprecisely speaking, if the sequential limits and some other ones from certain directions yield the same results, the result is likely to be the multiple limit.

In some applications, sometimes the two statements can be executed, however, the double limit does not exist. One must be very careful and try as many directions as possible, to see whether they all yield the same results.

If the sequential limit in a particular direction is different from those from other directions, this necessarily indicates that the multiple limit does not exist.

Example 3.27. Compute the multiple limit $\lim\limits_{(x,y)\to(0,0)} x\sin(1/y) + y\sin(1/x)$.

Solutions. Try the limit computation from three different directions, e. g., let $y = kx$, $y \to x^2$, or $x \to y^2$. Since all the limits are $L_1 = L_2 = L_3 = 0$, it is probably safe to say that the multiple limit is zero.

```
>> syms k x y;
   f(x,y)=x*sin(1/y)+y*sin(1/x);
   L1=limit(f(x,k*x),x,0)
   L2=limit(limit(f,x,y^2),y,0), L3=limit(limit(f,y,x^2),x,0)
```

Now, consider the function meshgrid(). In a small neighborhood of $(0,0)$, mesh grids can be generated. In order to avoid the inclusion of the lines $x = 0$ and $y = 0$, a small bias is intentionally introduced. The values of z at the mesh grid are generated, and 3D surface can be drawn, as shown in Figure 3.9. It can be seen that the surface around $(0,0)$ is relatively flat. Smaller neighborhood can also be tried, and similar results can be obtained. Therefore, it is safe to conclude that the multiple limit is zero.

```
>> [x0 y0]=meshgrid((-0.1+0.0001):0.002:0.1);
   z=double(f(x0,y0)); surf(x0,y0,z)
```

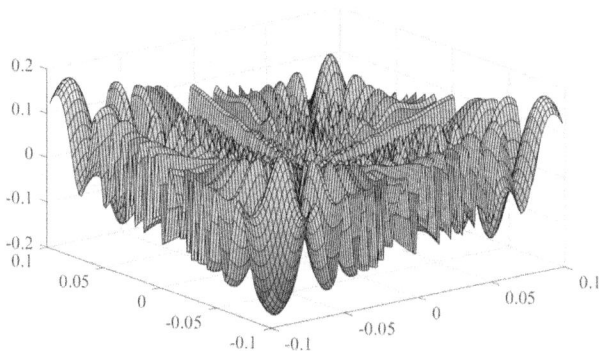

Figure 3.9: The surface in a neighborhood of $(0,0)$.

Example 3.28. Compute the multiple limit

$$\lim_{(x,y)\to(\infty,\infty)} \left(\frac{xy}{x^2+y^2}\right)^{x^2}.$$

Solutions. The two sequential limits can be found and they both equal to zero. Also, two extra directions $x \to y^2$ and $y \to x^2$ are probed, and identical results are achieved. Therefore, it might be safe to say that the multiple limit is zero.

```
>> syms x y; f(x,y)=(x*y/(x^2+y^2))^(x^2);   % 2D function
   L1=limit(limit(f,x,inf),y,inf), L2=limit(limit(f,y,inf),x,inf)
   L3=limit(limit(f,x,y^2),y,inf), L4=limit(limit(f,y,x^2),x,inf)
```

Example 3.29. Compute the multiple limit

$$\lim_{(x,y)\to(0,0)} \frac{xy}{x^2+y^2}.$$

Solutions. It is almost impossible to probe the sequential limits from any possible direction. In contrast, if one can find a particular direction such that the limit is different from others, it is sufficient to indicate that the multiple limit does not exist. For instance, let $y = rx$, where r is a symbolic variable, if the limit depends on r, it is sufficient to indicate that the multiple limit does not exist. The following commands can be tried:

```
>> syms r x y; f(x,y)=x*y/(x^2+y^2); L=limit(subs(f,y,r*x),x,0)
```

and it is found that $L = r/(r^2 + 1)$, which is r-dependent. Therefore, the multiple limit does not exist.

Suitable graphical display may better explain the results and the phenomenon in the example. Similar to the method used in Example 3.27, a small neighborhood of $(0,0)$ is selected, mesh grids can be generated, and the surface of the function can be obtained, as shown in Figure 3.10. It can be seen that when approaching the origin from different directions, the limit can be any value in the interval $(-0.5, 0.5)$. Therefore, the multiple limit does not exist.

```
>> [x0 y0]=meshgrid((-0.1+0.0001):0.002:0.1);
   z=double(f(x0,y0)); surf(x0,y0,z)
```

In fact, $L = r/(r^2 + 1)$ can also be drawn as shown in Figure 3.11, from which it can be seen that the limit is indeed in the interval of $(-0.5, 0.5)$.

```
>> r=-10:0.1:10; L=r./(1+r.^2); plot(r,L)
```

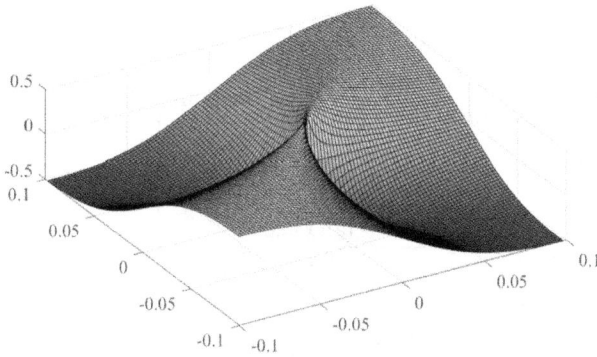

Figure 3.10: The surface around the origin $(0, 0)$.

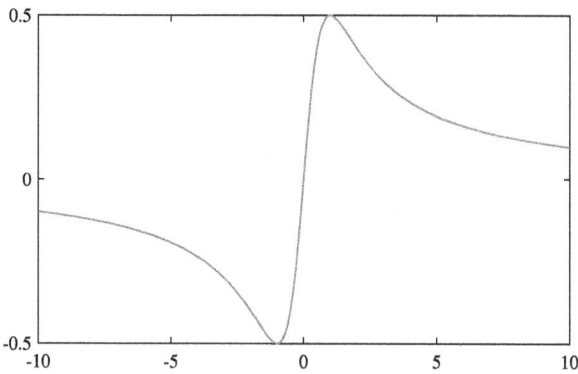

Figure 3.11: The curve of the function $L = r/(1 + r^2)$.

3.5 Exercises

3.1 Compute $1/2^{30}, 1/2^{60}, 1/2^{365}, 1/2^{10000 \times 365}$ (the scale of an atomic nucleus is around 10^{-15} m).

3.2 Compute the following limits:

(1) $\lim\limits_{x \to \infty} (3^x + 9^x)^{1/x}$, (2) $\lim\limits_{x \to \infty} \dfrac{(x + 2)^{x+2}(x + 3)^{x+3}}{(x + 5)^{2x+5}}$,

(3) $\lim\limits_{x \to a} \left(\dfrac{\tan x}{\tan a} \right)^{\cot(x-a)}$, (4) $\lim\limits_{x \to 0} \left[\dfrac{1}{\ln(x + \sqrt{1 + x^2})} - \dfrac{1}{\ln(1 + x)} \right]$.

3.3 Compute the limit $\lim\limits_{x \to \infty} \left[\sqrt[3]{x^3 + x^2 + x + 1} - \sqrt{x^2 + x + 1}\, \dfrac{\ln(e^x + x)}{x} \right]$.

3.4 Solve the following limit problems:

(1) $\lim\limits_{x \to 0} \dfrac{(\sqrt{1 + x^2} + x)^n - (\sqrt{1 + x^2} - x)^n}{x}$, $n \geqslant 0$,

(2) $\lim\limits_{x \to a} \dfrac{\sin(a + 2x) - 2\sin(a + x) + \sin a}{x^2}$.

3.5 Compute the limit $\lim\limits_{x \to 0} \dfrac{\text{arcsinh} \sinh x - \text{arcsinh} \sin x}{\sinh x - \sin x}$, where $\text{arcsinh}\, x = \ln(x + \sqrt{1 + x})$.

3.6 Compute the following limits:

(1) $\lim\limits_{x \to 0} \dfrac{4\sin x + x^2 \cos(4/x)}{(5 + 2\cos x)\ln(1 + x)}$,　(2) $\lim\limits_{x \to 0} \dfrac{(\sin x - \tan x)(\cos x - e^{x^2})}{x(1 + x^2/2 - \sqrt{1 + x^2})}$.

3.7 Compute the following sequence limits:

(1) $\lim\limits_{n \to \infty} \sqrt{n}(\sqrt{n + 1} - \sqrt{n})$,　(2) $\lim\limits_{n \to \infty} (\sqrt[n]{1} + \sqrt[n]{2} + \cdots + \sqrt[n]{10})$.

3.8 If $x_n = \dfrac{n - 1}{n + 1} \cos \dfrac{2n\pi}{3}$, find $\varliminf\limits_{n \to \infty} x_n$ and $\varlimsup\limits_{n \to \infty} x_n$.

3.9 Find from the given limits the values of a and b such that

(1) $\lim\limits_{x \to \infty} \left(ax + b - \dfrac{x^3 + 1}{x^2 + 1} \right) = 0$,　(2) $\lim\limits_{x \to \infty} (\sqrt{x^2 - x + 1} - ax - b) = 0$.

3.10 Compute the following single-sided limits:

(1) $\lim\limits_{x \to \pi/4^+} \left[\tan\left(\dfrac{\pi}{8} + x \right) \right]^{\tan 2x}$,　(2) $\lim\limits_{x \to 0^+} \ln(x \ln a) \ln\left(\dfrac{\ln ax}{\ln(x/a)} \right)$, $a > 1$.

3.11 Consider the following sequences, find the general term a_n, and compute $\lim\limits_{n \to \infty} a_n$ and $\lim\limits_{n \to \infty} a_{n+1}/a_n$:

$$S = \sum_{n=1}^{\infty} a_n = \dfrac{\cos 1!}{1 \times 2} + \dfrac{\cos 2!}{2 \times 3} + \dfrac{\cos 3!}{3 \times 4} + \dfrac{\cos 4!}{4 \times 5} + \dfrac{\cos 5!}{5 \times 6} + \cdots.$$

3.12 Find the roots of the equation $\sqrt{x}\sin(x^3 + 1/x) = 0$ in $x \in (-0.01, 0.01)$.

3.13 For the following functions, compute the limit $F(x) = \lim\limits_{h \to 0} [f(x + h) - f(x)]/h$.

(1) $f(x) = \sin \cos \sin 2x,$ [24]　(2) $f(x) = \ln \dfrac{1 + \sqrt{\sin x}}{1 - \sqrt{\sin x}} + 2\arctan \sqrt{\sin x}.$ [5]

3.14 Compute the sequential limits $\lim\limits_{x \to a}[\lim\limits_{y \to b} f(x, y)]$ and $\lim\limits_{y \to b}[\lim\limits_{x \to a} f(x, y)]$:

(1) $f(x, y) = \sin \dfrac{\pi x}{2x + y}$, $a = \infty$, $b = \infty$,

(2) $f(x, y) = \dfrac{1}{xy} \tan \dfrac{xy}{1 + xy}$, $a = 0$, $b = \infty$.

3.15 Compute the double limits:

(1) $\lim\limits_{(x,y) \to (-1,2)} \dfrac{x^2 y + xy^3}{(x + y)^3}$,　(2) $\lim\limits_{(x,y) \to (0,0)} \dfrac{xy}{\sqrt{xy + 1} - 1}$,

(3) $\lim\limits_{(x,y) \to (0,0)} \dfrac{1 - \cos(x^2 + y^2)}{(x^2 + y^2)e^{x^2 + y^2}}$.

3.16 Show that $f(x, y) = \dfrac{x^2 y^2}{x^2 y^2 + (x - y)^2}$ satisfies

$$\lim_{x \to 0}\left[\lim_{y \to 0} f(x, y)\right] = \lim_{y \to 0}\left[\lim_{x \to 0} f(x, y)\right] = 0,$$

however, the double limit $\lim\limits_{(x,y) \to (0,0)} f(x, y)$ does not exist.

3.17 Find the value of $f(0)$ such that $f(x) = \dfrac{\sqrt{1 + x} - 1}{\sqrt[3]{1 + x} - 1}$ is continuous.

3.18 Find the values of a and b such that the piecewise function is continuous.

$$f(x) = \begin{cases} x + 1, & x < 1, \\ ax + b, & 1 \leqslant x < 2, \\ 3x, & x \geqslant 2. \end{cases}$$

4 Derivatives and differentials

Derivatives and differentials are important topics in differential calculus. The study of derivatives originated from two practical examples.

Example 4.1. Assume that a particle is moving in a straight line. The displacement at t is described by a function $s(t)$. In a small period of time h, the displacement changes to $s(t + h)$. The total displacement in the period of length h is $s(t + h) - s(t)$. If h is small enough, the instant velocity is the average velocity in the period

$$v(t) = \lim_{h \to 0} \frac{s(t + h) - s(t)}{h}.$$

Example 4.2. For a given function $f(x)$, how can we determine the tangent line at point $x = x_0$? The function value at x_0 is known as $f(x_0)$. Find a nearby point x and the function value $f(x)$. If x approaches x_0 and is close enough, the tangent line can be constructed, with the slope being

$$k = \lim_{x \to x_0} \frac{f(x) - f(x_0)}{x - x_0}.$$

The two examples were considered respectively by the British scientist Sir Isaac Newton and German mathematician Gottfried Wilhelm Leibniz. Based their research background, they established the concept of derivative independently.

In this chapter, the concepts and computations in differential calculus are presented. In Section 4.1, the definition of a function derivative is proposed, and the computation of derivatives is also presented for univariate functions. Higher-order derivatives, derivatives of composite functions, as well as matrix derivatives, are also presented. In Section 4.2, the derivatives of parametric equations are proposed, followed by a universal MATLAB function, capable of finding high-order derivatives of parametric equations. In Section 4.3, the concept of partial derivatives of multivariate functions is also proposed, and the idea of total differential is addressed. In Section 4.4, essentials in the field theory are presented, including the concepts and computations of a gradient, divergence, curl, and potential. In Section 4.5, some of the commonly used derivative matrices, such as Jacobian and Hessian matrices, and Laplacian operators are presented. The concept and computation of partial derivatives of implicit functions are discussed in Section 4.6, where a universal high-order partial derivative solver is proposed. In Section 4.7, some of the applications regarding derivatives and differentials are addressed. The problems involving extreme values, Newton iteration method in equation solutions, and tangent plane construction for surfaces are demonstrated.

https://doi.org/10.1515/9783110666977-004

4.1 Derivatives and high-order derivatives

The definition of derivative is given first in the section, followed by an example of evaluating a derivative from the definition. Direct computation of derivatives and high-order derivatives is presented. MATLAB implementation of the well-established mathematical induction method is presented such that the obtained results are more reliable.

4.1.1 Derivatives and differentials

The above mentioned two examples are equivalent in mathematics, from which the definition of a derivative can be established. In fact, the derivatives of certain functions can easily be obtained with powerful tools such as MATLAB. Of course, easier and more concise approaches will be presented later to find the derivatives of given functions.

Definition 4.1. The first-order derivative of a function $y = f(x)$ with respect to the independent variable x is mathematically defined as

$$f'(x) = \frac{df(x)}{dx} = \lim_{\Delta x \to 0} \frac{f(x + \Delta x) - f(x)}{\Delta x}. \tag{4.1.1}$$

It can be seen from the definition that the slopes of the tangent lines of the given curves are derivative functions, and the instant velocity of a particle is also a derivative of the displacement. Further, instant acceleration can also be found by taking the derivative of the velocity. High-order derivatives can also be established in a similar manner.

Definition 4.2. If $\Delta x \to 0$, for the independent variable x, dx is its differential, while dy is the differential of the function $y(x)$.

The derivative of a given function can be computed directly with limits, and the computation of limits can be obtained with the approaches studied in the previous chapter. An example is used to demonstrate the derivative finding process.

Example 4.3. For a given function $f(x) = \dfrac{\sin x}{x^2 + 4x + 3}$, find its first-order derivative using the definition.

Solutions. The approaches in Chapter 3 can be used to find the derivative, via limits. The first-order derivative can be computed directly from

```
>> syms x h; f(x)=sin(x)/(x^2+4*x+3);
   F=limit((f(x+h)-f(x))/h,h,0)
```

and the result is

$$F(x) = \frac{\cos x}{x^2 + 4x + 3} - \frac{(2x + 4)\sin x}{(x^2 + 4x + 3)^2}.$$

With the powerful function `fplot()`, the curves of the original function and its first-order derivative can be obtained as shown in Figure 4.1.

```
>> fplot([f,F],[0,5]) % draw the function and its derivative
```

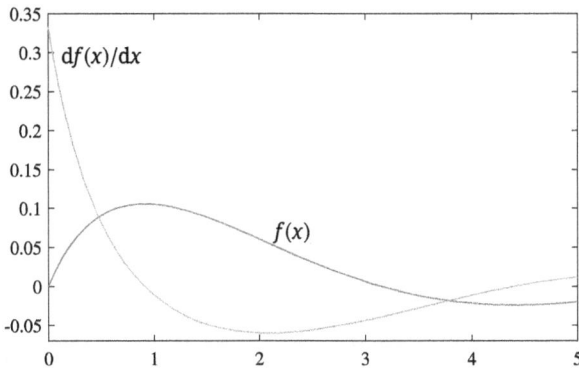

Figure 4.1: Curves of the function and its derivative.

4.1.2 Higher-order derivatives

The second-order derivative of $f(x)$ with respect to x is the derivative of $f'(x)$ with respect to x, simply denoted as $f''(x)$. The third-order derivative is denoted as $f'''(x)$. Similarly, the nth order derivative can be defined and denoted as $d^n f(x)/dx^n$, or simply as $f^{(n)}(x)$.

If the function and its independent variables are known, `diff()` function can be used to evaluate directly the derivatives of different orders. The syntax of `diff()` function is f_1=`diff`(f,x,n), where f is the given function, x is the independent variable, and they both must be symbolic variables. The argument n is the order, and it must be a given integer. In first-order derivative evaluation, n can be omitted. If f contains only one symbolic variable, x can be omitted.

Example 4.4. For the given function $f(x) = \dfrac{\sin x}{x^2 + 4x + 3}$, find $\dfrac{d^4 f(x)}{dx^4}$.

Solutions. To find the derivatives, three steps must be completed:
(1) Declare x as a symbolic variable;
(2) Describe the original function with MATLAB;
(3) Call the function `diff()` to find the derivatives.

The three steps can be carried out with the following statements, to compute the first-order derivative:

```
>> syms x; f(x)=sin(x)/(x^2+4*x+3); f1=diff(f)
```

The result is exactly the same as that obtained in Example 4.3.

The fourth-order derivative of the original function can be found with LaTeX.

```
>> f4=diff(f,x,4); latex(f4) % find 4th order derivative and convert to LaTeX
```

With LaTeX, better display of the result can be obtained as

$$\frac{d^4 f(t)}{dt^4} = \frac{\sin x}{x^2 + 4x + 3} + 4\frac{(2x + 4)\cos x}{(x^2 + 4x + 3)^2} - 12\frac{(2x + 4)^2 \sin x}{(x^2 + 4x + 3)^3} + 12\frac{\sin x}{(x^2 + 4x + 3)^2}$$
$$- 24\frac{(2x + 4)^3 \cos x}{(x^2 + 4x + 3)^4} + 48\frac{(2x + 4)\cos x}{(x^2 + 4x + 3)^3} + 24\frac{(2x + 4)^4 \sin x}{(x^2 + 4x + 3)^5}$$
$$- 72\frac{(2x + 4)^2 \sin x}{(x^2 + 4x + 3)^4} + 24\frac{\sin x}{(x^2 + 4x + 3)^3}.$$

It can be seen from the simplified expression that sometimes the results of function simplify() may not be satisfactory. Suitable simplification method should be adopted according to real situations. For instance, $\sin x$ and $\cos x$ terms can be collected directly, and possibly the simplest results can be obtained with

```
>> simplify(collect(simplify(f4),cos(x)))
```

in which case the concise result obtained is

$$\frac{d^4 f(x)}{dx^4} = f_1(x)\frac{\cos x}{(x^2 + 4x + 3)^5} + f_2(x)\frac{\sin x}{(x^2 + 4x + 3)^5},$$

where $f_1(x) = 8x^7 + 112x^6 + 552x^5 + 1040x^4 - 296x^3 - 4080x^2 - 5640x - 2448$ and $f_2(x) = x^8 + 16x^7 + 72x^6 - 32x^5 - 1094x^4 - 3120x^3 - 3120x^2 + 192x + 1581$.

The current MATLAB function diff() can be used to find higher-order derivatives. For instance, the 100th order derivative can be obtained within 4 seconds (in older versions of MATLAB, Maple was its symbolic engine, and under 1 second was needed to find the 100th order derivative).

```
>> tic, diff(f,x,100); toc % find the 100th order derivative and measure time
```

Example 4.5. Find the nth order derivative of $y(x) = (ax + b)/(cx + d)$.

Solutions. In the new version of diff() function, variable n cannot be used, and n must be a given positive integer. The following code can be tried:

```
>> syms x a b c d; f(x)=(a*x+b)/(c*x+d);
   f1=simplify(diff(f,x,1)), f2=simplify(diff(f,x,2)),
   f3=simplify(diff(f,x,3)), f4=simplify(diff(f,x,4)),
   f10=simplify(diff(f,x,10)), f11=simplify(diff(f,x,11))
```

and the results are obtained as

$$f_1 = \frac{ad - bc}{(d + cx)^2}, \quad f_2 = -\frac{2c(ad - bc)}{(d + cx)^3}, \quad f_3 = \frac{6c^2(ad - bc)}{(d + cx)^4}, \quad f_4 = -\frac{24c^3(ad - bc)}{(d + cx)^5}$$

$$f_{10} = \frac{-3\,628\,800c^9(ad - bc)}{(d + cx)^{11}}, \quad f_{11} = \frac{39\,916\,800c^{10}(ad - bc)}{(d + cx)^{12}}.$$

Based on these results, the following conclusion can be drawn:

$$\frac{d^n}{dx^n}\left(\frac{ax + b}{cx + d}\right) = \frac{(-1)^{n+1}n!\,c^{n-1}(ad - bc)}{(d + cx)^{n+1}}. \tag{4.1.2}$$

Normally, the conclusion summarizing a few known formulas may not be a strict mathematical result. Sometimes further proof is needed, for instance, by the mathematical induction method.

Definition 4.3. The most common and simplest mathematical induction method is to prove that, for any natural number n, a proposition holds. Two steps are usually carried out: (1) proving that, when $k = 1$, the proposition holds; (2) assuming that, when $k = n$, the proposition holds, proving that, when $k = n + 1$, the proposition also holds.

Example 4.6. Prove by the mathematical induction method that for any natural number n, (4.1.2) holds.

Solutions. It is obvious that when $k = 1$, (4.1.2) holds. Now assume that when $k = n$, (4.1.2) is satisfied. Taking the derivative of the right side of the equation, the $(n + 1)$th order derivative of $(ax + b)/(cx + d)$ can be obtained

```
>> syms a b c d x n
   f1=(-1)^(n+1)*factorial(n)*c^(n-1)*(a*d-b*c)/(d+c*x)^(n+1)
   simplify(diff(f1,x))
```

It can be seen that the new result is $(-1)^n c^n n!(ad - bc)(n + 1)/(d + cx)^{n+2}$. Since $(-1)^n = (-1)^{n+2}$, $n!(n + 1) = (n + 1)!$, it can be seen that when $k = n + 1$, (4.1.2) also holds. Hence, by the mathematical induction method, the equation holds for all natural numbers n.

Theorem 4.1 (Leibniz formula). *If functions $u(x)$ and $v(x)$ are given, then*

$$\frac{d^n}{dx^n}[u(x)v(x)] = \sum_{k=0}^{n} C_n^k u^{(n-k)}(x)v^{(k)}(x), \tag{4.1.3}$$

where the binomial coefficients are $C_n^k = n(n - 1) \cdots (n - k + 1)/k!$.

Example 4.7. Validate Theorem 4.1 for small integers n.

Solutions. The proof of the theorem cannot be carried out with MATLAB, for the nth order derivatives. Therefore, the theorem can only be validated for small values of n by

```
>> syms x u(x) v(x);
   for i=1:4, simplify(diff(u*v,x,i)), end
```

The results obtained are as follows:

$$u(x)v'(x) + v(x)u'(x),$$
$$u(x)v''(x) + v(x)u''(x) + 2u'(x)v'(x),$$
$$u(x)v'''(x) + v(x)u'''(x) + 3u'(x)v''(x)3v'(x)u''(x),$$
$$u(x)v^{(4)}(x) + v(x)u^{(4)}(x) + 4u'(x)v'''(x) + 4v'(x)u'''(x) + 6u''(x)v''(x).$$

It can be seen that for smaller n, the Leibniz formula is valid. It can be shown using mathematical induction that Leibniz formula is correct for all n. However, it is hard to show it with a computer, and simple to do so by hand.

4.1.3 Derivatives of composite functions

Theorem 4.2. *The derivative of a composite function $F(x) = f(g(x))$ is*

$$\frac{dF(x)}{dx} = \frac{df(g(x))}{dg(x)} \frac{dg(x)}{dx}. \tag{4.1.4}$$

Example 4.8. Prove Theorem 4.2 with MATLAB.

Solutions. It can be seen from the following statements that the nth order derivative is $D(f)(g(x))*\mathrm{diff}(g(x),x)$, which is the MATLAB representation of (4.1.4).

```
>> syms x g(x) f(x); diff(f(g(x)),x)
```

In fact, with the powerful tools provided in MATLAB, it is not even necessary to find the derivative through Theorem 4.2, since the rules in the related theorems are already embedded in the `diff()` function. Even though the reader does not know at all the theorem, the function `diff()` can still be used directly to find the derivatives of composite functions.

Example 4.9. For the given function $f(x) = u(x)^{v(x)}$, compute $f'(x)$ and $f''(x)$.

Solutions. With the composite function defined, the function `diff()` can be called directly to compute the derivative functions

```
>> syms x u(x) v(x); f=u^v; simplify(diff(f,x))
   simplify(diff(f,x,2))
```

The first-order derivative is

$$f'(x) = u(x)^{v(x)-1}[v(x)u'(x) + \ln u(x)u(x)v'(x)],$$

and the second-order derivative can be found as

$$f''(x) = u(x)^{v(x)-2}[v^2(x)(u'(x))^2 - v(x)(u'(x))^2 + \ln u(x)u^2(x)v''(x)$$
$$+ u(x)v(x)u''(x) + (\ln u(x))^2 u^2(x)(v'(x))^2$$
$$+ 2u(x)u'(x)v'(x) + 2\ln u(x)u(x)v(x)u'(x)v'(x)].$$

Example 4.10. Find the third-order derivative of the function $F(t) = t^2 f(t) \sin t$. If $f(t)$ is specified as $f(t) = e^{-t}$, compare the results with different methods.

Solutions. The symbolic function $f(t)$ can be declared directly with syms command. Then, the third-order derivative of $F(t)$ can be obtained directly with the following statements:

```
>> syms t f(t); F(t)=t^2*f*sin(t);
   G=simplify(diff(F,t,3)) % direct derivatives
```

and the result is

$$\frac{d^3 F(t)}{dt^3} = \left[\frac{d^3 f(t)}{dt^3}\sin t + 3\frac{d^2 f(t)}{dt^2}\cos t - 3\frac{df(t)}{dt}\sin t - f(t)\cos t\right]t^2$$
$$+ \left[6\frac{d^2 f(t)}{dt^2}\sin t + 12\frac{df(t)}{dt}\cos t - 6f(t)\sin t\right]t + 6\frac{df(t)}{dt}\sin t + 6f(t)\cos t.$$

With the following statements, the third-order derivative with $f(t) = e^{-t}$ embedded can immediately be obtained:

```
>> y1=simplify(subs(G,f,exp(-t)));
   simplify(diff(t^2*sin(t)*exp(-t),3)-y1)
```

and the same result is $y_1(t) = 2e^{-t}(t^2 \cos t + t^2 \sin t - 6t \cos t + 3 \cos t - 3 \sin t)$.

4.1.4 Derivatives of piecewise functions

Example 4.11. Find the derivative of the following piecewise function

$$f(x) = \begin{cases} x^2 + \sin x^2, & x \leqslant 0, \\ \ln(1+x), & x > 0. \end{cases}$$

Solutions. The following statements can be used to input the piecewise function into MATLAB environment, and then the derivative can be found with `diff()` function.

```
>> syms x;
   f(x)=piecewise(x<=0,x^2+sin(x^2), x>0,log(1+x));
   f1=diff(f)
```

The result obtained directly is

$$f_1(x) = \begin{cases} 1/(x+1), & 0 < x, \\ 2x + 2x\cos x^2, & x < 0. \end{cases}$$

It can be seen that when $x = 0$, the derivative function is not defined. The curves of the function and its first-order derivative can be obtained with the following statements, as shown in Figure 4.2. The original function is continuous, while the derivative function is not.

```
>> fplot([f, f1],[-2,2])
```

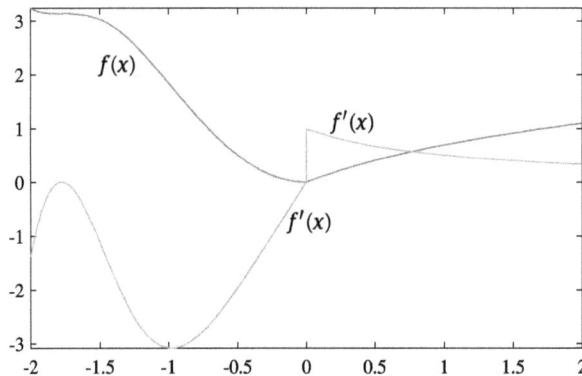

Figure 4.2: The curves of the function and its first-order derivative.

4.1.5 Derivatives of matrices

The so-called derivatives of matrices are, in fact, the derivatives of all individual elements $h_{i,j}(x)$ in the matrix. The function `diff()` can be used directly in finding the derivatives of matrices.

Example 4.12. Find the third-order derivative of the following matrix:

$$H(x) = \begin{bmatrix} 4\sin 5x & e^{-4x^2} \\ 3x^2 + 4x + 1 & \sqrt{4x^2 + 2} \end{bmatrix}.$$

Solutions. With the `diff()` function, the third-order derivative of the matrix $H(x)$ can be immediately found and stored in the new matrix $N(x)$:

```
>> syms x; H=[4*sin(5*x), exp(-4*x^2); 3*x^2+4*x+1, sqrt(4*x^2+2)]
   N=diff(H,x,3)   % input the matrix and find the 3rd-order derivative
```

The third-order derivative matrix can be found as

$$N(x) = \frac{d}{dx}H(x) = \begin{bmatrix} -500\cos 5x & 192x\,e^{-4x^2} - 512x^3\,e^{-4x^2} \\ 0 & 12\sqrt{2}(2x^3 - 1)/(2x^2 + 1)^{3/2} \end{bmatrix}.$$

4.2 Derivatives of parametric equations

Theorem 4.3. *For the parametric equations $y = f(t)$, $x = g(t)$, the nth order derivative $d^n y/dx^n$ can be iteratively derived as*

$$\frac{dy}{dx} = \frac{f'(t)}{g'(t)},$$
$$\frac{d^2 y}{dx^2} = \frac{d}{dt}\left(\frac{f'(t)}{g'(t)}\right)\frac{1}{g'(t)} = \frac{d}{dt}\left(\frac{dy}{dx}\right)\frac{1}{g'(t)}, \qquad (4.2.1)$$
$$\vdots$$
$$\frac{d^n y}{dx^n} = \frac{d}{dt}\left(\frac{d^{n-1} y}{dx^{n-1}}\right)\frac{1}{g'(t)}.$$

There is no existing MATLAB function capable of finding derivatives of parametric equations, nor of higher-order derivatives. A general purpose MATLAB function is needed for this task. A loop structure can be used to find high-order derivatives of parametric equations in MATLAB, and the efficiency is higher than the recursive algorithm used in [28].

```
function result=paradiff(y,x,t,n)
if mod(n,1)=0, error('n should positive integer, please correct')
else, g1=diff(x,t); result=diff(y,t)/g1;
   for i=2:n, result=diff(result,t)/g1; end
end
```

Example 4.13. For the given parametric equations, compute $d^3 y/dx^3$:

$$y(t) = \frac{\sin t}{(t+1)^3}, \quad x(t) = \frac{\cos t}{(t+1)^3}$$

Solutions. The function from the previous problem can be called, and high-order derivatives can be obtained via

```
>> syms t; y=sin(t)/(t+1)^3; x=cos(t)/(t+1)^3;
   f=simplify(paradiff(y,x,t,3))
```

and the result obtained is

$$\frac{d^3y}{dx^3} = \frac{3(t+1)^7}{(3\cos t + \sin t + t\sin t)^5}(23\cos t + 24\sin t - 6t^2\cos t - 4t^3\cos t - t^4\cos t$$
$$+ 12t^2\sin t + 4t^3\sin t - 4t\cos t + 32t\sin t).$$

If the $\sin t$ and $\cos t$ terms are collected, then, though manual processing, the following result can be found:

$$\frac{d^3y}{dx^3} = \frac{3(t+1)^7[(-t^4 - 4t^3 - 6t^2 - 4t + 23)\cos t + (4t^3 + 12t^2 + 32t + 24)\sin t]}{(3\cos t + \sin t + t\sin t)^5}.$$

Example 4.14. For the given parametric equations $x(t) = \ln t$, $y(t) = t^m$, compute d^ny/dx^n.[5]

Solutions. The function `paradiff()` written earlier can only be used for specific values of n. Variable n cannot be accepted in the function call. Several values of specific n can be used, and the nth order derivatives can be summarized. The following MATLAB commands can be used:

```
>> syms m t; x(t)=log(t); y(t)=t^m;
   f1=simplify(paradiff(y,x,t,1)), f2=simplify(paradiff(y,x,t,2))
   f3=simplify(paradiff(y,x,t,3)), f4=simplify(paradiff(y,x,t,4))
```

and it is found that $f_1 = mt^m$, $f_2 = m^2t^m$, $f_3 = m^3t^m$, and $f_4 = m^4t^m$. It can be seen that $d^ny/dx^n = m^nt^m$.

Mathematical induction can be used to prove the above conclusion. From f_1, it can be seen that for $k = 1$ the formula holds. Assume that for $k = n$, the formula also holds, i. e., $d^ny/dx^n = m^nt^m$. Since the expression is also a parametric equation, when $k = n + 1$, the $(n + 1)$th-order derivative $d^{n+1}y/dx^{n+1}$ can be found by

```
>> syms n; F=m^n*t^m; F1=simplify(diff(F,t)/diff(x,t))
```

with the conclusion that $F_1 = m^{n+1}t^m$, so that, when $k = n + 1$, the formula also holds. Hence by mathematical induction $d^ny/dx^n = m^nt^m$ holds for all integers n.

If one wants to find the high-order derivatives of $x(t)$ with respect to $y(t)$, the following statements can be used:

```
>> f1=simplify(paradiff(x,y,t,1)), f2=simplify(paradiff(x,y,t,2))
   f3=simplify(paradiff(x,y,t,3)), f4=simplify(paradiff(x,y,t,4))
```

The obtained solutions are $f_1 = 1/(mt^m)$, $f_2 = -1/(mt^{2m})$, $f_3 = 2/(mt^{3m})$, and $f_4 = -6/(mt^{4m})$, so we conclude that $d^n x/dy^n = (-1)^{n+1}(n-1)!/(mt^{nm})$.

The proof of the formula can also be carried out by mathematical induction. It is known that for $k = 1$, the proposition is satisfied. Assume that, when $k = n$, the proposition is also true. Taking further a first-order derivative of the formula, it can be seen that, when $k = n+1$, the result is $(-1)^n n(n-1)!/(mt^{m(n+1)})$. Hence, when $k = n+1$, the proposition also holds. Therefore, by mathematical induction we have established that the formula is correct for all n.

```
>> F1=(-1)^(n+1)*factorial(n-1)/m/t^(n*m); diff(F1,t)/diff(y,t)
```

By the way, careful readers may notice that for this particular example, $d^n y/dx^n \neq d^n x/dy^n$. This is a normal phenomenon.

4.3 Partial derivatives of multivariate functions

With the concepts and fundamental knowledge of multivariate functions and limits, it is relatively simple to understand the partial derivatives of multivariate functions. In this section, the definitions and computation of partial derivatives are given, followed by the concept of total differential.

4.3.1 Partial derivatives

The partial derivatives are, in fact, the derivatives of a multivariate function with respect to a specific independent variable. The definition of partial derivatives of a 2D function is presented below.

Definition 4.4. For a 2D function $z = f(x, y)$, the partial derivative of z with respect to x is defined as

$$\frac{\partial f(x,y)}{\partial x} = \lim_{\Delta x \to 0} \frac{f(x + \Delta x, y) - f(x, y)}{\Delta x}. \tag{4.3.1}$$

Moreover, although multivariate functions are related to different independent variables, they are always considered as constants when the partial derivative with respect to x is computed. The original function can just be regarded as a univariate function of x alone.

Similarly, the partial derivative $\partial f(x, y)/\partial y$ can also be defined, and high-order partial derivatives with respect to x and y can be defined as well. For a multivariate function $f(x_1, x_2, \ldots, x_n)$, a similar definition of partial derivatives can be written.

There is no dedicated MATLAB function to compute the partial derivatives of given multivariate functions. The partial derivatives can be evaluated with the `diff()` function.

From the given 2D function $f(x, y)$, function `diff()` can be called in a nested format to find the partial derivatives $\partial^{m+n} f / (\partial x^m \partial y^n)$

f_1=diff(diff(f,x,m),y,n) or f_1=diff(diff(f,y,n),x,m)

The command f_1=diff(f,$\underbrace{x,\dots,x}_{m\text{ terms}}$, $\underbrace{y,\dots,y}_{n\text{ terms}}$) can also be used.

Example 4.15. For the 2D function $z = (x^2 - 2x)e^{-x^2 - y^2 - xy}$, find the first-order partial derivatives. Draw plots of the results.

Solutions. The following statements can be used to compute $\partial z / \partial x$ and $\partial z / \partial y$:

```
>> syms x y; clear z
   z(x,y)=(x^2-2*x)*exp(-x^2-y^2-x*y);
   zx=simplify(diff(z,x)), zy=diff(z,y)
```

whose mathematical forms (also known as gradients) are

$$\frac{\partial z(x,y)}{\partial x} = -e^{-x^2-y^2-xy}(-2x + 2 + 2x^3 + x^2 y - 4x^2 - 2xy),$$

$$\frac{\partial z(x,y)}{\partial y} = -x(x-2)(2y+x)e^{-x^2-y^2-xy}.$$

Using the mesh grids generated for $x \in (-3, 2)$ and $y \in (-2, 2)$, numerical expressions of the function and its partial derivatives can be evaluated, and the surface plot is shown in Figure 4.3.

```
>> [x0,y0]=meshgrid(-3:.2:2,-2:.2:2); z0=double(z(x0,y0));
   surf(x0,y0,z0), zlim([-0.7 1.5]) % 3D surface with z limits
```

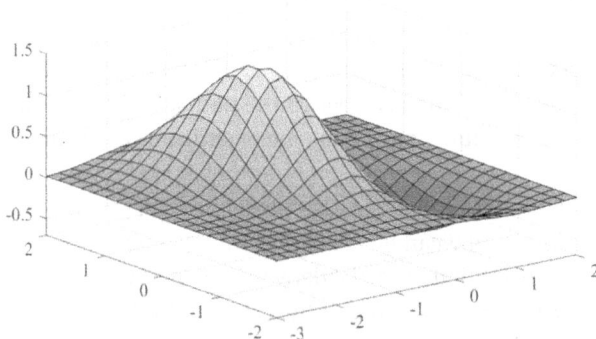

Figure 4.3: Surface of the 2D function.

From the first-order partial derivatives, the quiver() function can be used to show the gradients. The gradients are superimposed on the contours with contour() function, as shown in Figure 4.4. If a ball is placed on top of the surface, it will roll down in the direction shown by the gradients, with the speed shown in the length of arrows.

```
>> contour(x0,y0,z0,30), hold on;  % contours
   zx0=double(zx(x0,y0)); zy0=double(zy(x0,y0));
   quiver(x0,y0,-zx0,-zy0) % negative gradients
```

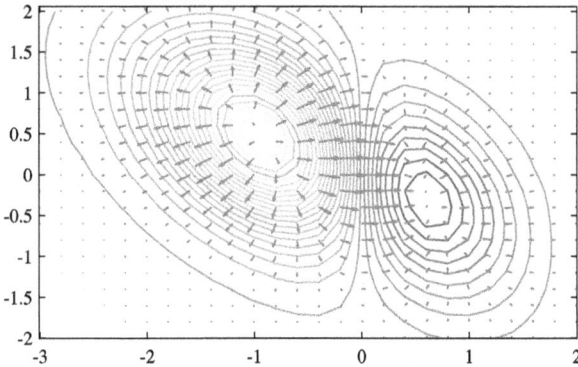

Figure 4.4: Contours with gradients.

Example 4.16. For the 3D function $f(x,y,z) = \sin(x^2 y)e^{-x^2 y - z^2}$, compute $\dfrac{\partial^4 f(x,y,z)}{\partial x^2 \partial y \partial z}$.

Solutions. Symbolic variable can be declared, then MATLAB statements can be used to compute the needed partial derivatives as

```
>> syms x y z;
   f(x,y,z)=sin(x^2*y)*exp(-x^2*y-z^2);  % original function
   F=diff(f,x,x,y,z)                     % high-order derivatives
```

The mathematical expression of the result is

$$F = -4ze^{-x^2 y - z^2}(\cos x^2 y - 10yx^2 \cos x^2 y + 4x^4 y^2 \sin x^2 y + x^4 y^2 4 \cos x^2 y - \sin x^2 y).$$

Example 4.17. Assume that x_0, y_0, z_0, and a are constants, show that the function

$$u(x,y,z,t) = \frac{1}{(2a\sqrt{\pi t})^3} e^{-[(x-x_0)^2 + (y-y_0)^2 + (z-z_0)^2]/(4a^2 t)}$$

satisfies the heat conduction equation[5]

$$\frac{\partial u}{\partial t} = a^2 \left(\frac{\partial^2 u}{\partial x^2} + \frac{\partial^2 u}{\partial y^2} + \frac{\partial^2 u}{\partial z^2} \right).$$

Solutions. The symbolic variables and constants can be declared first, and the function $u(x,y,z,t)$ can be defined. The difference of the left- and right-hand sides of the heat conduction equation can be obtained, and simplified as zero. It can be shown that function $u(x,y,z,t)$ satisfies the equation of heat conduction.

```
>> syms x0 y0 z0 x y z t a
   u(x,y,z,t)=1/(2*a*sqrt(pi*t))^3*...
       exp(-((x-x0)^2+(y-y0)^2+(z-z0)^2)/(4*a^2*t));
   diff(u,t)-a^2*(diff(u,x,x)+diff(u,y,y)+diff(u,z,z));
   simplify(ans)
```

4.3.2 Total differential

The concept of partial derivatives was introduced based on the assumption that a certain independent variable is changed. If all the independent variables change, the concept of total differential is proposed.

Definition 4.5. If $z = z(x,y)$, the total differential of z is defined as

$$dz = \frac{\partial z(x,y)}{\partial x}dx + \frac{\partial z(x,y)}{\partial y}dy. \tag{4.3.2}$$

Definition 4.6. Similarly, second-order total differential of a 2D function $z = z(x,y)$ is defined as

$$d^2z = d(dz) = \frac{\partial^2 z}{\partial x^2}dx^2 + 2\frac{\partial^2 z}{\partial x \partial y}dxdy + \frac{\partial^2 z}{\partial y^2}dy^2. \tag{4.3.3}$$

Second-order total differentiation is simply denoted as

$$d^2z = \left(dx\frac{\partial}{\partial x} + dy\frac{\partial}{\partial y}\right)^2 z. \tag{4.3.4}$$

Definition 4.7. More generally, for a multivariate function $z = z(x_1, x_2, \ldots, x_m)$, the nth order total differential is defined as

$$d^n z = d(d^{n-1}z) = \left(dx_1\frac{\partial}{\partial x_1} + dx_2\frac{\partial}{\partial x_2} + \cdots + dx_m\frac{\partial}{\partial x_m}\right)^n z. \tag{4.3.5}$$

4.3.3 Derivatives of composite functions

Multivariate composite functions are also important research objects in multivariate calculus. Derivatives of composite functions with two variables are presented in this section, together with the solutions using MATLAB.

Definition 4.8. If $u = f(x,y)$, and $x = x(s,t)$, $y = y(s,t)$, then u is a composite function of s and t.

Theorem 4.4. *The two-dimensional composite function in Definition 4.8 can be computed with*

$$\frac{\partial u}{\partial s} = \frac{\partial u}{\partial x}\frac{\partial x}{\partial s} + \frac{\partial u}{\partial y}\frac{\partial y}{\partial s}, \quad \frac{\partial u}{\partial t} = \frac{\partial u}{\partial x}\frac{\partial x}{\partial t} + \frac{\partial u}{\partial y}\frac{\partial y}{\partial t}. \tag{4.3.6}$$

Example 4.18. Given $u = e^x \sin y$ and $x = 2st$, $y = t+s^2$, compute the partial derivatives of u with respect to s and t.

Solutions. Before computing the partial derivatives, the composite function should be entered into MATLAB environment. Observe the way the function is stored!

```
>> syms s t; x(s,t)=2*s*t; y(s,t)=t+s^2; u=exp(x)*sin(y)
```

The obtained result is $u = \sin(s^2 + t)e^{2st}$. It can be seen that function u can be substituted automatically as a function of s and t. Therefore there is no need to evaluate partial derivatives according to Theorem 4.4. Direct derivative evaluations with respect to s and t are sufficient. If one wants to evaluate the expected partial derivatives, the following commands can be used directly

```
>> simplify(diff(u,s)), simplify(diff(u,s)),
   simplify(diff(u,s,t)), simplify(diff(u,s,t,t)).
```

The results obtained are as follows:

$$\frac{\partial u}{\partial s} = 2e^{2st}(s\cos(s^2 + t) + t\sin(s^2 + t)),$$

$$\frac{\partial u}{\partial t} = 2e^{2st}(s\cos(s^2 + t) + t\sin(s^2 + t)),$$

$$\frac{\partial^2 u}{\partial s \partial t} = 2e^{2st}(\sin(s^2 + t) + 2s^2\cos(s^2 + t) + t\cos(s^2 + t) - s\sin(s^2 + t) + 2st\sin(s^2 + t)),$$

$$\frac{\partial u^3}{\partial s \partial t^2} = 2e^{2st}(2\cos(s^2 + t) + 4s^3\cos(s^2 + t) - 4s^2\sin(s^2 + t) - s\cos(s^2 + t)$$
$$+ 4s\sin(s^2 + t) - t\sin(s^2 + t) + 4st\cos(s^2 + t) + 4s^2t\sin(s^2 + t)).$$

4.4 Introduction to fields

With the concept of partial derivatives of multivariate functions, specific 3D functions (or even 2D functions) can be established. In this section, the concept of "field" in physics is introduced, and the concepts and computations of gradient, divergence, curl, and potentials will be presented.

4.4.1 Scalar and vector fields

In physics, a physical quantity in spacial distribution can be expressed by fields. Fields are classified into scalar and vector fields.

Definition 4.9. A scalar field can be expressed as $\varphi(x,y,z)$, while a vector field is expressed as

$$v(x,y,z) = [X(x,y,z), Y(x,y,z), Z(x,y,z)].\tag{4.4.1}$$

For example, for a moving particle in a 3D space, its real-time position can be expressed as a scalar field, while its instantaneous velocity can be expressed as a vector field. In MATLAB, scalar fields can be modeled by scalar functions, while vector fields can be expressed as vector functions.

The fields studied in physics are all 3D fields. In some particular areas, 2D or even higher-dimensional fields may also be encountered. These fields can also be handled directly with MATLAB symbolic functions.

4.4.2 Gradient, divergence, and curl

An example was used in the previous section to present the concept and physical interpretation of gradients. In this section, the definition and computation of gradients are addressed.

Definition 4.10. The gradient of a scalar field is defined as

$$\text{grad }\varphi(x,y,z) = \left[\frac{\partial\varphi(x,y,z)}{\partial x}, \frac{\partial\varphi(x,y,z)}{\partial y}, \frac{\partial\varphi(x,y,z)}{\partial z}\right].\tag{4.4.2}$$

Gradient is also denoted as $\nabla\varphi(x,y,z)$. It can be seen that the gradient can be used to convert a scalar field into a vector field. The gradient can be evaluated directly with the MATLAB function g=gradient(φ), and the function applies to 2D or even higher-dimensional fields.

Definition 4.11. The divergence and curl of a vector field $v(x,y,z)$ are respectively defined as

$$\text{div }v(x,y,z) = \frac{\partial X(x,y,z)}{\partial x} + \frac{\partial Y(x,y,z)}{\partial y} + \frac{\partial Z(x,y,z)}{\partial z},\tag{4.4.3}$$

$$\text{curl }v(x,y,z) = \left[\left(\frac{\partial Z}{\partial y} - \frac{\partial Y}{\partial z}\right), \left(\frac{\partial X}{\partial z} - \frac{\partial Z}{\partial x}\right), \left(\frac{\partial Y}{\partial x} - \frac{\partial X}{\partial y}\right)\right]^{\text{T}}.\tag{4.4.4}$$

Curl is sometimes also denoted as rot $v(x,y,z)$.

Definition 4.12. For a 2D vector field $v(x,y) = [X(x,y), Y(x,y)]$, the curl is defined as

$$\text{curl } v(x,y) = \frac{\partial X(x,y)}{\partial y} - \frac{\partial Y(x,y)}{\partial x}. \tag{4.4.5}$$

Function d=divergence(v, [x,y,z]) can be used to compute the divergence of a vector v. It can also be used to handle 2D and multidimensional vector fields.

The curl of a 3D vector field can be evaluated from c=curl(v, [x,y,z]). The divergence is a scalar function, while the curl is a vector function.

The curl of a 2D vector field can be evaluated with c=curl2(v, [x,y]), whose code is given by

```
function C=curl2(v,vxy)
x=vxy(1); y=vxy(2); C=diff(v(1),y)-diff(v(2),x);
```

Example 4.19. For the 2D vector field $v(x,y) = [5x^2y - 4xy, 3x^2 - 2y]$, compute its divergence and curl.

Solutions. The following statements can be used to compute directly the divergence and curl of the 2D vector field. Note that the function curl() cannot be used for a 2D field. The new function curl2() should be used instead.

```
>> syms x y; v=[5*x^2*y-4*x*y, 3*x^2-2*y];
   divergence(v,[x,y]), curl2(v,[x,y])
```

The result obtained is div $v(x,y) = 10xy - 4y - 2$, curl $v(x,y) = 5x^2 - 10x$.

Example 4.20. For the vector field $v(x,y,z) = [X(x,y,z), Y(x,y,z), Z(x,y,z)]$, where

$$X(x,y,z) = x^2 \sin y, \quad Y(x,y,z) = y^2 \sin xz, \quad Z(x,y,z) = xy \sin \cos z,$$

compute its divergence and curl.

Solutions. The original vector field should be entered first, and the following statements can be used to compute directly:

```
>> syms x y z;
   v=[(x^2)*sin(y), (y^2)*sin(x*z), x*y*sin(cos(z))]; % vector field
   d=divergence(v,[x,y,z]), c=curl(v,[x,y,z]) % divergence and curl
```

Then the divergence and curl are respectively

$$d = 2y \sin xz + 2x \sin y - xy \cos \cos z \sin z,$$

$$c = [\, x \sin \cos z - xy^2 \cos xz, \; -y \sin \cos z, \; y^2 z \cos xz - x^2 \cos y\,]^T.$$

Definition 4.13. If the curl of a vector field $v(x,y,z)$ is zero, the field is referred to as a conservative field.

Example 4.21. Prove curl [grad $u(x, y, z)$] = **0**.

Solutions. The scalar field can be expressed first in MATLAB, and with the following statements, a zero vector is obtained, which proves the proposition.

```
>> syms x y z u(x,y,z);
   v=gradient(u,[x,y,z]); simplify(curl(v,[x,y,z]))
```

It can be seen from the example that the gradient vector of a scalar field forms a conservative vector field.

4.4.3 Potentials of a vector field

Definition 4.14. If $F(x, y, z)$ is a vector field and $F(x, y, z) = \nabla f(x, y, z)$, then scalar function $f(x, y, z)$ is referred to as the potential function.

If the MATLAB expression $F(x, y, z)$ is known, the potential function can be obtained directly with the command f=potential(F, [x,y,z]). In fact, if $f(x, y, z)$ is a potential function, $f(x, y, z) + C$ is also a potential function, where C is an arbitrary constant.

Example 4.22. For the vector field given below, compute its potential function[21]

$$v(x, y, z) = \left[\frac{y}{1 + x^2 y^2}, \frac{x}{1 + x^2 y^2} + \frac{z}{\sqrt{1 - y^2 z^2}}, \frac{y}{\sqrt{1 - y^2 z^2}} + \frac{1}{z}\right].$$

Solutions. The field should be entered into MATLAB environment, and the potential function can be obtained directly with the following statements:

```
>> syms x y z; a=1+x^2*y^2; b=sqrt(1-y^2*z^2);
   F=[y/a, x/a+z/b, y/b+1/z]; P=potential(F,[x y z])
```

The result $P(x, y, z) = \text{atan } xy + \ln z + z \text{ asinh } y\sqrt{-z^2}/\sqrt{-z^2}$ is obtained directly, and due to the appearance of the $\sqrt{-z^2}$ and $\ln z$ terms, z must be declared as a positive variable, with

```
>> syms z positive; P=potential(F,[x y z])
```

The simplified potential function is

$$P(x, y, z) = \text{atan } xy + \ln z + z \text{ asin } y + C.$$

4.5 Derivative matrices

From the previous section, it can be seen that the gradient involves the first-order derivatives of a scalar function. If the scalar function is extended to a vector function, Jacobian matrix can be defined. If the derivatives are taken to be second-order, Hessian matrix can be defined. Laplacian operator can also be defined.

4.5.1 Jacobian matrix

Definition 4.15. Assume that for m functions with n independent variables

$$
\begin{cases}
y_1 = f_1(x_1, x_2, \ldots, x_n), \\
y_2 = f_2(x_1, x_2, \ldots, x_n), \\
\vdots \\
y_m = f_m(x_1, x_2, \ldots, x_n).
\end{cases}
\tag{4.5.1}
$$

Taking partial derivatives of y_i with respective to x_j, the Jacobian matrix can be defined as

$$
J = \begin{bmatrix}
\partial y_1/\partial x_1 & \partial y_1/\partial x_2 & \cdots & \partial y_1/\partial x_n \\
\partial y_2/\partial x_1 & \partial y_2/\partial x_2 & \cdots & \partial y_2/\partial x_n \\
\vdots & \vdots & \ddots & \vdots \\
\partial y_m/\partial x_1 & \partial y_m/\partial x_2 & \cdots & \partial y_m/\partial x_n
\end{bmatrix}.
\tag{4.5.2}
$$

Jacobian matrices are useful in certain areas, such as robotics and image processing. Jacobian matrix is evaluated from the MATLAB function `jacobian()`, with the syntax J=`jacobian(y,x)`, where x is the vector of independent variables, and y is the vector of the functions.

Example 4.23. It is known that the coordinate conversion formula is given by $x = r\sin\theta\cos\phi$, $y = r\sin\theta\sin\phi$, $z = r\cos\theta$, compute the Jacobian matrix of the three functions $[x, y, z]$ with respect to the independent variables $[r, \theta, \phi]$.

Solutions. Symbolic variables should be declared first, then the three functions can be entered. The following statements can be used to compute the Jacobian matrix:

```
>> syms r theta phi;
   x=r*sin(theta)*cos(phi); y=r*sin(theta)*sin(phi); z=r*cos(theta);
   J=jacobian([x; y; z],[r theta phi]) % compute Jacobian matrix
```

The Jacobian matrix obtained is

$$
J = \begin{bmatrix}
\sin\theta\cos\phi & r\cos\theta\cos\phi & -r\sin\theta\sin\phi \\
\sin\theta\sin\phi & r\cos\theta\sin\phi & r\sin\theta\cos\phi \\
\cos\theta & -r\sin\theta & 0
\end{bmatrix}.
$$

4.5.2 Hessian matrix

Definition 4.16. For an n-dimensional scalar function $f(x_1, x_2, \ldots, x_n)$, the Hessian matrix is defined as

$$
H = \begin{bmatrix}
\partial^2 f/\partial x_1^2 & \partial^2 f/\partial x_1 \partial x_2 & \cdots & \partial^2 f/\partial x_1 \partial x_n \\
\partial^2 f/\partial x_2 \partial x_1 & \partial^2 f/\partial x_2^2 & \cdots & \partial^2 f/\partial x_2 \partial x_n \\
\vdots & \vdots & \ddots & \vdots \\
\partial^2 f/\partial x_n \partial x_1 & \partial^2 f/\partial x_n \partial x_2 & \cdots & \partial^2 f/\partial x_n^2
\end{bmatrix}.
\tag{4.5.3}
$$

It can be seen that the Hessian matrix is a second-order partial derivative matrix for a 2D scalar function $f(x, y)$. Function hessian() is provided in MATLAB to construct Hessian matrix, with the syntax H=hessian(f, x), with the vector $x = [x_1, x_2, \ldots, x_n]$.

Example 4.24. Consider again the 2D function in Example 4.15. Compute the Hessian matrix.

Solutions. The Hessian matrix can be obtained with the following statements:

```
>> syms x y; f=(x^2-2*x)*exp(-x^2-y^2-x*y);
   H=simplify(hessian(f,[x,y]))
   H1=simplify(hessian(f,[x,y])/exp(-x^2-y^2-x*y)) % simplify
```

The result obtained is

$$
H_1 = e^{-x^2-y^2-xy} \begin{bmatrix}
4x - 2(2x - 2)(2x + y) - 2x^2 - (2x - x^2)(2x + y)^2 + 2 \\
2x - (2x - 2)(x + 2y) - x^2 - (2x - x^2)(x + 2y)(2x + y)
\end{bmatrix}
$$

$$
\begin{aligned}
&2x - (2x - 2)(x + 2y) - x^2 - (2x - x^2)(x + 2y)(2x + y) \\
&x(x - 2)(x^2 + 4xy + 4y^2 - 2)
\end{aligned} \Bigg].
$$

4.5.3 Laplacian operators for scalar fields

Definition 4.17. For a scalar field $f(x_1, x_2, \ldots, x_n)$, the Laplacian operator is defined as

$$
\Delta f(x_1, x_2, \ldots, x_n) = \left[\frac{\partial^2}{\partial x_1^2} + \frac{\partial^2}{\partial x_2^2} + \cdots + \frac{\partial^2}{\partial x_n^2} \right] f(x_1, x_2, \ldots, x_n).
\tag{4.5.4}
$$

The Laplacian operator is the sum of the diagonal elements of the Hessian matrix, or the trace of the Hessian matrix.

MATLAB function L=laplacian$(f, [x_1, x_2, \ldots, x_n])$ can be used directly in computing the Laplacian of a given function.

Example 4.25. Consider the 2D function in Example 4.24 and find its Laplacian operator.

Solutions. With the following statements, the Laplacian operator of the function can be found as

```
>> syms x y; f(x,y)=(x^2-2*x)*exp(-x^2-y^2-x*y);
   F=laplacian(f,[x y])
```

and the result obtained is

$$F(x,y) = e^{-x^2-xy-y^2}\left(5x^4 + 8x^3y - 10x^3 + 5x^2y^2 - 16x^2y - 12x^2 - 10xy^2 - 4xy + 16x + 4y + 2\right).$$

4.6 Partial derivatives of implicit functions

Since usually the explicit form of the implicit function $y = f(x_1, x_2, \ldots, x_n)$ cannot be found, the function `diff()` cannot be used directly. Practical approaches and their MATLAB implementations should be considered and introduced to evaluate the partial derivatives of an implicit function. In this section, the problem will be studied.

4.6.1 First-order derivative of an implicit function

Consider now the 2D implicit function $f(x,y) = 0$. If there exists an explicit relationship between x and y, denoted as $y = y(x)$, a composite function $f[x, y(x)] = 0$ can be constructed. Taking the total differential of it, one has

$$df(x,y) = \frac{\partial f(x,y)}{\partial x}dx + \frac{\partial f(x,y)}{\partial y}dy \equiv 0, \tag{4.6.1}$$

from which it is found immediately that

$$\frac{dy}{dx} = -\frac{\partial f(x,y)/\partial x}{\partial f(x,y)/\partial y}. \tag{4.6.2}$$

The partial derivatives $\partial f(x,y)/\partial x$ can be simply denoted as $f_x(x,y)$, or even simpler, f_x. Equation (4.6.2) can be extended to the case of partial derivatives of multivariate implicit functions.

Theorem 4.5. *If the mathematical form of a multivariate implicit function is given by* $f(x_1, x_2, \ldots, x_n) = 0$, *the partial derivatives among the related variables can be obtained directly. For instance,*

$$\frac{\partial x_i}{\partial x_j} = -\frac{\partial f(x_1, x_2, \ldots, x_n)/\partial x_j}{\partial f(x_1, x_2, \ldots, x_n)/\partial x_i} = -\frac{f_{x_j}}{f_{x_i}}. \tag{4.6.3}$$

Since the partial derivatives of f with respect to x_i and x_j can all be evaluated with the MATLAB function `diff()`, the partial derivatives among the variables can be computed directly from $F_1 = -\text{diff}(f, x_j)/\text{diff}(f, x_i)$.

Example 4.26. Consider the function in Example 4.15, and construct an implicit function $(x^2 - 2x)e^{-x^2-y^2-xy} = 0$. Compute dy/dx.

Solutions. It can be seen from (4.6.3) that the required partial derivative dy/dx can be obtained directly with

```
>> syms x y; f(x,y)=(x^2-2*x)*exp(-x^2-y^2-x*y);
   F=-diff(f,x)/diff(f,y), F=simplify(F)
```

and the result is

$$\frac{dy}{dx} = F(x, y) = \frac{2x + 2xy - x^2y + 4x^2 - 2x^3 - 2}{x(x + 2y)(x - 2)}.$$

4.6.2 Higher-order derivatives of implicit functions

Theorem 4.6. For a 2D implicit function $f(x, y) = 0$, if $\partial y/\partial x = F_1(x, y)$ is available, the second-order derivative can be obtained directly from

$$F_2(x, y) = \frac{\partial^2 y}{\partial x^2} = \frac{\partial F_1(x, y)}{\partial x} + \frac{\partial F_1(x, y)}{\partial y} F_1(x, y). \tag{4.6.4}$$

Higher-order partial derivatives can also be obtained recursively from

$$F_n(x, y) = \frac{\partial^n y}{\partial x^n} = \frac{\partial F_{n-1}(x, y)}{\partial x} + \frac{\partial F_{n-1}(x, y)}{\partial y} F_1(x, y). \tag{4.6.5}$$

The above method can easily be extended to higher-order derivatives of multivariate implicit functions.

The higher-order derivative method in Theorem 4.6 can easily be implemented in MATLAB with

```
function dy=impldiff(f,x,y,n)
if mod(n,1)=0,
   error('n should positive integer, please correct')
else, F1=-simplify(diff(f,x)/diff(f,y)); dy=F1;         % 1st order
   for i=2:n, dy=simplify(diff(dy,x)+diff(dy,y)*F1); % loop for (4.6.5)
end, end
```

With the recursive algorithm embedded in the MATLAB function, the nth order partial derivatives $\partial^n y/\partial x^n$ of a given implicit function f can be obtained directly, with $f_1 = \text{impldiff}(f, x, y, n)$.

Example 4.27. Consider again the problem in Example 4.26. Compute $\partial^3 y/\partial x^3$.

Solutions. With (4.6.3), the partial derivative $\partial^3 y/\partial x^3$ can be obtained directly with the following statements:

```
>> syms x y; f(x,y)=(x^2-2*x)*exp(-x^2-y^2-x*y);
   F2=simplify(impldiff(f,x,y,2)), F3=impldiff(f,x,y,2),
   [n,d]=numden(F3), collect(n)
```

and the higher-order derivatives of y are

$$\frac{\partial^2 y}{\partial x^2} = F_2 = -\frac{3x^4 - 12x^3 + 16x^2 - 8x + 8}{2x^2(x + 2y)(x - 2)^2} - \frac{(-3x^3 + 6x^2 + 4x - 4)^2}{2x^2(x + 2y)^3(x - 2)^2},$$

$$\frac{\partial^3 y}{\partial x^3} = F_3 = -\frac{1}{x^3(x + 2y)^5(x - 2)^3}\left[-6x^6 + (24 - 6y)x^5\right.$$
$$+ (-6y^2 + 24y - 14)x^4 + (24y^2 - 32y - 32)x^3$$
$$+ (-32y^2 + 16y + 12)x^2 + (16y^2 - 16y + 16)x - 16y^2 - 8\left.\right].$$

Example 4.28. For an implicit function $x^2 \sin y + y^2 z + z^2 \cos y - 4z = 0$, compute the partial derivatives $\partial z/\partial x$ and $\partial^2 z/\partial x^2$.

Solutions. For the given 3D implicit function $f(x, y, z) = 0$, if the partial derivative of x with respect to z is expected, variable y can be regarded as a constant. Therefore, the following statements can be employed:

```
>> syms x y z; f(x,y,z)=x^2*sin(y)+y^2*z+z^2*cos(y)-4*z;
   F1=simplify(impldiff(f,x,z,1))
   F2=simplify(impldiff(f,x,z,2))
```

and the results are

$$\frac{\partial z}{\partial x} = F_1 = \frac{-2x \sin y}{2z \cos y + y^2 - 4},$$

$$\frac{\partial^2 z}{\partial x^2} = F_2 = \frac{-2 \sin y}{2z \cos y + y^2 - 4} - \frac{8x^2 \cos y \sin^2 y}{(2z \cos y + y^2 - 4)^3}.$$

Example 4.29. Find the partial derivatives of the implicit function[5]

$$x^2 + xy + y^2 = 3.$$

Solutions. Partial derivatives of different orders can be obtained with the following statements. Besides, since $x^2 + xy + y^2 = 3$, variable substitution can further be used to simplify the results.

```
>> syms x y z; f=x^2+x*y+y^2-3; F1=impldiff(f,x,y,1)
   f2=impldiff(f,x,y,2); F2=subs(f2,x^2+x*y+y^2,3)
   f3=impldiff(f,x,y,3); F3=subs(f3,x^2+x*y+y^2,3)
   f4=impldiff(f,x,y,4); F4=subs(f4,x^2+x*y+y^2,3)
```

It is found from the above statements that

$$F_1 = -\frac{2x+y}{x+2y}, \qquad\qquad F_2 = -\frac{18}{(x+2y)^3},$$

$$F_3 = -\frac{162x}{(x+2y)^5}, \qquad\qquad F_4 = -\frac{648(4x^2+xy+y^2)}{(x+2y)^7},$$

where substitution using subs() function may not be completed sometimes; F_4 can be manually simplified as

$$F_4 = \frac{-1\,944(x^2+1)}{(x+2y)^7}.$$

4.6.3 Partial derivatives of simultaneous implicit functions

Consider the simultaneous implicit functions

$$\begin{cases} F(x,y,z) = 0, \\ G(x,y,z) = 0. \end{cases} \tag{4.6.6}$$

One implicit function represents a surface, while the simultaneous implicit functions represent the intersection curves of the two surfaces.

Taking first-order derivatives with respect to x, it is found that

$$F_x + F_y y'(x) + F_z z'(x) = 0, \quad G_x + G_y y'(x) + G_z z'(x) = 0. \tag{4.6.7}$$

If the functions $F(x,y,z)$ and $G(x,y,z)$ are known, F_x and other related derivatives can easily be obtained with function diff(), and can be regarded as known quantities. The undetermined quantities are $y'(x)$ and $z'(x)$. It can be seen from (4.5.2) that the two unknowns satisfy a linear algebraic equation, and the solution can be found as

$$y'(x) = -\frac{J_{x,z}(F,G)}{J_{y,z}(F,G)}, \quad z'(x) = -\frac{J_{y,x}(F,G)}{J_{y,z}(F,G)}, \tag{4.6.8}$$

where the J operator is the simplified notation of determinant of the Jacobian matrix, with

$$J_{y,z}(F,G) = \begin{vmatrix} F_y & F_z \\ G_y & G_z \end{vmatrix}, \quad J_{x,z}(F,G) = \begin{vmatrix} F_x & F_z \\ G_x & G_z \end{vmatrix}, \quad J_{y,x}(F,G) = \begin{vmatrix} F_y & F_x \\ G_y & G_x \end{vmatrix}. \tag{4.6.9}$$

Since the obtained $y'(x)$ is a function of x and y, $y''(x)$ can be expressed as

$$y''(x) = \frac{dy'(x)}{dx} + \frac{dy'(x)}{dy}y'(x). \tag{4.6.10}$$

Similarly, it is found that

$$z''(x) = \frac{dz'(x)}{dx} + \frac{dz'(x)}{dz}z'(x). \tag{4.6.11}$$

The formulas are used to find recursively the nth order derivatives y or z with respect to x. Also, the nth order derivatives among arbitrary combinations of the independent variables can be found. The idea can also be extended to the multivariate simultaneous implicit functions.

Example 4.30. Consider the implicit functions $x^2 + y^2 - z^2 = 0$ and $x^2 + 2y^2 + 3z^2 = 4$, with x as their independent variable. Compute dy/dx, dz/dx, d^2y/dx^2, and d^2z/dx^2.

Solutions. With (4.6.6) in mind, denote

$$F(x,y,z) = x^2 + y^2 - z^2, \quad G(x,y,z) = x^2 + 2y^2 + 3z^2 - 4,$$

so that the necessary determinants of Jacobian matrices in (4.6.9) can be computed. Then, (4.6.8) can be used to compute dy/dx and dz/dx.

How can we find the second-order derivatives? Equations (4.6.10) and (4.6.11) can be used to find directly the second-order derivatives, and the following MATLAB statements can be written:

```
>> syms x y z;
   F(x,y,z)=x^2+y^2-z^2; G(x,y,z)=x^2+2*y^2+3*z^2-4;
   Fyz=det(jacobian([F,G],[y,z])); Fxz=det(jacobian([F,G],[x,z]));
   Fyx=det(jacobian([F,G],[y,x])); yx=-Fxz/Fyz, zx=-Fyx/Fyz
   y2x=diff(yx,x)+diff(yx,y)*yx, z2x=diff(zx,x)+diff(zx,z)*zx
```

yielding the following results:

$$\frac{dy}{dx} = -\frac{4x}{5y}, \quad \frac{dz}{dx} = \frac{x}{5z}, \quad \frac{d^2y}{dx^2} = -\frac{16x^2}{25y^3} - \frac{4}{5y}, \quad \frac{d^2z}{dx^2} = -\frac{x^2}{25z^3} + \frac{1}{5z}.$$

Further, the third- and fourth-order derivatives can also be found via

```
>> z3x=simplify(diff(z2x,x)+diff(z2x,z)*zx)
   z4x=simplify(diff(z3x,x)+diff(z3x,z)*zx)
```

and the results are

$$\frac{d^3z}{dx^3} = \frac{3x(x^2 - 5z^2)}{125z^5}, \quad \frac{d^4z}{dx^4} = -\frac{3(x^4 - 6x^2z^2 + 5z^4)}{125z^7}.$$

In fact, based on the above ideas, the higher-order partial derivatives of and variables u and v can be obtained, where u and v are any two from x, y, and z, and w is the remaining one.

Theorem 4.7. *For given implicit functions $F(x,y,z) = 0$, $G(x,y,z) = 0$, the nth order partial derivative of u with respect to v can be evaluated recursively as*

$$\frac{\partial u}{\partial v} = -\frac{J_{w,v}(F,G)}{J_{w,u}(F,G)}, \quad \frac{\partial^n u}{\partial v^n} = \frac{\partial}{\partial v}\left(\frac{\partial^{n-1} u}{\partial v^{n-1}}\right) + \frac{\partial}{\partial u}\left(\frac{\partial^{n-1} u}{\partial v^{n-1}}\right)\frac{\partial u}{\partial v}. \tag{4.6.12}$$

With this theorem, the following universal MATLAB function can be written:

```
function d=impldiff2(F,G,x,y,z,u,v,n)
if nargin==7, n=1; end
if mod(n,1)~=0,    % check whether integer
    error('n should be a positive integer, please correct')
else, w=setdiff([x y z],[u v]); % the third variable
    if length(w)~=1, error('wrong u v specifications'); end
    d=-det(jacobian([F,G],[w,v]))/det(jacobian([F,G],[w,u])); d0=d;
    for i=2:n, d=simplify(diff(d,v)+diff(d,u)*d0); end
end
```

with the syntax d=impldiff2(F,G,x,y,z,u,v,n), where the arguments are the same as those in mathematics. Arguments u and v are any of the two arguments x, y, and z, n is the order, with the default value of 1. The returned variable d is $\partial^n u/\partial v^n$.

Example 4.31. Solve again the problem in Example 4.30, with the impldiff2() function, and find other partial derivatives.

Solutions. The partial derivative of z with respect to x in Example 4.30 can be evaluated directly with the following statements. It can be seen that the results are the same, with much simpler commands.

```
>> syms x y z;
   F(x,y,z)=x^2+y^2-z^2; G(x,y,z)=x^2+2*y^2+3*z^2-4;
   impldiff2(F,G,x,y,z,z,x,3), impldiff2(F,G,x,y,z,z,x,4)
```

If one wants to find high-order partial derivative of x with respect to z, with the following statements:

```
>> F1=impldiff2(F,G,x,y,z,x,z,3), F2=impldiff2(F,G,x,y,z,x,z,4)
   F3=impldiff2(F,G,x,y,z,x,z,5)
```

the three partial derivatives can be obtained as

$$-\frac{75z(x^2 - 5z^2)}{x^5}, \quad -\frac{75(x^4 - 30x^2z^2 + 125z^4)}{x^7}, \quad \frac{1875z(3x^4 - 50x^2z^2 + 175z^4)}{x^9}.$$

4.7 Applications of derivatives and differentials

The concepts and theory were originated from real applications. For instance, the instantaneous velocity and tangent lines were extracted from problems in applications. In this section, some other applications will be presented, where the extreme value problem, equation solving problem, and tangent plane problem will be studied with the ideas of derivatives and differentials.

4.7.1 Extreme value problem

An extreme value problem is one of the earliest studied research topics in calculus. The first paper of Leibniz's on calculus started from extreme values and tangent lines. The concept and theorems of extreme values are presented first, and then direct solutions with MATLAB are given.

Definition 4.18. If $f(x)$ is continuous in an interval (a, b), and in a neighborhood $(x_0 - \delta, x_0 + \delta)$ of point x_0, where $\delta > 0$, satisfies $f(x) \leqslant f(x_0)$, then $x = x_0$ is referred to as a maximum point, while if in some neighborhood of x_0, $f(x) \geqslant f(x_0)$, then $x = x_0$ is referred to as a minimum point.

Theorem 4.8. *The two conditions for the possible existence of extreme points are:* (1) $f'(x_0) = 0$; (2) $f'(x)$ is not defined at x_0.

Theorem 4.9 (Extreme value theorem). *If $f(x)$ is continuous in the closed interval $[a, b]$, then there exist points c and d such that all the points x satisfy $f(d) \leqslant f(x) \leqslant f(c)$.*

Theorem 4.10 (Lagrange mean value theorem). *If $f(x)$ is continuous in the closed interval $[a, b]$, and differentiable in the open interval (a, b), there exists at least one point ξ in (a, b) such that*

$$f'(\xi) = \frac{f(b) - f(a)}{b - a}. \tag{4.7.1}$$

Example 4.32. Find the extreme points in the interval $(-1, 1/2)$ for the function $f(x) = (x - 1)\sqrt[3]{x^2}$.

Solutions. The function $f(x)$ should be entered first, then its first-order derivative $F(x)$ can be obtained via

```
>> syms x; f(x)=(x-1)*x^(2/3), F(x)=diff(f)
```

It can be seen that the first-order derivative is $F(x) = 2(x-1)3x^{-1/3} + x^{2/3}$. There are two possible extreme points, $x = 0$ (where the derivative is not defined) and $x = 2/5$ (where the first-order derivative equals 0).

Example 4.33. Find the extreme values for the piecewise function[5]

$$f(x) = \begin{cases} e^{-1/|x|}(\sqrt{2} + \sin 1/x), & x \neq 0, \\ 0 & x = 0. \end{cases}$$

Solutions. The piecewise function should be entered first, and then the first-order derivative of the function can be found as shown in Figure 4.5. It can be seen from the curves that the equation $f'(x) = 0$ has infinitely many solutions. Therefore, such functions are not suitable for the extreme point analysis tasks.

```
>> syms x; L=2*pi;
   f(x)=piecewise(x==0,0,...
           x~=0,exp(-1/abs(x))*sin(sqrt(2)*x^3+sin(1/x)));
   F=diff(f,x); fplot(F)
```

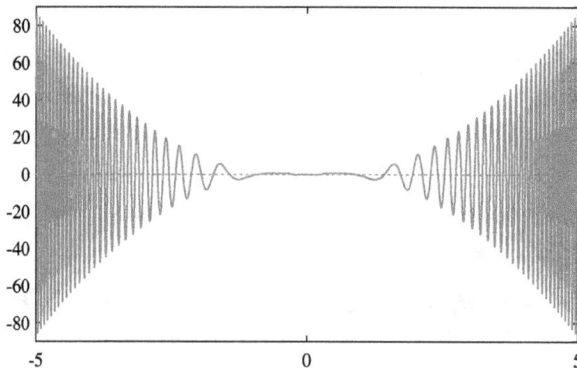

Figure 4.5: The curve of $f'(x)$.

In fact, command fplot (f) can be used to draw directly the curve of the original function, as shown in Figure 4.6. It can be seen that there are indeed too many extreme points, which cannot be found one by one.

The existence conditions for extreme points were studied earlier. However, a more important issue was intentional neglected. How to find out whether an extreme point is a maximum point or a minimum point? To answer the question, second-order derivatives are needed. If the value of the second-order derivative is positive, then the extreme point is a minimum point, otherwise, it is a maximum point.

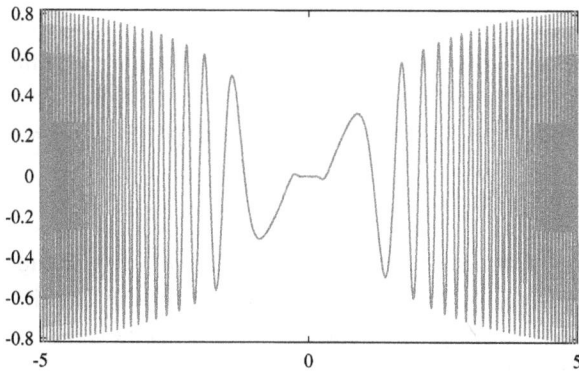

Figure 4.6: The curve of function $f(x)$.

Example 4.34. Consider again the extreme point problem studied in Example 4.32. Judge whether the extreme points are minima or maxima.

Solutions. Two extreme points, $x = 0$ and $x = 2/5$, were found in Example 4.32. Since at $x = 0$, the first-order derivative does not exist, the above method cannot be used. Now let us consider the point $x = 2/5$. The second-order derivative can be found using

```
>> syms x; f(x)=(x-1)*(x^2)^(1/3),
   F1(x)=simplify(diff(f,2)), F1(2/5), double(ans)
```

and it can be seen that $F_1(x) = (10x + 2)/\sqrt[3]{x^4}$, whose value at $x = 2/5$ is

$$F_1(2/5) = 5\sqrt[3]{2^2}\sqrt[3]{5}/6 = 2.2620 > 0.$$

Therefore, the extreme point at $x = 2/5$ is a minimum point. The curve of the function $f(x)$ can also be drawn, as shown in Figure 4.7. It can be seen from the curve that

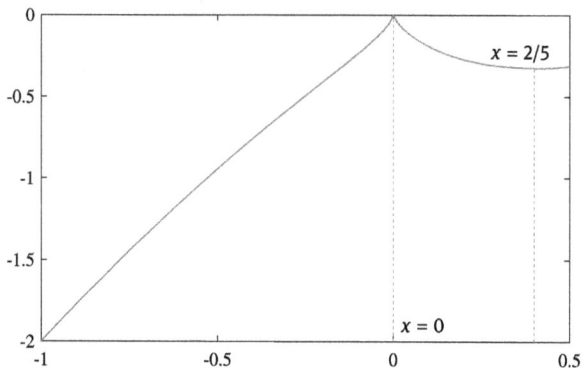

Figure 4.7: Properties of the extreme points of $f(x)$.

$x = 0$ is a maximum point, also the global maximum point in the interval $x \in (-1, 1/2)$, with maximum function value of 0. The point $x = -1$ is the global minimum point, with the function value of -2. The extreme point at $x = 2/5$ is neither a global maximum point, nor a global minimum point. It is merely a local extreme point for $f(x) = 0$.

```
>> x0=-1:0.01:(1/2); f1=double(f(x0)); plot(x0,f1)
```

In real applications, the global maximum and minimum points are often of interest, not the extreme points. Therefore, more emphasis should be put on the optimization problems, rather than on the extreme point problems. Volume IV of the series will concentrate on the problems in optimization with MATLAB.

4.7.2 Newton–Raphson iterative algorithm

For a given equation $f(x) = 0$, it is known from the definition of the derivative that the tangent equation at $x = x_n$ can be formulated as

$$y = f(x_n) + f'(x_n)(x_{n+1} - x_n), \quad n = 0, 1, 2, \dots \tag{4.7.2}$$

Letting $y = 0$, the Newton–Raphson iterative algorithm can be established.

Theorem 4.11. *For the equation $f(x) = 0$, if an initial point x_0 is selected, the equation can be solved iteratively with the following formula:*

$$x_{n+1} = x_n - \frac{f(x_n)}{f'(x_n)}, \quad n = 0, 1, 2, \dots \tag{4.7.3}$$

In real applications, it is, of course, not necessary to let the iteration process go on and on. Termination conditions must be set up. For instance, two small quantities $\epsilon_1, \epsilon_2 > 0$ can be selected. If $|x_{n+1} - x_n| < \epsilon_1$ or $|f(x_n)| < \epsilon$, the iteration process can be terminated, and the solution can be considered to be at $x = x_n$.

Based on the above Newton–Raphson iterative algorithm, the following MATLAB function can be written to find numerically the equation solutions:

```
function [xn,k]=fnonleq(f,x0,e1,e2)
symvar(f), df(symvar(f))=diff(f); k=0; f1=1; x1=1000;
while (abs(f1)>e2 | abs(xn-x0)>e2)
    xn=x0-double(f(x0)/df(x0)); x0=xn; k=k+1; f1=f(xn);
end
```

Example 4.35. Select $x_0 = 0.1$, find a solution to the equation in Example 3.20 with Newton–Raphson iterative algorithm.

Solutions. For the given equation $\sin(x^3 + 1/x) = 0$, selecting an initial search point $x_0 = 0.1$, and rather tough error tolerances $\epsilon_1 = \epsilon_2 = 10^{-15}$, the following MATLAB statements can be used to find the solution which is $x_1 = 0.1601$, with $k = 4$ and $e = -3.6513 \times 10^{-16}$.

```
>> syms x; e1=1e-15; e2=e1; f(x)=sin(x^3+1./x);
   [x1 k]=fnonleq(f,0.1,e1,e2), e=double(f(x1))
```

It can be seen that only with four iteration steps an accurate solution to the given equation can be found. Compared with the curve in Figure 3.7, it is found that at $x = 0.1$, the first-order derivative is larger than 0, therefore, an outbound direction is searched such that an $x > 0.1$ can finally be found, which is not the same as that shown in Example 3.20.

If a tougher error tolerance of eps is selected, the code would be very time consuming, since the convergence conditions may not be satisfied. It is concluded that, if iterative algorithms are used, it is not appropriate to select eps as the error tolerance. A slightly larger one, say, 10^{-15}, should be used instead.

The equation solving facilities illustrated here is only one application of derivatives. In Volume IV of the series, more practical algorithms will be presented for equation solution processes.

4.7.3 Tangent planes and normal lines

Theorem 4.12. *For a 2D function $z = F(x, y)$, the tangent plane at a point $p_0 = (x_0, y_0, z_0)$ can be expressed as*

$$\left.\frac{\partial z}{\partial x}\right|_{p_0}(x - x_0) + \left.\frac{\partial z}{\partial y}\right|_{p_0}(y - y_0) - (z - z_0) = 0. \tag{4.7.4}$$

In fact, a simpler explicit form of the tangent plane can be written as

$$z = z_0 + \left.\frac{\partial z}{\partial x}\right|_{p_0}(x - x_0) + \left.\frac{\partial z}{\partial y}\right|_{p_0}(y - y_0). \tag{4.7.5}$$

Theorem 4.13. *Considering the 2D function $z = F(x, y)$, the normal line at $p_0 = (x_0, y_0, z_0)$ can be expressed as*

$$z - z_0 = -\frac{x - x_0}{F_x(x_0, y_0)} = -\frac{y - y_0}{F_y(x_0, y_0)}. \tag{4.7.6}$$

Letting the equation equal t, the parametric equations of the normal line equation can be established, when $t \in (0, t_M)$ is selected.

Example 4.36. Draw the tangent plane at $(-1, 0.4, 1.403)$, for the function $z = (x^2 - 2x)e^{-x^2-y^2-xy}$ in Example 4.15.

Solutions. For the 2D explicit function $z = z(x, y)$, the following statements can be used to draw the surface of the function. We also select the point and establish the tangent plane. With the following statements, the function surface and the tangent plane can both be obtained as shown in Figure 4.8

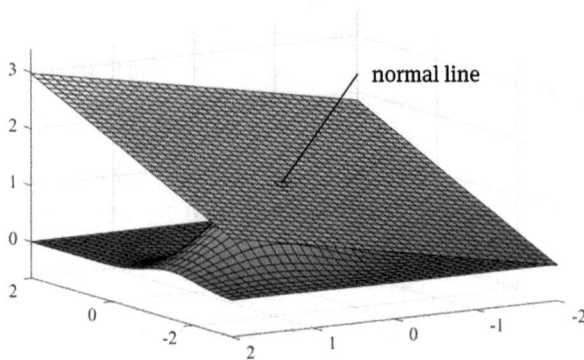

Figure 4.8: The tangent plane and the surface (after rotation).

```
>> syms x y; z(x,y)=(x^2-2*x)*exp(-x^2-y^2-x*y);
   zx=diff(z,x); zy=diff(z,y); x0=-1; y0=0.4; z0=1.403;
   fsurf(z,[-3 2,-2 2]); hold on
   zx0=double(zx(x0,y0)), zy0=double(zy(x0,y0))
   zt(x,y)=z0+zx0*(x-x0)+zy0*(y-y0), fsurf(zt,[-3 2 -2 2])
```

with $F_x = 0.3741$, $F_y = 0.2806$. The mathematical form of the tangent plane can be obtained as

$$z_t(x, y) = 1.403 + 0.3741(x + 1) + 0.2806(y - 0.4) = 0.3741x + 0.2806y + 1.6649.$$

The equation of the normal line is $x = -1 - 0.3741t$, $y = 0.4 - 0.2806t$, and $z = 1.403 + t$. The following statements can be used to superimpose normal lines on the tangent plane:

```
>> t=[0 2]; plot3(-1,0.4,1.403,'o')
   plot3(-1-0.3741*t,0.4-0.2806*t,1.403+t)
```

4.8 Exercises

4.1 Compute the fourth-order derivative of

$$y(t) = \sqrt{(x-1)(x-2)/[(x-3)(x-4)]}.$$

4.2 Compute the derivatives:

(1) $y(x) = \sqrt{x \sin x \sqrt{1 - e^x}}$, (2) $y = \dfrac{1 - \sqrt{\cos ax}}{x(1 - \cos \sqrt{ax})}$,

(3) $\operatorname{atan} \dfrac{y}{x} = \ln(x^2 + y^2)$, (4) $y(x) = -\dfrac{1}{na} \ln \dfrac{x^n + a}{x^n}$, $n > 0$.

4.3 Find the first-order derivatives to the following functions:[5]

(1) $y(t) = \arccos^2 x + (\ln^2 \arccos x - \ln \arccos x + 1/2)$,

(2) $y(t) = \dfrac{1}{2} \arctan \sqrt[4]{1 + x^4} + \dfrac{1}{4} \ln \dfrac{\sqrt[4]{1 + x^4} + 1}{\sqrt[4]{1 + x^4} - 1}$,

(3) $y(x) = \dfrac{e^{-x^2} \arcsin e^{-x^2}}{\sqrt{1 - e^{-2x^2}}} + \dfrac{1}{2} \ln(1 - e^{-2x^2})$.

4.4 Find the tenth-order derivative of

$$y(x) = \frac{1 - \sqrt{\cos ax}}{x(1 - \cos \sqrt{ax})}.$$

4.5 In calculus textbooks, the limit of fractions, where the numerator and denominator tend to zero or infinity simultaneously, L'Hôpital's rule can be used, i. e., we can take derivatives of the numerator and denominator separately, and then find the limit from their ratio. Compare the results from L'Hôpital's rule and the direct solutions for

$$\lim_{x \to 0} \frac{\ln(1 + x) \ln(1 - x) - \ln(1 - x^2)}{x^4}.$$

4.6 For the given parametric equations, compute $\dfrac{dy}{dx}$ and $\dfrac{d^2y}{dx^2}\Big|_{t=\pi/3}$:

$$\begin{cases} x = \ln \cos t, \\ y = \cos t - t \sin t. \end{cases}$$

4.7 Find first- and second-order derivatives for the parametric functions:

(1) $\begin{cases} x(t) = a(\ln \tan t/2 + \cos t - \sin t), \\ y(t) = a(\sin t + \cos t), \end{cases}$ (2) $\begin{cases} x(t) = 2at/(1 + t^3), \\ y = a(3at^2)/(1 + t^3). \end{cases}$

4.8 For the function $f(x) = x^2 a^x$ $(a > 0)$ derive and prove the formula of $f^{(n)}(x)$.

4.9 If $u = \arccos \sqrt{x/y}$, show that $\partial^2 u/(\partial x \partial y) = \partial^2 u/(\partial y \partial x)$.

4.10 Assuming that $u = xyze^{x+y+z}$, compute $\partial^{p+q+r}u/(\partial x^p \partial y^q \partial z^r)$. (Hints: diff() function cannot be used to find the pth order derivatives. Select different values for p, q, r, and summarize from the results the final solution.)

4.11 If $\begin{cases} xu + yv = 0 \\ yu + xv = 1, \end{cases}$ compute $\dfrac{\partial^2 u}{\partial x \partial y}$.

4.12 For the given $f(x,y) = \displaystyle\int_0^{xy} e^{-t^2} dt$, compute $\dfrac{x}{y}\dfrac{\partial^2 f}{\partial x^2} - 2\dfrac{\partial^2 f}{\partial x \partial y} + \dfrac{\partial^2 f}{\partial y^2}$.

4.13 Compute dy/dx, d^2y/dx^2, and d^3y/dx^3 from the following parametric functions:

(1) $x = e^{2t} \cos^2 t$, $y = e^{2t} \sin^2 t$,

(2) $x = \arcsin t/\sqrt{1+t^2}$, $y = \arccos t/\sqrt{1+t^2}$.

4.14 If $u(x,y) = x - y + x^2 + 2xy + y^2 + x^3 - 3x^2y - y^3 + x^4 - 4x^2y^2 + y^4$, compute

$$\dfrac{\partial^4 u(x,y)}{\partial x^4}, \quad \dfrac{\partial^4 u(x,y)}{\partial x^3 \partial y}, \quad \dfrac{\partial^4 u(x,y)}{\partial x^2 \partial y^2}.$$

4.15 Compute the divergence and curl of the following vector functions:

(1) $v(x,y) = [5x^2y - 4xy, 3x^2 - 2y]$, (2) $v(x,y,z) = [x^2y^2, 1, z]$,

(3) $v(x,y,z) = [2xyz^2, x^2z^2 + z\cos yz, 2x^2yz + y\cos yz]$.

4.16 Show that $\text{div}[\text{curl } v(x,y,z)] = 0$.

4.17 If $x^2 - xy + 2y^2 + x - y - 1 = 0$, compute the values of dy/dx, d^2y/dx^2, and d^3y/dx^3 at $x = 0$ and $y = 1$.

4.18 Compute the Jacobian matrix for

$$f(x,y,z) = \begin{bmatrix} 3x + e^y z \\ x^3 + y^2 \sin z \end{bmatrix}.$$

4.19 Compute the Laplacian operator for the function $u(x,y)$ in Example 4.14.

4.20 If $u = \ln\dfrac{1}{\sqrt{(x-\xi)^2 + (y-\eta)^2}}$, compute $\dfrac{\partial^4 u}{\partial x \partial y \partial \xi \partial \eta}$.

4.21 If $z = \psi(x^2 + y^2)$, compute $y\dfrac{\partial z}{\partial x} - x\dfrac{\partial z}{\partial y}$.

4.22 If $u = x\phi(x+y) + y\psi(x+y)$, compute $\dfrac{\partial^2 u}{\partial x^2} - 2\dfrac{\partial^2 u}{\partial x \partial y} + \dfrac{\partial^2 u}{\partial y^2}$.

4.23 If $z = F(r,\theta)$, where r and θ are functions of x and y, $x = r\cos\theta$, $y = r\sin\theta$, compute $\partial z/\partial x$ and $\partial z/\partial y$.

4.24 Consider the implicit function studied in Example 4.30. Compute $z^{(8)}(x)$, and also $x^{(8)}(z)$.

5 Integrals

Integral problems are inverse problems to those of differentiation. If a function $F(x)$ is known, the approaches studied in Chapter 4 can be used to find the derivative function $f(x)$. On the contrary, if a function $f(x)$ is known, how can we find $F(x)$? This is the essential problem to be studied in integral calculus.

In 1675, German mathematician Gottfried Wilhelm Leibniz in his manuscript used the sign $\int$ for the first time, since he thought integrals are the sums of an infinite number of infinitesimals. He used the stretched S (from the Latin word *summa* for sum) sign to denote integrals.

In classical integral calculus courses, computing an indefinite integral is a very challenging task. The students are required to memorize many formulas, and some of the approaches, such as variable substitution, integration by parts, and so on, should be mastered. Some improper and multiple integral problems may not be suitable to solve by hand. For a student, the solvability of certain integral problems may still depend upon the experience and skills, and, to some extent, luck. A more objective method will be introduced here, to guide the readers when sending the integral problem directly to computers. With the standard use of MATLAB, various integral problems can be solved.

In Section 5.1, the indefinite integral problems are addressed, including the concepts and solution methods. All the integral problems are sent to computers. In Section 5.2, definite integral problems are presented, and solutions to improper and infinite integral problems are considered. In Section 5.3, multiple integrals of multivariate functions are presented. In Section 5.4, some applications of definite integrals, including the arc length, volume, probability and Laplace transform computation, are demonstrated. In Sections 5.5 and 5.6, path and surface integrals will be thoroughly discussed. Since there were no relevant MATLAB functions, universal MATLAB functions will be written, so that path and surface integral problems can be solved directly with the universal functions by the author.

Although the integral computing facilities of MATLAB are very powerful, there is are indeed a significant number of functions whose integrals do not have analytical expressions. The numerical approach may be the only choice. These problems will be fully discussed in Chapter 8.

5.1 Indefinite integrals of univariate functions

If a function $F(x)$ is the derivative of a function $f(x)$, how can we restore $f(x)$ from $F(x)$? This is the so-called integral problem. In this section, indefinite integral is discussed. More importantly, computing integrals and even high-order integrals with MATLAB is addressed.

https://doi.org/10.1515/9783110666977-005

Definition 5.1. The indefinite integral of a given function $f(x)$ is expressed as

$$F(x) = \int f(x)dx, \tag{5.1.1}$$

where $f(x)$ is called the integrand, while $F(x)$ is termed the primitive function.

In calculus textbooks, it is a hard job to find indefinite integrals, since too many tactics and approaches are needed, and many skills are expected. For instance, the variable substitution method, the integration by parts method and many others are needed.

A MATLAB function `int()` is provided in the Symbolic Math Toolbox, and it can be used to find indefinite integrals for a given function, in the syntax of `F=int(f,x)`. If there is only one variable in the symbolic expression f, x in the statement can be omitted.

It should be noted that if $F(x)$ is the primitive function of $f(x)$, $F(x) + C$ forms a family of primitive functions, where C is an arbitrary constant.

Definition 5.2. Multiple integrals can be expressed as

$$F(x_1, \ldots, x_n) = \int \cdots \int f(x_1, \ldots, x_n)\, dx_n \cdots dx_1. \tag{5.1.2}$$

Multiple indefinite integrals can also be obtained with the nested use of the `int()` function. The format in `diff()` cannot be used directly by merely specifying the order of integrals. If many integrals are needed, loops can be used to implement the `int()` function. Based on different orders in the integration process, combinations of constants C_i and polynomials are needed to construct the family of primitive functions.

For integrable functions, a universal MATLAB Symbolic Math Toolbox function `int()` can be used to solve the trivial and heavy tasks involving indefinite integrals. The solutions to the integral problems can be obtained by the direct use of computers. For non-integrable functions, MATLAB cannot be used to find analytical solutions. Examples will be given in this section to demonstrate the solutions for indefinite integral problems.

Example 5.1. Consider the problem in Example 4.4. The first-order derivative of the function $f(x)$ can be obtained by `diff()` function. Take the indefinite integral of the result and see whether the original function can be restored.

Solutions. The original function should be entered into MATLAB first. The derivative and integral actions can be carried out with

```
>> syms x; y(x)=sin(x)/(x^2+4*x+3);
   y1=diff(y); y0=int(y1) % derivative then integral
```

It is found that the solution is $y_0(x) = \sin x/[2(x+1)] - \sin x/[2(x+3)]$. If fact, the actual primitive function is a family of functions given by

$$y_1(x) = \frac{\sin x}{2(x+1)} - \frac{\sin x}{2(x+3)} + C,$$

where C is an arbitrary constant, also known as undetermined constant.

If a point in the primitive function is known, e. g., when $x = 0$, $y_1 = 1$, then the undetermined constant can be determined uniquely such that the given point fits the equation. Substituting $x = 0$ into $y_1(x)$, it can be found that $C = 1$.

Taking the fourth-order derivative of the original function, and then computing the integral four times to get the result, the primitive function can be restored, namely, $\sin x/[(x+1)(x+3)]$, which is exactly the same as the original function. It can be seen that the result automatically generated with MATLAB is valid.

```
>> y4=diff(y,4); y0=int(int(int(int(y4))));
   simplify(y0) % 4th order integrals
```

If the "arbitrary constants" are involved, the proper family of primitive functions should be written as

$$F(x) = \frac{\sin x}{(x+1)(x+3)} + C_1 + C_2x + C_3x^2 + C_4x^3.$$

Now, the "undetermined constants" are not merely constants, an undetermined polynomial should be constructed. If the four undetermined coefficients are to be found, four known points should be provided.

Example 5.2. Show that

$$\int x^3 \cos^2 ax \, dx = \frac{x^4}{8} + \left(\frac{x^3}{4a} - \frac{3x}{8a^3}\right)\sin 2ax + \left(\frac{3x^2}{8a^2} - \frac{3}{16a^4}\right)\cos 2ax + C.$$

Solutions. The following MATLAB commands can be used to compute and simplify the integral

```
>> syms a x; f=simplify(int(x^3*cos(a*x)^2,x)) % direct integration
```

and the result is

$$f = \frac{1}{8a^4}(3\sin^2 ax + 2a^3x^3 \sin 2ax - 6a^2x^2 \sin^2 ax + 3a^2x^2 - 3ax \sin 2ax) + \frac{x^4}{8}.$$

Unfortunately, this result is different from that given on the right-hand side of the stated expression. Inputting the latter into MATLAB and simplifying the difference between them, it can be seen that the difference is $-3/(16a^4)$.

```
>> f1=x^4/8+(x^3/(4*a)-3*x/(8*a^3))*sin(2*a*x)+...
      (3*x^2/(8*a^2)-3/(16*a^4))*cos(2*a*x); % right side
   simplify(f-f1)                           % simplified difference
```

Although the two sides are not exactly equal, fortunately the difference is a constant $3/(16a^4)$. Since in forming the primitive function family, an arbitrary constant C is added, the difference can be included in C, hence the equation is proved.

Example 5.3. Compute the indefinite integral

$$I = \int \sin^3(x^2 + 1)^4 \cos(x^2 + 1)^4 (x^2 + 1)^3 x \, dx.$$

Solutions. If manual formulation is required, usually variable substitution method should be adopted.[24] However, if computers are involved, it is not necessary to specify the methods. No other tactics no skills are required. What is required is the direct use of the function int(). The following statements can be used:

```
>> syms x; F(x)=sin((x^2+1)^4)^3*cos((x^2+1)^4)*(x^2+1)^3*x;
   I=int(F,x)
```

and the results obtained is

$$I = \frac{\cos 4(x^2 + 1)^4}{256} - \frac{\cos 2(x^2 + 1)^4}{64} + C.$$

Example 5.4. Compute the indefinite integral

$$F = \int \frac{(2r - 1) \cos \sqrt{3(2r - 1)^2 + 6}}{\sqrt{3(2r - 1)^2 + 6}} \, dr.$$

Solutions. The following commands are sufficient for the problem:

```
>> syms r
   f(r)=(2*r-1)*cos(sqrt(3*(2*r-1)^2+6))/sqrt(3*(2*r-1)^2+6)
   F=int(f,r)
```

with the result being $F(r) = \frac{1}{6} \sin \sqrt{3(2r - 1)^2 + 6} + C.$

If a different constant C is selected, the family of primitive functions can be obtained as shown in Figure 5.1. It can be seen that the primitive functions are just translated up and down.

```
>> r=0:0.01:6; F1=double(F(r)); F2=[];
   for C=1:6, F2=[F2; F1+C]; end, plot(r,F2)
```

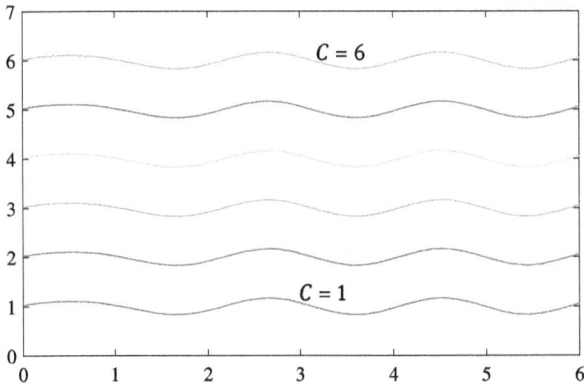

Figure 5.1: A family of primitive functions.

Example 5.5. Compute the complex integral $\int e^{-x-jx} \sin(7 + j2)x dx$.

Solutions. Complex integral problems can be solved in exactly the same way as are real integrals.

```
>> syms x; f(x)=exp(-x-1i*x)*sin((7+2i)*x); I=int(f,x)
```

The result obtained is

$$I = -e^{(-3+6j)x+4-14j}\left(\frac{4}{65} - \frac{1j}{130}\right) + \frac{1}{15} - \frac{1j}{30} + C.$$

Example 5.6. Compute the indefinite integral of $f(x) = e^{-x^2/2}$.

Solutions. In calculus textbooks, the integrand $f(x) = e^{-x^2/2}$ is always considered non-integrable in closed form. With MATLAB, the following commands can be used:

```
>> syms x; int(exp(-x^2/2)) % direct method tried
```

Although the original integrand is non-integrable in closed form, mathematicians invented a special function, known as the error function

$$\text{erf}(x) = \frac{2}{\sqrt{\pi}} \int_0^x e^{-t^2} dt,$$

such that the indefinite integral can be obtained as $\sqrt{\pi/2}\,\text{erf}(x/\sqrt{2})$.

Example 5.7. Try to solve the integral problem for $g(x) = x \sin(ax^4)e^{x^2/2}$.

Solutions. The integral problem can be tackled directly with MATLAB. Unfortunately, exactly the same command is returned, indicating nothing is done to solve the problem, which means that there is no solution to the original problem.

```
>> syms a x; int(x*sin(a*x^4)*exp(x^2/2))
```

It can be seen from the examples that the function `int()` can be tried directly to any integral problem, without bothering the low-level problems, such as which particular method or which tactic to use. The computers can be fully relied on. If solutions cannot be found when proper MATLAB commands are employed, it may probably mean that there is no analytical solution to the original problem. In this case, numerical ideas should be introduced.

5.2 Definite and improper integrals

In the indefinite integral problems, an integral may lead to a family of primitive functions. If an interval or integration region is specified, a dedicated primitive function is obtained. This type of integral problem is regarded as a definite integral problem. In this section, improper and multiple integral problems are also presented.

5.2.1 Definite integrals

Definition 5.3. The mathematical description of a definite integral is

$$I = \int_a^b f(x) \, dx. \tag{5.2.1}$$

Theorem 5.1 (Newton–Leibniz formula). *If a function $f(x)$ is continuous in the closed interval $[a, b]$, and the indefinite integral of $f(x)$ is $F(x) + C$, the definite integral can be obtained directly as*

$$\int_a^b f(x)dx = F(b) - F(a). \tag{5.2.2}$$

In real applications, sometimes the indefinite integrals may not exist, however, specific definite integrals or even infinite integrals can be found. For instance, special functions or numerical algorithms can be used to find the solutions.

Again, the function `int()` can be used to compute the definite or even infinite integrals, with the syntax I=`int`(f,x,a,b), where x is the independent variable, (a, b) is the integral for the definite integral, where a and b are allowed to be set to `-Inf` or `Inf`, if infinite integrals are expected. Function `vpa()` may sometimes be used to evaluate definite or even infinite integrals with high precision algorithms..

It should be noted that, although Newton–Leibniz formula was introduced in the section, it is purely due to the requirement for the completeness of mathematical formulation. If function `int()` is used in finding the definite integrals, there is no need to consider the use of Newton–Leibniz formula, since it is already embedded in the MATLAB solvers, and MATLAB may consider it for you.

Example 5.8. Compute the definite integral

$$I = \int_0^1 (y^3 + 6y^2 - 12y + 9)^{-1/2}(y^2 + 4y - 4)dy.$$

Solutions. No other tactic is needed, just type in the command

```
>> syms y; f(y)=(y^3+6*y^2-12*y+9)^(-1/2)*(y^2+4*y-4);
   I=int(f,y,0,1), vpa(I)
```

and the result will be obtained for you. However, for this particular example, the result is very complicated and cannot be simplified by the computer. Luckily, the function `vpa()` can be used, and it can be seen that the result is infinitely approaching $-2/3$. The curve of the integrand can also be obtained, as shown in Figure 5.2. The integral can physically be interpreted as the area enclosed by the integrand and the horizontal axis.

```
>> x0=0:0.01:1; f1=double(f(x0));
   x0=[0 x0 1]; f1=[0 f1 0]; fill(x0,f1,'g')
```

Since there exists a zero-crossing point in the integrand, the final area is a sum of the "positive area" and the "negative area". Also, due to the existence of the zero inside the radical, which is in fact in the denominator albeit not written, the integral is, in

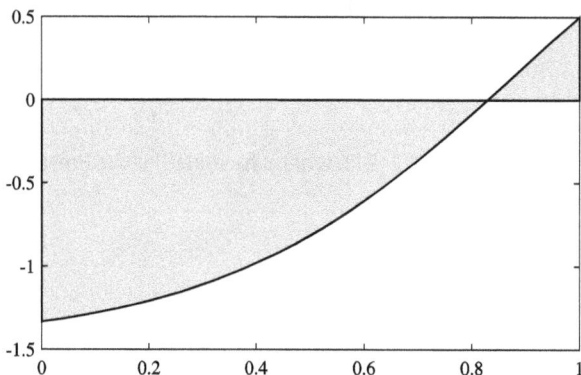

Figure 5.2: The area enclosed.

fact, an improper integral, although it may not be noticed if the integrand curve is not drawn. For this particular example, since the integrand is continuous, Newton–Leibniz formula can be used directly, and the accurate result of $-2/3$ is obtained.

```
>> I1=int(f,y); I1(1)-I1(0)
```

Example 5.9. Consider the definite integral of $f(x) = e^{-x^2/2}$. Compute the definite integral when $a = 0$, and $b = 1.5$ or ∞.

Solutions. The following statements can be used directly:

```
>> syms x; I1=int(exp(-x^2/2),x,0,1.5), I1a=vpa(I1),
   I2=int(exp(-x^2/2),x,0,inf)
```

with the solutions

$$I_1 = \sqrt{\pi/2}\,\mathrm{erf}(3\sqrt{2}/4), \quad I_{1a} = 1.0858533176660165697024.$$

The analytical value of the infinite integral is $I_2 = \sqrt{\pi/2}$.

Example 5.10. For the piecewise function in Example 4.11, which is provided below, compute $\int_{-1}^{1} f(x)dx$:

$$f(x) = \begin{cases} x^2 + \sin x^2, & x \leq 0, \\ \ln(1+x), & x > 0. \end{cases}$$

Solutions. The piecewise function should be fed into MATLAB environment, then the `int()` function can be called to compute the definite integral

```
>> syms x;
   f(x)=piecewise(x<=0,x^2+sin(x^2), x>0,log(1+x));
   I=int(f,x,-1,1)
```

The result obtained is $\ln 4 + \sqrt{2\pi}\,\mathrm{fresnels}(\sqrt{2/\pi})/2 - 2/3$, where, fresnels(·) is the Fresnels sinusoidal special function, defined as

$$\mathrm{fresnels}(x) = \int_0^x \sin\left(\frac{1}{2}\pi\tau^2\right)d\tau.$$

An illustration of the area enclosed by the integrand is shown in Figure 5.3, where the total area is the definite integral of the piecewise function.

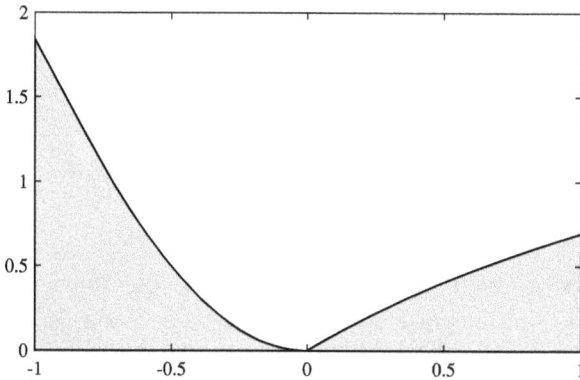

Figure 5.3: The area enclosed by the integrand and x axis.

5.2.2 Infinite and improper integrals

It was always assumed in the previous section that the integrand function is continuous in the interval (a, b) or the interval is finite, therefore, Newton–Leibniz formula applies. Here, improper integral problems will be studied.

Definition 5.4. If the boundary a or b in (5.2.1) is infinite, this kind of definite integral is referred to as an infinite integral, also known as an improper integral with infinite boundary.

In mathematics, for the upper infinite boundary problems, one has to assume that the upper boundary is b. When the integral is obtained, the following limit is taken

$$\int_a^\infty f(x)\, dx = \lim_{b\to\infty} \int_a^b f(x)\, dx, \tag{5.2.3}$$

however, if MATLAB is used, it is not necessary to do so.

Definition 5.5. If a function $f(x)$ is continuous in the interval (a, b), except for point c, and $\lim_{x\to c} |f(c)| = \infty$, its integral in this interval is referred to as an improper integral.

The total interval can be divided into two subintervals excluding the singularity c, namely (a, c_1) and (c_2, b). Therefore, the improper integral can be computed through

$$\int_a^b f(x)\, dx = \lim_{c_1\to c^-} \int_a^{c_1} f(x)\, dx + \lim_{c_2\to c^+} \int_{c_2}^b f(x)\, dx. \tag{5.2.4}$$

These rules have already been embedded into MATLAB function int(). It is not necessary to consider the low-level issues such as singularities and single-sided limits. The integral capability in MATLAB is significantly simplified.

Example 5.11. Compute the improper integral $\displaystyle\int_1^\infty \frac{1}{x^p}dx$.

Solutions. For the simplicity of presentation, p can be assumed to be real, therefore, the following statements can be used to solve the integral problem:

```
>> syms p real, syms x; I=int(1/x^p,x,1,inf)
```

The result obtained is a piecewise function, which can be interpreted as follows. If $p \leqslant 1$, the integral is ∞; if $p > 1$, the integral is $I = 1/(p-1)$. It can be seen from the example that the function int() can be used directly, and low-level trivial issues can be completely ignored. Computers can solve this problem for you.

Example 5.12. Compute the improper integral $\displaystyle\int_1^{2e} \frac{1}{x\sqrt{1-\ln^2 x}}dx$.

Solutions. It can be seen that at $x = e$, the integrand is not continuous, therefore, the integral is an improper integral. This problem can be solved directly, with the result being $\arcsin(\ln 2 + 1) \approx 1.5708 - 1.1182j$.

```
>> syms x; f(x)=1/x/sqrt(1-log(x)^2);
   I=int(f,x,1,2*exp(sym(1))), vpa(I) %direct computation
```

Again, it is not necessary to consider low-level issues, for instance, the complex quantities in the interval $x \in (e, 2e)$. An ordinary call to int() function is adequate.

In fact, if a set of samples is selected, the curves of the integrand and integral can both be obtained, as shown in Figure 5.4. It can be seen that at $x = e$, there are sudden jumps in the real and imaginary parts of the integrand, while the integral has conver-

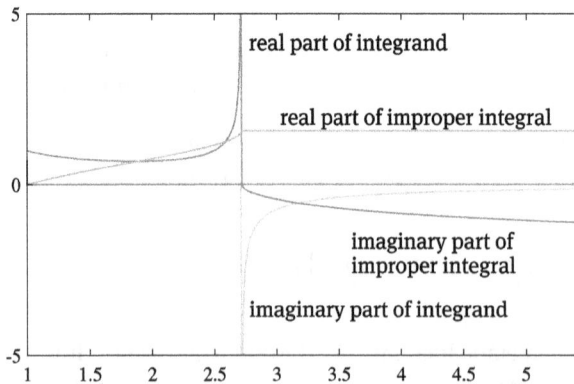

Figure 5.4: The curves of improper integral and primitive function.

gent real and imaginary parts. Therefore, correct results for the improper integral can be obtained.

```
>> syms t; x0=1:0.005:2*exp(2); f1=f(x0);
   I1(t)=int(f,x,1,t); f2=I1(x0); f1=double(f1); f2=double(f2);
   plot(x0,real(f1),x0,real(f2),x0,imag(f1),x0,imag(f2))
   axis([1 2*exp(1) -5 5])
```

Example 5.13. Compute the definite integral $I(t) = \displaystyle\int_{\cos t}^{e^{-2t}} \dfrac{-2x^2 + 1}{(2x^2 - 3x + 1)^2}\,dx$ with functional boundaries.

Solutions. The function int() can be used directly to solve definite integral problems, even though one has functional boundaries. The following MATLAB commands can be used:

```
>> syms x t; f=(-2*x^2+1)/(2*x^2-3*x+1)^2;
   I0=simplify(int(f,x,cos(t),exp(-2*t)))
```

and the result obtained is a piecewise function. The most important conclusion in the result is: when the conditions

$$\left(1 < \cos t \text{ or } -1 \leqslant e^{-2t}\right) \quad \text{and} \quad \left(\frac{1}{2} < \cos t \text{ or } -\frac{1}{2} \leqslant e^{-2t}\right)$$

are satisfied, there is an analytical expression for the integral given by

$$I_0 = -\frac{(e^{-2t} - \cos t)(2e^{-2t}\cos t - 1)}{(\cos t - 1)(2\cos t - 1)(e^{-2t} - 1)(2e^{-2t} - 1)}.$$

It seems that the conclusion obtained is rather complicated. In older versions of MATLAB, I_0 is returned directly, and no conditions are given. In fact, if the conditions are not given, the solution obtained is not complete. Since the boundaries are functions of t, and there exist poles in the integrand, at $x = -1/2$ and $x = -1$, the relationship of the boundary functions and poles should be considered. If the poles are enclosed, the integral problem is in fact an improper one, and Newton–Leibniz formula cannot be applied directly.

For this particular problem, the internal strategy taken by the computer is: Find out first the intervals where Newton–Leibniz formula holds. The above conditions can then be established automatically. Within the intervals, Newton–Leibniz formula can then be used to find the solution. Hence, the results obtained above can be generated, and they are essentially due to Newton–Leibniz formula.

5.3 Multiple integrals

Multiple integrals can also be computed directly, with the sophisticated `int()` function. The order of integrals should be selected first, and normally, the easier integrable part can be placed into the inner integrals, while complicated parts can be placed as outer integrals. In each of the integrals, `int()` function is applied. If there is no analytical solution found, even though proper order of integrals is arranged, there is probably no analytical solution to the original problem. In this case, numerical techniques may be introduced.

5.3.1 Multiple indefinite integrals

The mathematical form of multiple integrals has been presented in Definition 5.2. The nested use of `int()` can be applied to compute multiple indefinite integrals. MATLAB-based solution strategy is introduced in this section for multiple integral problems.

Example 5.14. Consider the 3D function $F(x, y, z)$ as given below

$$F(x, y, z) = -4ze^{-x^2y-z^2}(\cos x^2y - 10yx^2 \cos x^2y$$
$$+ 4x^4y^2 \sin x^2y + 4x^4y^2 \cos x^2y - \sin x^2y).$$

Compute $\iiint F(x, y, z)\mathrm{d}x^2\mathrm{d}y\mathrm{d}z$.

Solutions. As a matter of fact, this integrand was the result in Example 4.16, when partial derivatives were taken. The inverse process can be introduced, to see whether the original function $f(x, y, z)$ can be restored.

Now let us start the integration process. To compute multiple integrals, the order of integrals should be considered first. An integral with respect to z can be computed once, then with respect to y, and finally, twice with respect to x, the result through simplification can be obtained as $f_1 = e^{-x^2y-z^2} \sin x^2y$, and it can be seen that the result is exactly the same as the original function given in Example 4.16.

```
>> syms x y z C1 C2 C3 C4;
   f0=-4*z*exp(-x^2*y-z^2)*(cos(x^2*y)-10*cos(x^2*y)*y*x^2+...
      4*sin(x^2*y)*x^4*y^2+4*cos(x^2*y)*x^4*y^2-sin(x^2*y)); % integrand
   f1=int(f0,z)+C1; f1=int(f1,y)+C2; f1=int(f1,x)+C3;
   f1=simplify(int(f1,x)+C4) % compute multiple integral
```

It can also be seen that, when the order of integration is considered, the family of the primitive functions can be written as

$$f_1 = e^{-x^2y-z^2} \sin x^2y + C_1x^2y/2 + C_2x^2/2 + C_3x + C_4.$$

If the order of integration is changed to $z \to x \to x \to y$, the same result can also be obtained

```
>> f2=int(f0,z)+C1; f2=int(f2,x)+C2; f2=int(f2,x)+C3;
   f2=simplify(int(f2,y)+C4) % different orders in the integral
```

If the new order is applied, the family of primitive functions becomes

$$f_2 = e^{-x^2 y - z^2} \sin x^2 y + C_1 x^2 y/2 + C_2 xy + C_3 y + C_4.$$

Although the main primitive is returned exactly the same, the undetermined polynomials are completely different.

Example 5.15. Compute the genuine multiple integral given bellow

$$\iiiint\int xy^2 zuw^4 e^{-x-2y-3z-4u-5w} \, dw \, du \, dz \, dy \, dx.$$

Solutions. It seems that this kind of integral cannot be computed easily by hand. We can send the problem to a computer directly and let it solve for us.

```
>> syms w u x y z
   f=x*y^2*z*u*w^4*exp(-x-2*y-3*z-4*u-5*w);
   I=int(int(int(int(int(f,w),u),z),y),x)
```

The result obtained by the computer is

$$I = -e^{-x-2y-3z-4u-5w}(4u+1)(3z+1)(x+1)$$
$$(2y^2 + 2y + 1)(625w^4 + 500w^3 + 300w^2 + 120w + 24)/1800000.$$

If undetermined polynomials are expected, an order for the integration process can be selected. Then the following MATLAB commands can be given:

```
>> syms C1 C2 C3 C4 C5
   I=int(int(int(int(int(f,w)+C1,u)+C2,z)+C3,y)+C4,x)+C5
```

Under the selected integration order, the undetermined polynomial is $C_5 + x(C_4 + C_3 y + C_2 yz + C_1 uyz)$, i. e., $C_1 uxyz + C_2 xyz + C_3 xy + C_4 x + C_5$. Different integration orders may lead to different undetermined polynomials.

5.3.2 Constructions of undetermined polynomials

It has been indicated in the previous examples that, when different integration orders are selected, the undetermined polynomials obtained may be different. For instance, two different undetermined polynomials, f_1 and f_2, are obtained in Example 5.14. The new question to be considered is which of f_1 and f_2 is correct?

How can we construct the complete set of undetermined polynomials? There seems to be no discussion in the calculus textbooks, and for a multiple integral problem, this question seems inevitable, since the integrals should be independent of the order of integration. We shall try to answer this question now.

Example 5.16. Consider again the integral of a 2D function $f(x, y)$. Construct a correct undetermined polynomial and propose guidelines for similar multiple integral problems.

Solutions. If a 2D function $F(x, y)$ is defined, the integral with respect to first x then y may lead to an undetermined polynomial $C_1 x + C_2$, while taken first with respect to y and then with respect to x leads to $C_1 y + C_2$. Neither of them is complete. The complete one should be $C_1 x + C_2 y + C_3$ since the derivative $\partial^2 F / \partial z \partial y$ eliminates the undetermined polynomial completely.

For multivariate multiple integral problems, a similar method can be used to construct the complete undetermined polynomial. For instance, for the problem in Example 5.14, two integrals with respect to x are computed, and one with respect to y and z, respectively. From the four quantities x, x, y, and z, all third-, second-, and first-order combinations should be considered. Finally, the complete undetermined polynomial can be constructed as

$$\underbrace{C_1 x^2 y + C_2 x^2 z + C_3 xyz}_{\text{all 3rd-order terms}} + \underbrace{C_4 x^2 + C_5 xy + C_6 xz + C_7 yz}_{\text{all 2nd-order terms}} + \underbrace{C_8 x + C_9 y + C_{10} z}_{\text{all 1st-order terms}} + C_{11}.$$

Now, we are able to answer the previous question. Both f_1 and f_2 are correct, however, they are both incomplete families. The undetermined polynomial constructed above provides a complete family of primitive functions.

It is also noticed that in Example 5.14, in order to find the undetermined constants, four known points are required to find them, however, the solution is not unique. For instance, in $C_1 x^2 y / 2 + C_2 xy + C_3 y + C_4$, the other six undetermined constants have already been set to zero. If one wants to uniquely establish a complete undetermined polynomial, ten known points should be provided.

5.3.3 Computation of multiple integrals

Similar to the cases in univariate integral problems, if the region of integration is specified, the multiple indefinite integral problems may become definite integral problems. Here double, triple, and even multiple definite integrals are demonstrated.

Definition 5.6. The standard form of a double definite integral is

$$I = \int_{x_m}^{x_M} \int_{y_m(x)}^{y_M(x)} f(x,y)\,dydx. \qquad (5.3.1)$$

Definition 5.7. Ordinary triple integral is described as

$$I = \int_{x_m}^{x_M} \int_{y_m(x)}^{y_M(x)} \int_{z_m(x,y)}^{z_M(x,y)} f(x,y,z)\,dzdydx. \qquad (5.3.2)$$

In the mathematical presentations it can be seen that, even if the analytical forms to inner integrals exist, difficulties may be encountered when computing outer integrals, so that the original integral problem may not be solvable. The integration functions in MATLAB can be tried. If there is no solution found, pure numerical algorithms should be used. These topics will be further explored in Chapter 8.

Example 5.17. Compute the double integral

$$J = \int_{-1}^{1} \int_{-\sqrt{1-y^2}}^{\sqrt{1-y^2}} e^{-x^2/2} \sinh(x^2 + y)\,dxdy.$$

Solutions. The conventional method can be tried with MATLAB. Care must be taken to match the integration order and integral boundaries. If all these are assigned properly, you can just wait for the final results.

```
>> syms x y; f(x,y)=exp(-x^2/2)*sinh(x^2+y);
   I=int(int(f,x,-sqrt(1-y^2),sqrt(1-y^2)),y,-1,1),
   tic, I1=vpa(I), toc
```

Unfortunately, there is no analytical solution to the problem. The function vpa() can further be used to try to find a high precision solution, and the result is $I_1 = 0.70412133490335689947800312022517$. The variable precision solutions are obtained from the symbolic mechanism inside the Symbolic Math Toolbox, and sometimes it is very time-consuming. For this particular example, 354 seconds are needed to find the above result.

Example 5.18. Compute the triple integral $\int_0^2 \int_0^\pi \int_0^\pi 4xze^{-x^2y-z^2}\,dzdydx$.

Solutions. The following command can be used directly:

```
>> syms x y z;
   I=int(int(int(4*x*z*exp(-x^2*y-z^2),x,0,2),y,0,pi),z,0,pi)
```

It is seen that the result is $-(\mathrm{e}^{-\pi^2} - 1)(\gamma + \ln(4\pi) - \mathrm{Ei}(-4\pi))$, where γ is the Euler constant, $\mathrm{Ei}(z)$ is a special function, defined as $\mathrm{Ei}(z) = \displaystyle\int_{-\infty}^{z} \mathrm{e}^{-t} t^{-1}\,\mathrm{d}t$. Although the function is non-integrable in closed-form, high precision numerical result is obtained with vpa(ans), as 3.10807940208541272283461464767l4.

Example 5.19. Compute the following multiple integral:

$$\int_0^1 \int_0^2 \int_{-3}^3 \int_0^4 \int_0^5 xy^2zuw^4 \mathrm{e}^{-x-2y-3z-4u-5w}\,\mathrm{d}w\mathrm{d}u\mathrm{d}z\mathrm{d}y\mathrm{d}x.$$

Solutions. The solution can be found directly with the commands

```
>> syms w u x y z
   f=x*y^2*z*u*w^4*exp(-x-2*y-3*z-4*u-5*w);
   I=int(int(int(int(int(f,w,0,5),u,0,4),z,-3,3),y,0,2),x,0,1)
   I1=vpa(I)
```

and the result obtained is

$$
\begin{aligned}
I = \;& \mathrm{e}^{-9}(2\mathrm{e}^{-1} - 1)/7\,500 - 17\mathrm{e}^{-7}(2\mathrm{e}^{-1} - 1)/9\,375 - 13\mathrm{e}^{5}(2\mathrm{e}^{-1} - 1)/9\,375 \\
& + \mathrm{e}^{9}(2\mathrm{e}^{-1} - 1)/9\,375 + 221\exp^{-11}(2\mathrm{e}^{-1} - 1)/9\,375 \\
& - 13\mathrm{e}^{-13}(2\mathrm{e}^{-1} - 1)/7\,500 - 461\,249\mathrm{e}^{-16}(2\mathrm{e}^{-1} - 1)/225\,000 \\
& + 5\,996\,237\mathrm{e}^{-20}(2\mathrm{e}^{-1} - 1)/225\,000 - 17\mathrm{e}^{-25}(2\mathrm{e}^{-1} - 1)/7\,500 \\
& + 221\mathrm{e}^{-29}(2\mathrm{e}^{-1} - 1)/7\,500 + 7\,841\,233\mathrm{e}^{-32}(2\mathrm{e}^{-1} - 1)/225\,000 \\
& - 461\,249\mathrm{e}^{-34}(2\mathrm{e}^{-1} - 1)/180\,000 - 101\,936\,029\mathrm{e}^{-36}(2\mathrm{e}^{-1} - 1)/225\,000 \\
& + 5\,996\,237\mathrm{e}^{-38}(2\mathrm{e}^{-1} - 1)/180\,000 + 7\,841\,233\mathrm{e}^{-50}(2\mathrm{e}^{-1} - 1)/180\,000 \\
& - 101\,936\,029\mathrm{e}^{-54}(2\mathrm{e}^{-1} - 1)/180\,000.
\end{aligned}
$$

High-precision solution $I_1 = -0.17401016114421936736883805l8$ can also be found.

5.3.4 Conversions of integration regions

If the integration region bounds of a multiple integral can be expressed using expressions in (5.3.1), the integral problem can easily be solved with the nested use of function int(). Integration region conversion problems are solved with examples.

Example 5.20. Solve the following double integral problem:

$$I = \int\int_{|x|+|y|\leq 1} (|x| + |y|)\mathrm{d}x\mathrm{d}y.$$

Solutions. The condition $|x| + |y| \leqslant 1$ can be converted to $|y| \leqslant 1 - |x|$ and $-1 \leqslant x \leqslant 1$, while $|y| \leqslant 1 - |x|$ condition can further be converted as $y \leqslant 1 - |x|$ and $y > -(1 - |x|)$. Therefore, the original integral problem can further be converted to

$$I = \int_{-1}^{1} \int_{-(1-|x|)}^{1-|x|} (|x| + |y|) \, dy \, dx.$$

From the standard form of the double integral, the analytical solution can be obtained with the following statements, with the result being $I = 4/3$:

```
>> syms x y; f(x,y)=abs(x)+abs(y);    % symbolic variables and integrand
   I=int(int(f,y,-(1-abs(x)),1-abs(x)),x,-1,1) % double integral
```

5.4 Applications of definite integrals

In this section, several applications of definite integrals will be presented. The computation of arc length of curves will be presented first, followed by the volume and probability computation. Finally, an introduction to integral transforms is presented.

5.4.1 Computation of arc length

There are several ways to describe a curve in Cartesian coordinates, for example, using parametric or polar equations. The arc length under these representations is summarized below.

Theorem 5.2. *If the curve is described by $x = f(y)$, with $y_m \leqslant y \leqslant y_M$, the arc length can be evaluated from*

$$s = \int_{y_m}^{y_M} \sqrt{1 + (df(y)/dy)^2} \, dy. \tag{5.4.1}$$

Theorem 5.3. *If the curve is described by $y = f(x)$, with $x_m \leqslant x \leqslant x_M$, the arc length can be evaluated from*

$$s = \int_{x_m}^{x_M} \sqrt{1 + (df(x)/dx)^2} \, dx. \tag{5.4.2}$$

Theorem 5.4. *If parametric equations $x = x(t)$, $y = y(t)$, $z = z(t)$ are given, with $t_m \leqslant t \leqslant t_M$, the arc length can be computed from*

$$s = \int_{t_m}^{t_M} \sqrt{(dx/dt)^2 + (dy/dt)^2 + (dz/dt)^2} \, dt. \tag{5.4.3}$$

Theorem 5.5. *If the curve is represented using a polar equation $\rho = \rho(\theta)$, with $\theta_m \leqslant \theta \leqslant \theta_M$, the arc length can be computed from*

$$s = \int_{\theta_m}^{\theta_M} \sqrt{\rho^2 + (d\rho/d\theta)^2} \, d\theta. \qquad (5.4.4)$$

Example 5.21. Assume that a function $y = 4\sqrt{2}x^{3/2}/3 - 1$ is given, with $-1 \leqslant x \leqslant 1$. Compute the arc length.[24]

Solutions. Using the formula in Theorem 5.3, the explicit function of the curve should be entered first. Then, derivatives and integrals can be computed, and finally, the arc length can be found as $L = 13/6$.

```
>> syms x positive; y(x)=4*sqrt(2)*x^(3/2)/3-1;
   L=int(sqrt(1+diff(y,x)^2),x,0,1)
```

In fact, the curve can be obtained as shown in Figure 5.5. It can be seen that at point x, a differential dx is created, with the differential of the function being d$y = f'(x)$dx. Using the Pythagorean theorem, the differential of the arc is computed as

$$ds = \sqrt{(dx)^2 + (dy)^2} = \sqrt{1 + (f'(x))^2}\,dx.$$

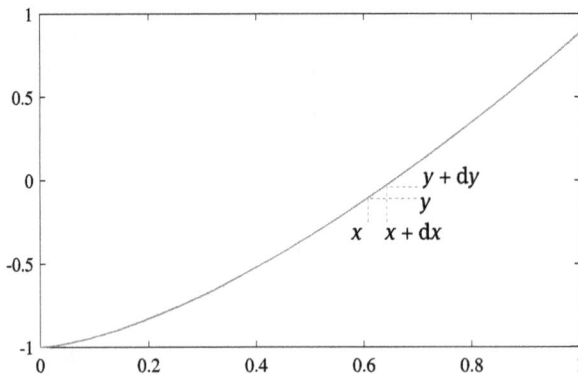

Figure 5.5: The curve of the given function.

If the differential of the arc is added up, the total arc length can be computed from (5.4.2).

```
>> x0=0:0.01:1; y0=y(x0); plot(x0,y0)
```

Example 5.22. Assume that the curve is defined by the polar function $\rho = a\sin^3\theta/5$, $0 \leqslant \theta \leqslant 3\pi$, $a > 0$. Compute the arc length.

Solutions. The variable t_0 is used to denote θ, and it should be set as a real symbolic variable. The following statements can be used to evaluate the arc length, found to be $s = 3a\pi/2$:

```
>> syms t0 real; syms a positive; r=a*sin(t0/3)^3;
   F=simplify(sqrt(r^2+diff(r,t0)^2)); s=int(F,t0,0,3*pi)
```

5.4.2 Computation of volume

Theorem 5.6. *For a given function* $y = f(x)$, *with* $x \in (a, b)$, *the volume spanned by the curve rotated around the x axis can be computed from*

$$V = \int_a^b \pi f^2(x)\,dx. \qquad (5.4.5)$$

Example 5.23. For a given function $y = f(x) = 1 + x\sin 4/x$, with $0 \leqslant x \leqslant \pi$. Rotate the curve $360°$ around the x axis. Compute the volume enclosed.

Solutions. The curve of the function can be drawn first, as shown in Figure 5.6. We can try to compute the volume enclosed through (5.4.5). Unfortunately, the analytical solution cannot be obtained with int(), not even the variable precision solution with vpa().

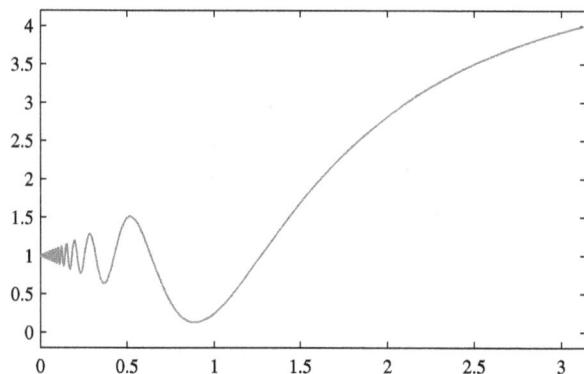

Figure 5.6: The curve of the given function.

```
>> syms x; f(x)=1+x*sin(4/x); fplot(f,[0,pi])
   V=int(pi*f^2,x,0,pi), vpa(V)
```

Finding a numerical solution to this problem is left for Chapter 8, where numerical integral computation problems are fully addressed. Here, only the commands and results are given, without further explanations, yielding $V \approx 57.5928$.

```
>> f=@(x)pi*(1+x.*sin(4./x)).^2; V=integral(f,0,pi,'RelTol',1e-10)
```

MATLAB function `cylinder()` can be used to draw a cylindrical surface. The default surface is obtained by rotating a curve about the z axis. The range of z is set in the interval $(0,1)$, and the x axis is regarded as the function, i.e., x is physically the z axis. Necessary data can be generated for `cylinder()` function, then data processing will be carried out, and finally, the surface plot for the problem can be constructed, as shown in Figure 5.7. It can be seen that the required volume is enclosed by the surface.

```
>> z=0:0.03:pi; x=1+z.*sin(4./z);      % create data for the function
   [x0,y0,z0]=cylinder(x); z0=pi*z0;  % create cylinder data and stretch
   x1=z0; z1=x0; surf(x1,y0,z1)        % swap and draw surface
```

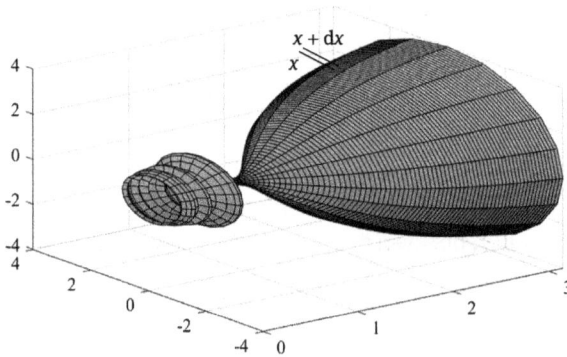

Figure 5.7: The 3D representation of a surface.

It can be seen from the 3D surface that a very thin slice at x, with a height of dx can be generated, and the slice can be regarded as a cylinder, with radius of $f(x)$. The differential of the volume of the slice is $\pi f^2(x)dx$. Adding up such volume, it can be seen that the whole volume can be evaluated from (5.4.5). This is the physical explanation of Theorem 5.6.

5.4.3 Volume and mass computation

The volume to be studied here is not the same as that presented in Section 5.4.2, where the volume was spanned by a curve rotated around a certain axis. The volume here is enclosed by three-dimensional surfaces.

Definition 5.8. The mathematical description of the volume is

$$I = \iiint_V dxdydz. \tag{5.4.6}$$

The key point here is how to convert the triple integral into standard form of the triple integral with explicit bounds for the three independent variables. The difficulties also lie in the problem of how to convert V into corresponding bounds. A simple example is given here to demonstrate the computation method.

Example 5.24. Consider a sphere centered at the origin, with radius r. Compute the volume of the hemisphere if $z \geqslant 0$.

Solutions. For this particular example, the integration region can be manually rewritten as

$$-\sqrt{r^2 - y^2 - z^2} \leqslant x \leqslant \sqrt{r^2 - y^2 - z^2}, \quad -\sqrt{r^2 - z^2} \leqslant y \leqslant \sqrt{r^2 - z^2}, \quad 0 \leqslant z \leqslant r,$$

where $r > 0$. Letting the integrand be 1, the volume can be expressed as

$$V = \int_0^r \int_{-\sqrt{r^2-z^2}}^{\sqrt{r^2-z^2}} \int_{-\sqrt{r^2-y^2-z^2}}^{\sqrt{r^2-y^2-z^2}} dx dy dz. \tag{5.4.7}$$

From the standard triple integral formula, the volume of the hemisphere can be evaluated with the following commands:

```
>> syms x y z; syms r positive
   int(int(int(sym(1),x,-sqrt(r^2-y^2-z^2),sqrt(r^2-y^2-z^2)),...
         y,-sqrt(r^2-z^2),sqrt(r^2-z^2)),z,0,r)
```

The volume is $V = 2\pi r^3/3$. Since it is known that the volume is $4\pi r^3/3$, it can be seen that the MATLAB solution is correct.

Example 5.25. Assume that in the hemisphere, the density at point (x, y, z) is $d(x, y, z) = |xyz|e^{-(x^2+y^2+z^2)}$, compute the mass of the hemisphere.

Solutions. The formula in (5.4.7) is the volume of a hemisphere, which is also the mass, if the density at any point inside the hemisphere is 1. If the density it not 1 but a given three-dimensional function $d(x, y, z)$, the mass of the hemisphere can be computed from

$$M = \int_0^r \int_{-\sqrt{r^2-z^2}}^{\sqrt{r^2-z^2}} \int_{-\sqrt{r^2-y^2-z^2}}^{\sqrt{r^2-y^2-z^2}} |xyz|e^{-(x^2+y^2+z^2)} dx dy dz. \tag{5.4.8}$$

With the standard triple integral presented above, the mass can easily be evaluated with the following statements:

```
>> syms x y z; syms r positive
   d(x,y,z)=abs(x*y*z)*exp(-(x^2+y^2+z^2));
   int(int(int(d,x,-sqrt(r^2-y^2-z^2),sqrt(r^2-y^2-z^2)),...
       y,-sqrt(r^2-z^2),sqrt(r^2-z^2)),z,0,r)
```

The mass obtained is $M = -e^{-r^2}(2r^2 - 2e^{r^2} + r^4 + 2)/4$.

5.4.4 Computation of probability distribution

Definition 5.9. If the probability density function $p(x)$ of a stochastic variable is known, the probability of the variable ξ falling into the interval $(-\infty < \xi \leqslant x)$ can be evaluated by the integral

$$P(\xi \leqslant x) = \int_{-\infty}^{x} p(y)dy. \tag{5.4.9}$$

The interval can also be simply denoted as $\xi \leqslant x$.

Definition 5.10. If the joint probability density function for a multivariate stochastic variable ξ is given by $p(x_1, x_2, \ldots, x_n)$, the probability of a stochastic variable falling into the region $(-\infty < \xi_i \leqslant x_i)$ can be computed from

$$P(\xi_1 \leqslant x_1, \ldots, \xi_n \leqslant x_n) = \int_{-\infty}^{x_1} \cdots \int_{-\infty}^{x_n} p(y_1, \ldots, y_n)\, dy_n \cdots dy_1. \tag{5.4.10}$$

Example 5.26. Given the joint probability density function of a 2D stochastic variable (ξ, η) defined as follows, compute $P(\xi < 1/2, \eta < 1/2)$:

$$p(x,y) = \begin{cases} x^2 + \dfrac{xy}{3}, & 0 \leqslant x \leqslant 1, 0 \leqslant y \leqslant 2, \\ 0, & \text{otherwise.} \end{cases}$$

Solutions. If the probability density function $p(x,y)$ is given, definite integrals can be computed, and the probability $P(\xi < 1/2, \eta < 1/2)$ can be found as $5/192$.

```
>> syms x y; f(x,y)=x^2+x*y/3;
   P=int(int(f,x,0,1/2),y,0,1/2)  % compute from the joint pdf
```

In theory, the lower bounds should be $-\infty$, rather than 0. Since the value of the integrand is zero when x or y is zero, these parts of the integral are omitted and the final integral can be found as above.

5.4.5 An introduction to integral transforms

Laplace transform is a useful tool in a variety of fields. The ideas of this integral transform will be demonstrated by an example.

Definition 5.11. For a time domain function $f(t)$, its Laplace transform is defined as

$$F(s) = \int_0^\infty f(t)e^{-st}dt. \qquad (5.4.11)$$

A more formal exposition of Laplace and other integral transforms will be presented in Chapter 9. Here, only one example is used to show Laplace transform techniques.

Example 5.27. For a given function $f(t) = t^3 \sin\left(4t + \frac{\pi}{4}\right)$, compute

$$F(s) = \int_0^\infty f(t)e^{-st}dt.$$

Solutions. Assuming s is a positive symbolic variable, the following commands can be used to compute the Laplace transform for function $f(t)$:

```
>> syms t; syms s positive; f(t)=t^3*sin(4*t+pi/4)
   F=int(f*exp(-s*t),t,0,inf)
```

The result obtained is

$$F = \frac{3\sqrt{2}(s^4 + 16s^3 - 96s^2 - 256s + 256)}{(s^2 + 16)^4}.$$

In fact, there is no need to constrain s in Laplace transforms as a positive variable. However, there are problems in computing Laplace transforms from its definition, since the results cannot be found if otherwise stated.

5.5 Path integrals

Extending the arc length and volume computation problems, the concepts of path and surface integrals can be proposed. In the following two sections, the related problems are presented, and the methods to convert them into ordinary integral problems are also provided. There is no built-in path or surface integral solver in MATLAB. Attempts will be made to propose universal solvers for these problems.

5.5.1 Type I path integral

Path integral problems are usually classified as type I and II path integral problems. Type I path integral problem is originated from the total mass computation problem of a spatial arc, with variable density.[3]

Definition 5.12. Assume that the density of a spatial arc l can be expressed as $f(x,y,z)$, then the total mass can be formulated as

$$I_1 = \int_l f(x,y,z)ds, \qquad (5.5.1)$$

where ds is the differential of arc length at a certain point. This type of path integral is also known as the integral with respect to arc length.

Theorem 5.7. *If the curve is described by parametric equations $x = x(t)$, $y = y(t)$, $z = z(t)$, they can be substituted in the integrand $f(\cdot)$. Also the differential of the arc length can be represented as*

$$ds = \sqrt{(dx/dt)^2 + (dy/dt)^2 + (dz/dt)^2}\, dt, \qquad (5.5.2)$$

which is simply denoted as $ds = \sqrt{x_t^2 + y_t^2 + z_t^2}\, dt$. The path integral problem can be converted into an ordinary definite integral with respect to t, namely

$$I = \int_{t_m}^{t_M} f(x(t),y(t),z(t))\sqrt{x_t^2 + y_t^2 + z_t^2}\, dt. \qquad (5.5.3)$$

It can be seen that the arc length formula in Theorem 5.2 is only a special case of type I path integral, where the density of the curve equals 1 at any point.

If the integrand is a 2D function $f(x,y)$, a similar method can be applied to convert the problem into an ordinary definite integral problem. Therefore, it is concluded that type I path integral problem can be solved easily with MATLAB. Based on the above algorithms, a universal MATLAB function below can be written for the tasks.

```
function I=path_integral(F,vars,t,a,b)
if length(F)==1,
    I=int(F*sqrt(sum(diff(vars,t).^2)),t,a,b); % type I integral
else, F=F(:).';
    vars=vars(:); I=int(F*diff(vars,t),t,a,b); % type II
end
```

The syntaxes of the function for type I path integral are

I=path_integral(f,[x,y],t,t_m,t_M), %2D integrand

I=path_integral$(f,[x,y,z],t,t_m,t_M)$, %3D integrand
I=path_integral(f,v,t,t_m,t_M), %integrand in any dimension

where $[x,y]$ or $[x,y,z]$ can be used to describe the parametric equations of the curve. If an explicit function $y = f(x)$ is given, the vector can be expressed as $[x,y]$. If the integrand is $f = 1$, the arc length can be evaluated directly.

It can be seen that only one line of code is sufficient to compute type I path integral in any dimension. The function can also be used to evaluate type II path integral, and the syntaxes will be presented later.

Example 5.28. Compute $\displaystyle\int_l \frac{z^2}{x^2 + y^2}\,ds$, where l is a spiral curve, whose mathematical form is given by

$$x = a\cos t, \quad y = a\sin t, \quad z = at \quad (0 \leqslant t \leqslant 2\pi \text{ and } a > 0).$$

Solutions. The parametric equations of the curve should be entered into MATLAB first, then the solver can be called, and the result can be obtained as $I = 8\sqrt{2}\pi^3 a/3$.

```
>> syms t; syms a positive;
   x=a*cos(t); y=a*sin(t); z=a*t; f=z^2/(x^2+y^2);% integrand
   I=path_integral(f,[x,y,z],t,0,2*pi) % compute path integral
```

Example 5.29. The arc length in Example 5.21 can be evaluated directly by setting the integrand to 1, and the result is the same, $L_1 = 13/6$.

```
>> syms x positive; y(x)=4*sqrt(2)*x^(3/2)/3-1;
   L1=path_integral(sym(1),[x,y],x,0,1)
```

Example 5.30. Compute $\displaystyle\int_l (x^2+y^2)\,ds$, where l is the positive curve formed by the curves $y = x$ and $y = x^2$.

Solutions. The above two curves can be drawn as shown in Figure 5.8. The positive curve is labeled by arrows on the curves.

```
>> x=0:0.001:1.2; y1=x; y2=x.^2; plot(x,y1,x,y2)
   hold on, ii=find(x<=1); xx=[x(ii),x(ii(end):-1:1)];
   yy=[y2(ii), y1(ii(end):-1:1)]; fill(xx,yy,'g')
```

Two intersections can be found, at $x = 0$ and $x = 1$. The original integral can be divided into two parts, and the sum of them is the integral which we need to find.

```
>> syms x; y=x; f=(x^2+y^2); I1=path_integral(f,[x,y],x,1,0)
   y=x^2; f=(x^2+y^2); I2=path_integral(f,[x,y],x,0,1),
   I=I1+I2, vpa(I)
```

The result obtained is

$$I = -\frac{2}{3}\sqrt{2} + \frac{349}{768}\sqrt{5} - \frac{7}{512}\ln(2 + \sqrt{5}) \approx 0.05358.$$

5.5.2 Type II path integral

Type II path integral is also known as the path integral with respect to the coordinates. It was originated from the study of the work done when a variable force $\vec{f}(x,y,z)$ acts along a path l.

Definition 5.13. The mathematical form of type II path integral is

$$I_2 = \int_l \vec{f}(x,y,z) \cdot \mathrm{d}\vec{s}, \tag{5.5.4}$$

where $\vec{f}(x,y,z) = [P(x,y,z), Q(x,y,z), R(x,y,z)]$ is a row vector, and path $\mathrm{d}\vec{s}$ is a column vector, such that their product yields a scalar.

Theorem 5.8. *If the path is expressed using parametric equations in t, denoted as $x(t)$, $y(t), z(t)$ ($a \leqslant t \leqslant b$), $\mathrm{d}\vec{s}$ can be expressed as*

$$\mathrm{d}\vec{s} = \left[\frac{\mathrm{d}x}{\mathrm{d}t}, \frac{\mathrm{d}y}{\mathrm{d}t}, \frac{\mathrm{d}z}{\mathrm{d}t}\right]^{\mathrm{T}} \mathrm{d}t = [x'(t), y'(t), z'(t)]^{\mathrm{T}} \mathrm{d}t. \tag{5.5.5}$$

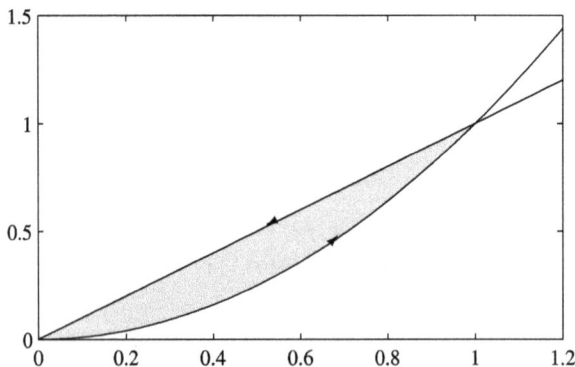

Figure 5.8: Illustration of the integration path.

The type II path integral problem can then be converted into an ordinary definite integral problem

$$I_2 = \int_a^b \left[P(x,y,z)x'(t) + Q(x,y,z)y'(t) + R(x,y,z)z'(t) \right] dt. \qquad (5.5.6)$$

With the new function `path_integral()` presented earlier, the solver for type II path integral is also be presented, with the syntaxes

I=path_integral([P,Q],[x,y],t,a,b), %2D integrand
I=path_integral([P,Q,R],[x,y,z],t,a,b), %3D integrand
I=path_integral(F,v,t,a,b), %any-D integrand

Example 5.31. Compute the path integral

$$\int_l \frac{x+y}{x^2+y^2}\, dx - \frac{x-y}{x^2+y^2}\, dy,$$

where l is the positive circle given by $x^2 + y^2 = a^2$.

Solutions. The original problem is a 2D path integral problem. From the circular equation $x^2 + y^2 = a^2$, $a > 0$, the parametric equations can be obtained as

$$x = a\cos t, \quad y = a\sin t \quad (0 \leqslant t \leqslant 2\pi).$$

The "positive" direction often mentioned is in fact the counterclockwise direction of the curve. For the parametric equation representation of the circle, it means t is changing from 2π to 0. Therefore, the following statements can be used, and the result is 2π.

```
>> syms t; syms a positive; x=a*cos(t); y=a*sin(t);   % curve equation
   F=[(x+y)/(x^2+y^2),-(x-y)/(x^2+y^2)];              % integrand vector
   I=path_integral(F,[x,y],t,2*pi,0)                  % path integral
```

Example 5.32. If the path l is a parabolic curve $y = x^2$ $(-1 \leqslant x \leqslant 1)$, compute the path integral

$$I = \int_l (x^2 - 2xy)dx + (y^2 - 2xy)dy.$$

Solutions. For the parabolic equation, x can be used as the parameter, such that $x = x$, $y = x^2$, and $x \in (-1, 1)$. Then the original path integral problem can be solved directly, and the result is $-14/15$.

```
>> syms x; y=x^2; F=[x^2-2*x*y,y^2-2*x*y];   % vector integrand
   I=path_integral(F,[x,y],x,-1,1)           % path integral
```

5.6 Surface integrals

5.6.1 Type I surface integrals

Definition 5.14. Type I surface integral is expressed as

$$I = \iint_S \phi(x,y,z)dS, \tag{5.6.1}$$

where dS is the differential of the surface area. Therefore, this type of integral is also known as surface integral with respect to area.

Theorem 5.9. *If the surface S can be expressed as an explicit function* $z = f(x,y)$, *the integral can be converted into the double integral in the xy plane*

$$I = \iint_{\sigma_{xy}} \phi[x,y,f(x,y)] \sqrt{1 + f_x^2 + f_y^2} \, dxdy, \tag{5.6.2}$$

where σ_{xy} *describes the integration domain in the xy plane.*

From the algorithm discussed earlier, a universal MATLAB function can be written for the computation of surface integrals. The name of the function is surf_integral(), whose listing is given below

```
function I=surf_integral(f,xx,uu,um,vm)
if length(f)==1       % type I surface integral - scalar integrand
    if length(xx)==1 % case 1: surface described by explicit function
        I=int(int(f*sqrt(1+diff(xx,uu(1))^2+diff(xx,uu(2))^2),...
            uu(2),um(1),um(2)),uu(1),vm(1),vm(2));   % (5.6.2) model
    else              % case 2 - parametric surface
        xx=[xx(:).' 1]; x=xx(1); y=xx(2); z=xx(3); u=uu(1); v=uu(2);
        E=diff(x,u)^2+diff(y,u)^2+diff(z,u)^2;
        F=diff(x,u)*diff(x,v)+diff(y,u)*diff(y,v)+diff(z,u)*diff(z,v);
        G=diff(x,v)^2+diff(y,v)^2+diff(z,v)^2;           % (5.6.4)
        I=int(int(f*sqrt(E*G-F^2),u,um(1),um(2)),v,vm(1),vm(2)); % (5.6.5)
    end
else                        % type II surface integral - vector integrand
    if length(xx)==1 % case 3 - explicit surface
        syms x y z; ua=sqrt(1+diff(xx,x)^2+diff(xx,y)^2);
        cA=-diff(xx,x)/ua; cB=-diff(xx,y)/ua; cC=1/ua;     % (5.6.8)
        I=surf_integral(f(:).'*[cA; cB; cC],xx,uu,um,vm); % (5.6.7)
    else,                   % case 4 - parametric surface
        x=xx(1); y=xx(2); z=xx(3); u=uu(1); v=uu(2);
        A=diff(y,u)*diff(z,v)-diff(z,u)*diff(y,v);
```

```
        B=diff(z,u)*diff(x,v)-diff(x,u)*diff(z,v);
        C=diff(x,u)*diff(y,v)-diff(y,u)*diff(x,v);
        F=A*f(1)+B*f(2)+C*f(3);                                     % (5.6.10)
        I=int(int(F,uu(1),um(1),um(2)),uu(2),vm(1),vm(2)); % (5.6.11)
end, end
```

The syntax of the function is

I=surf_integral$(f,z,[x,y],[y_m,y_M],[x_m,x_M])$

Four cases are considered separately in the function. If length(f)==1, type I surface integral is considered, where the integrand is a scalar function. If a vector integrand is involved, type II surface integral can be processed, which will be presented later.

For each type of surface integral, two cases are considered. Case 1 deals with the problem where the surface is described by an explicit function $z = f(x, y)$, while case 2 deals with parametric equations to be discussed later.

Example 5.33. Compute $\iint_S xyz\,dS$, where the surface S is composed of four planes, defined respectively as $x = 0$, $y = 0$, $z = 0$, and $x + y + z = a$, and $a > 0$.

Solutions. Denoting the four planes as S_1, S_2, S_3, and S_4, the original integral can be written as

$$\iint_S = \iint_{S_1} + \iint_{S_2} + \iint_{S_3} + \iint_{S_4}.$$

Since in the planes S_1, S_2 and S_3, the integrand is zero, the integrals are also zero. Therefore, only plane S_4 is considered. The S_4 plane can be expressed mathematically as $z = a - x - y$, and the boundaries can be expressed as $0 \leqslant y \leqslant a - x$, $0 \leqslant x \leqslant a$. Therefore, the following statements can be written and the result is $I = \sqrt{3}a^5/120$.

```
>> syms x y; syms a positive; z=a-x-y; f=x*y*z; % integrand and surface
   I=surf_integral(f,z,[x,y],[0,a-x],[0,a])      % direct integral
```

Theorem 5.10. *If the surface is expressed in parametric equations*

$$x = x(u, v), \quad y = y(u, v), \quad z = z(u, v), \tag{5.6.3}$$

then the surface integral can be evaluated directly with

$$I = \iint_\Sigma \phi[x(u, v), y(u, v), z(u, v)]\sqrt{EG - F^2}\,du\,dv, \tag{5.6.4}$$

where

$$E = x_u^2 + y_u^2 + z_u^2, \quad F = x_u x_v + y_u y_v + z_u z_v, \quad G = x_v^2 + y_v^2 + z_v^2. \tag{5.6.5}$$

With the function surf_integral(), this type of surface integral can be evaluated easily. In the listing of the function presented earlier, the "case 2" paragraph is involved. The syntax of the function is

```
I=surf_integral(f,[x,y,z],[u,v],[uₘ,uₘ],[vₘ,vₘ])
```

Example 5.34. Compute the surface integral $\iint\limits_S (x^2y + zy^2)\mathrm{d}S$, where S is the spiral surface expressed using parametric equations $x = u\cos v$, $y = u\sin v$, $z = v$, with $0 \leqslant u \leqslant a$ and $0 \leqslant v \leqslant 2\pi$.

Solutions. The following statements can be employed directly:

```
>> syms u v; syms a positive;
   x=u*cos(v); y=u*sin(v); z=v; % parametric equation of surface
   f=x^2*y+z*y^2;                % input the integrand
   I=surf_integral(f,[x,y,z],[u,v],[0,a],[0,2*pi]) % solution
```

and the result is

$$I = \frac{\pi^2}{8}(2a(a^2+1)^{3/2} - a\sqrt{a^2+1} - \operatorname{arcsinh} a).$$

5.6.2 Type II surface integrals

Type II surface integrals are also known as surface integrals with coordinates.

Definition 5.15. The mathematical form of a type II surface integral is

$$I = \iint\limits_{S^+} \vec{r} \cdot \mathrm{d}\vec{V}$$

$$= \iint\limits_{S^+} P(x,y,z)\mathrm{d}y\mathrm{d}z + Q(x,y,z)\mathrm{d}x\mathrm{d}z + R(x,y,z)\mathrm{d}x\mathrm{d}y, \qquad (5.6.6)$$

where the surface S^+ is expressed as an explicit function $z = f(x,y)$, while the integrand is a row vector $\vec{r} = [P, Q, R]$, with $\mathrm{d}\vec{V} = [\mathrm{d}y\mathrm{d}z, \mathrm{d}x\mathrm{d}z, \mathrm{d}x\mathrm{d}y]^{\mathrm{T}}$ being a column vector.

Theorem 5.11. *Type II surface integral can be converted into type I surface integral, with*

$$I = \iint\limits_{S^+} [P(x,y,z)\cos\alpha + Q(x,y,z)\cos\beta + R(x,y,z)\cos y]\mathrm{d}S, \qquad (5.6.7)$$

where z can be substituted by $f(x,y)$, and

$$\cos\alpha = \frac{-f_x}{\sqrt{1+f_x^2+f_y^2}}, \quad \cos\beta = \frac{-f_y}{\sqrt{1+f_x^2+f_y^2}}, \quad \cos y = \frac{1}{\sqrt{1+f_x^2+f_y^2}}. \qquad (5.6.8)$$

The term $\sqrt{1+f_x^2+f_y^2}$ in the denominator cancels the term in (5.6.2), so the whole integral can be rewritten as

$$I = \iint_{\sigma_{xy}} (-Pf_x - Qf_y + R)\,dS. \tag{5.6.9}$$

The algorithm is implemented in the surf_integral() function in the paragraph labeled "case 3". The syntax of the function is

I=surf_integral([P,Q,R],z,[u,v],[u_m,u_M],[v_m,v_M])

Theorem 5.12. *If the surface is described by parametric equations in (5.6.3), then (5.6.8) can further be expressed as*

$$\cos\alpha = \frac{A}{\sqrt{A^2+B^2+C^2}}, \quad \cos\beta = \frac{B}{\sqrt{A^2+B^2+C^2}}, \quad \cos\gamma = \frac{C}{\sqrt{A^2+B^2+C^2}}, \tag{5.6.10}$$

where $A = y_u z_v - z_u y_v$, $B = z_u x_v - x_u z_v$, $C = x_u y_v - y_u x_v$; A, B and C can also be described by the determinants of Jacobian matrices. It can be seen by comparing the denominator in (5.6.10) that it cancels the $\sqrt{EG-F^2}$ term in (5.6.4), and the whole integral can be simplified as

$$I = \iint_{S^+} [AP(u,v) + BQ(u,v) + CR(u,v)]\,du\,dv, \tag{5.6.11}$$

where S^+ is the surface formed by varying parameters u and v.

The algorithm is implemented in the paragraph "case 4" in the function surf_integral(). The syntax of the function is

I=surf_integral([P,Q,R],[x,y,z],[u,v],[u_m,u_M],[v_m,v_M])

Example 5.35. Compute the surface integral $\iint_S (xy+z)\,dy\,dz$, where S is the top surface of an ellipsoid, $\dfrac{x^2}{a^2} + \dfrac{y^2}{b^2} + \dfrac{z^2}{c^2} = 1$.

Solutions. Parametric equations can be introduced as

$$x = a\sin u\cos v, \quad y = b\sin u\sin v, \quad z = c\cos u, \quad 0\leqslant u\leqslant \pi/2,\ 0\leqslant v\leqslant 2\pi.$$

Then the original surface integral problem can be converted into that of a double integral

$$\int_0^{2\pi}\!\!\int_0^{\pi} CR\,du\,dv, \quad \text{where } R = xy+z,\ C = x_u y_v - y_u x_v.$$

The problem can also be solved with the following statements, and the solution is $2abc\,\pi/3$.

```
>> syms u v; syms a b c positive;                    % symbolic variables
   x=a*sin(u)*cos(v); y=b*sin(u)*sin(v); z=c*cos(u); % integrand
   I=surf_integral([0,0,x*y+z],[x,y,z],[u,v],[0,pi/2],[0,2*pi])
```

5.7 Exercises

5.1 Compute the following indefinite integrals:

(1) $I(x) = -\int \dfrac{3x^2 + a}{x^2(x^2 + a)^2}dx$, (2) $I(x) = \int \dfrac{\sqrt{x(x+1)}}{\sqrt{x} + \sqrt{1+x}}dx$,

(3) $I(x) = \int xe^{ax}\cos bx\,dx$, (4) $I(x) = \int e^{ax}\sin bx \sin cx\,dx$,

(5) $I(t) = \int (7t^2 - 2)3^{5t+1}dt$.

5.2 Explain the necessity of the sentence x0=[0 x0 1]; f1=[0 f1 0] used in Example 5.8. What is the physical meaning of such a sentence?

5.3 Compute the following definite or improper integrals:

(1) $I = \displaystyle\int_0^{\infty} \dfrac{\cos x}{\sqrt{x}}dx$, (2) $I = \displaystyle\int_0^1 \dfrac{1+x^2}{1+x^4}dx$, (3) $\displaystyle\int_{e^{-2n\pi}}^1 \left|\cos\left(\ln\dfrac{1}{x}\right)\right|dx$.

5.4 Compute the following definite integrals:

(1) $\displaystyle\int_0^{0.75} \dfrac{1}{(x+1)\sqrt{x^2+1}}dx$, (2) $\displaystyle\int_0^1 \dfrac{\arcsin\sqrt{x}}{\sqrt{x(1-x)}}dx$,

(3) $\displaystyle\int_0^{\pi/4} \left(\dfrac{\sin x - \cos x}{\sin x + \cos x}\right)^{2n+1}dx$.

5.5 Compute the following indefinite integrals:

(1) $\displaystyle\int \dfrac{\sin^2 x - 4\sin x \cos x + 3\cos^2 x}{\sin x + \cos x}dx$,

(2) $\displaystyle\int \dfrac{\sin^2 x - \sin x \cos x + 2\cos^2 x}{\sin x + 2\cos x}dx$.

5.6 Compute the integral $I(s) = \displaystyle\int_0^s \dfrac{e^x \sqrt{e^x - 1}}{e^x + 3}dx$.

5.7 It is known that the Laplace transform of a given function $f(t)$ is defined as

$$F(s) = \int_0^{\infty} e^{-st}f(t)dt,$$ find the Laplace transform of the following functions:

(1) $f(t) = 1$, (2) $f(t) = e^{\beta t}$, (3) $f(t) = \sin \alpha t$, (4) $f(t) = t^m$.

5.8 If $f(x) = e^{-5x} \sin(3x + \pi/3)$, compute the integral $R(t) = \int_0^t f(x)f(t+x)dx$.

5.9 Compute the following multiple integrals:

(1) $\displaystyle\int_0^\pi \int_0^\pi |\cos(x+y)|dxdy$, (2) $\displaystyle\int_0^1 \int_{-1}^{1-x} \arcsin(x+y)dydx$.

5.10 Compute the following integrals:

(1) $\displaystyle\int\int_{|x|+|y|\leqslant 1} (|x|+|y|)dxdy$, (2) $\displaystyle\int\int_{\pi^2\leqslant x^2+y^2\leqslant 4\pi^2} \sin\sqrt{x^2+y^2}dxdy$.

5.11 Compute the triple integral $\displaystyle\iiint_V x^3y^2zdxdydz$, where V is the region defined by

$0 \leqslant x \leqslant 1, 0 \leqslant y \leqslant x, 0 \leqslant z \leqslant xy$.

5.12 For different values of a, compute $I = \displaystyle\int_0^\infty \frac{\cos ax}{1+x^2}dx$.

5.13 Show that for any function $f(t)$, $\displaystyle\int_a^b f(t)\,dt = -\int_b^a f(t)\,dt$.

5.14 Compute the following multiple integrals:

(1) $\displaystyle\int_0^2 \int_0^{\sqrt{4-x^2}} \sqrt{4-x^2-y^2}\,dydx$, (2) $\displaystyle\int_0^3 \int_0^{3-x} \int_0^{3-x-y} xyz\,dzdydx$,

(3) $\displaystyle\int_0^2 \int_0^{\sqrt{4-x^2}} \int_0^{\sqrt{4-x^2-y^2}} z(x^2+y^2)\,dzdydx$.

5.15 Compute the following multiple integrals:

(1) $\displaystyle\int_0^1 \int_0^x \int_0^y \int_0^z xyzue^{6-x^2-y^2-z^2-u^2}\,dudzdydx$,

(2) $\displaystyle\int_0^{7/10} \int_0^{4/5} \int_0^{9/10} \int_0^1 \int_0^{11/10} \sqrt{6-x^2-y^2-z^2-w^2-u^2}\,dwdudzdydx$.

5.16 Construct a complete undetermined polynomial for Example 5.15.

5.17 Compute the following path integrals:

(1) $\displaystyle\int_l (x^2+y^2)ds$, l is the curve $x = a(\cos t + t\sin t), y = a(\sin t - t\cos t)$ $(0 \leqslant t \leqslant 2\pi)$;

(2) $\displaystyle\int_l (yx^3 + e^y)dx + (xy^3 + xe^y - 2y)dy$, l is the top half of $a^2x^2 + b^2y^2 = c^2$;

(3) $\int_l y dx - x dy + (x^2 + y^2) dz$, l is the curve $x = e^t$, $y = e^{-t}$, $z = at$, $0 \leqslant t \leqslant 1$, $a > 0$;

(4) $\int_l (e^x \sin y - my) dx + (e^x \cos y - m) dy$, l is the positive path from point $(a, 0)$ to

point $(0, 0)$, then closed by the half circle of $x^2 + y^2 = ax$;

5.18 If a curve can be described by a polar function $r = \rho(\theta)$, and $\theta \in (\theta_m, \theta_M)$, the arc length can be computed from

$$L = \int_{\theta_m}^{\theta_M} \sqrt{\rho^2(\theta) + [d\rho(\theta)/d\theta]^2} \, d\theta.$$

Find the arc length of $\rho = a \sin^2 \theta/3$, $\theta \in (0, 3\pi)$.

5.19 If the surface S is the bottom of a hemisphere, $z = \sqrt{R^2 - x^2 - y^2}$, compute the following surface integrals:

(1) $\int_S xyz^3 \, ds$, (2) $\int_S (x + yz^3) \, dxdy$.

6 Series and function fitting

Series problems are probably the oldest problems studied in calculus. Dating back to several centuries BCE, ancient Greek were starting the study on series problems. Chinese philosopher Zhuang Zhou (also known as Chuang Tzu, c369BCE–286BCE) stated that "For a foot-long stick, if you take half from the remaining part each day, you will never exhaust it in a million years". It can also be regarded as a series problem.

The first objective in this chapter is to explore series sums and convergence problems. In Section 6.1, finite and infinite sums of series will be explored. Number and functional series will be studied, and numerical and symbolic computations will all be presented for a sum of series. For infinite series, the convergence issue is very important, and in Section 6.2, convergence tests for infinite series are carried out with MATLAB implementation. In Section 6.3, the product of sequences will be presented, as well as convergence tests.

In the other part of the chapter, function approximations are addressed. For a given function $f(x)$, sometimes one may want to have a simpler function to approximate it. Normally, Taylor and Fourier series are considered. Also, continued fraction expansion and Padé approximation can be used to approximate given functions. In Section 6.4, Taylor series approximations to univariate as well as multivariate functions are presented, and fitting behaviors are also assessed. In Section 6.5, Fourier series approximation to periodic functions is presented. A universal MATLAB function is developed for finding Fourier series. In Section 6.6, continued fraction expansion and Padé approximation to given functions are presented. For a complex function $f(z)$, Laurent series expansion is also presented in Section 6.7, and a MATLAB tool for Laurent series expansion to rational functions is presented.

6.1 Series sums

Series sum problems are studied first with numerical and symbolic algorithms. The sum of a functional series is also studied. Finally, similar problems are also explored.

6.1.1 Number series sums

Definition 6.1. The mathematical form of a series is given by

$$S = a_{k_0} + a_{k_0+1} + \cdots + a_{k_n} = \sum_{k=k_0}^{k_n} a_k, \tag{6.1.1}$$

where a_k is known as the general term, k is the independent variable, k_0 and k_n are respectively the starting and terminating terms.

https://doi.org/10.1515/9783110666977-006

With the Symbolic Math Toolbox function symsum(), the sum in Definition 6.1 can be found. The syntax of the function is S=symsum(a_k, k, k_0, k_n), where the starting term can be set to $-\infty$, and the terminating term can be set to ∞, however, they cannot both be infinite at the same time.

If there is only one symbolic variable in the general term a_k, the independent variable k can be omitted from the function call.

Example 6.1. Find the sum of the finite series

$$S = 2^0 + 2^1 + 2^2 + 2^3 + 2^4 + \cdots + 2^{62} + 2^{63} = \sum_{i=0}^{63} 2^i.$$

Solutions. Numerical data type can be tried directly and it can be seen that the solution is $1.844674407370955 \times 10^{19}$.

```
>> format long; sum(2.^[0:63]) % display all the digits
```

Since double precision data type is used as a default, only at most 16 decimal digits can be reserved. The result under such a data type is not accurate enough. The symbolic function symsum() can be used, and accurate solution can be found, namely 18 446 744 073 709 551 615. If value 2 is converted into symbolic form, the simple command sum() function can be used, and the accurate solution can be found.

```
>> syms k; symsum(2^k,0,63) % or sum(sym(2).^[0:63])
```

Moreover, up to 201 terms can be added up with the following statements:

```
>> syms k; S=symsum(2^k,0,200) % or sum(sym(2).^[0:200])
```

and the result is

S = 3 213 876 088 517 980 551 083 924 184 682 325 205 044 405 987 565 585 670 602 751.

This is the result impossible in any other numerical data types.

Example 6.2. Find the sum of the following infinite series:

$$S = \frac{1}{1 \cdot 4} + \frac{1}{4 \cdot 7} + \frac{1}{7 \cdot 10} + \cdots + \frac{1}{(3n-2)(3n+1)} + \cdots.$$

Solutions. Manual solution skills may be used to find the sum. Each term can be split into a difference of two terms. The middle terms can be canceled, and only the first term can be retained:

$$\frac{1}{3}\left(\frac{1}{1} - \frac{1}{4}\right) + \frac{1}{3}\left(\frac{1}{4} - \frac{1}{7}\right) + \frac{1}{3}\left(\frac{1}{7} - \frac{1}{11}\right) + \cdots \rightarrow \frac{1}{3}.$$

It is not the intention of the book to guide the readers in the search for tricks. A more general-purpose solution pattern is suggested. That is, the intent is to send the problem to computers and let them do the low-level work for you. With the help of MATLAB, the same result can be obtained directly.

```
>> syms n; s=symsum(1/((3*n-2)*(3*n+1)),n,1,inf) % symbolic sum
```

Since the series contains only numbers, numerical methods can be tried. Suppose a total of first 10 000 000 terms are to be summed up, with the following commands:

```
>> m=1:10000000; s1=sum(1./((3*m-2).*(3*m+1)));
   format long; s1 % sum taken in double precision
```

Then the sum can be found, with the result 0.33333332222165. It can be seen that although significantly many numbers are added up, the accuracy cannot be ensured in the numerical framework.

Although many terms are added, and it is time-consuming, the error is still relatively large, about 10^{-6}. It can be seen from the general term that, when $m = 10^7$, the value of the general term is about 10^{-15}. It seems that the total error should not be so large. As a matter of fact, since double precision data type is used, some of the terms cannot be well added to the final sum.

With the use of symsum() function, the sum of the first k terms can also be found as $s = k/(3k + 1)$. It can be seen that, when $k \to \infty$, the limit of s is 1/3.

```
>> syms k; s=symsum(1/((3*n-2)*(3*n+1)),n,1,k)
```

There are various ways and skills for finding the sum of a series. However, skill-based solution methodology is not good enough in real applications. In MATLAB, a systematic tool, symsum(), is provided, and can be used to solve general problems when finding the sum of a series. With such a tool, most of the finite and infinite sum problems can be solved directly.

Example 6.3. Find the finite and infinite sum of the following series:

$$S = \frac{1}{2} - \frac{1}{3} + \frac{1}{4} + \frac{1}{9} + \frac{1}{8} - \frac{1}{27} + \frac{1}{16} + \frac{1}{81} + \cdots.$$

Solutions. The most important step in series-related problems is to find the general term of the series. If the general term cannot be properly found, the series sum cannot be well computed. It is hard to find general terms by observing the given terms alone, however, if the terms are grouped in pairs, it is not difficult to rewrite the series in the following form:

$$S = \left(\frac{1}{2} - \frac{1}{3}\right) + \left(\frac{1}{4} + \frac{1}{9}\right) + \left(\frac{1}{8} - \frac{1}{27}\right) + \left(\frac{1}{16} + \frac{1}{81}\right) + \cdots.$$

Now, the problem is clear. It is not hard to formulate the general term as

$$a_n = \frac{1}{2^n} + \frac{(-1)^n}{3^n}, \quad n = 1, 2, \dots$$

With the general term, the following statements can be used to describe the general term, and then the sum of the series can be computed directly. The result obtained is $S = 3/4$. Since n is the only symbolic variable in the expression, it can be omitted from the function call.

```
>> syms n; a=1/2^n+(-1)^n/3^n; S=symsum(a,n,1,inf)
```

Example 6.4. Find the finite and infinite sum for the following series:

$$S = \frac{1}{\sqrt{2}-1} - \frac{1}{\sqrt{2}+1} + \frac{1}{\sqrt{3}-1} - \frac{1}{\sqrt{3}+1} + \frac{1}{\sqrt{4}-1} - \frac{1}{\sqrt{4}+1} + \cdots.$$

Solutions. For this problem, if one tries to examine each term only, it is very difficult, if not impossible, to express the sequence. The terms in the sequence can be grouped in pairs

$$S = \left(\frac{1}{\sqrt{2}-1} - \frac{1}{\sqrt{2}+1}\right) + \left(\frac{1}{\sqrt{3}-1} - \frac{1}{\sqrt{3}+1}\right) + \left(\frac{1}{\sqrt{4}-1} - \frac{1}{\sqrt{4}+1}\right) + \cdots.$$

With the grouped series, it is easier to find the general term as

$$a_n = \frac{1}{\sqrt{n+1}-1} - \frac{1}{\sqrt{n+1}+1}, \quad n = 1, 2, \dots$$

The general term extraction may not be unique. For instance, one can write another general term for $n = 2, 3, \dots$, as

$$b_n = \frac{1}{\sqrt{n}-1} - \frac{1}{\sqrt{n}+1}.$$

It can be seen that although the general term of a series is important, finding its initial term is equally important. With this information, the following statement can be tried to find the sum of the series. Unfortunately, there is no analytical expression for the sum of the series.

```
>> syms n;
   S1=symsum(1/(sqrt(n+1)-1)-1/(sqrt(n+1)+1),n,1,inf)
   S2=symsum(1/(sqrt(n)-1)-1/(sqrt(n)+1),n,2,inf)
```

Is there an alternative way to find the sum of an infinite series? Since the series is a number series, the numerical method can be used to find the finite sum of its first n elements, and observe its trends. With the following MATLAB commands, we can see the relationship between the sum and the number of terms, as shown in Table 6.1. It can be seen that the sum is increasing without bounds as the number of terms increases. Therefore, the series may not be convergent.

Table 6.1: The relationship between term number and sum.

number of terms	10	100	1 000	10 000	100 000	1 000 000	10 000 000
partial sum	5.8579	10.375	14.971	19.575	24.18	28.785	33.391

```
>> NO=[10,100,1000,10000,100000,1000000 10000000]; T=[];
   for N=NO
      n=1:N; y=sum(1./(sqrt(n+1)-1)-1./(sqrt(n+1)+1));
      T=[T [N; y]];
   end
```

It can be seen from the example that convergence is also an important property in series. Convergence tests will be explored later.

6.1.2 Sum of infinite series

Although symsum() function can be used to find the sum of some infinite series, for some specific problems, sometimes it is hard, if not impossible, to find the sum with symsum() function. The original problem may need to be adjusted, such that infinite series problem may be changed to comprehensive problems of finite sums, followed by limits. This kind of problems will be demonstrated through examples.

Example 6.5. Solve the limit problem

$$\lim_{n\to\infty}\left[\left(1+\frac{1}{2}+\frac{1}{3}+\frac{1}{4}+\cdots+\frac{1}{n}\right)-\ln n\right].$$

Solutions. In the original problem, the infinite series can be evaluated with the statements

```
>> syms n; symsum(1/n,n,1,inf)
```

Unfortunately, the sum obtained is ∞. Therefore, the limit cannot be processed in this way. The infinite sum should not be evaluated alone, instead, the finite sum of the first n terms should be computed, the term $\ln n$ should be subtracted, and only then the limit can be taken.

For this specific example, the finite sum can be evaluated with the function symsum(1/m,1,n), and the following statements can be used to find the final results:

```
>> syms m;
   limit(symsum(1/m,m,1,n)-log(n),n,inf)
```

The result obtained is `eulergamma`, which is the Euler constant y, whose value can be evaluated with `vpa(ans,60)`

$$y \approx 0.577215664901532860606512090082402431042159335939923598805767.$$

Example 6.6. Compute the sum of the infinite series

$$S = \left(1 + \frac{1}{n^2}\right)\sin\frac{\pi}{n^2} + \left(1 + \frac{2}{n^2}\right)\sin\frac{2\pi}{n^2} + \cdots + \left(1 + \frac{n-1}{n^2}\right)\sin\frac{(n-1)\pi}{n^2} + \cdots.$$

Solutions. Consider the mathematical expression. It can be seen that the values in the numerator are changing from 1 to $n-1$, and it is quite difficult compute the infinite sum directly with `symsum()` function. One can set n as an integer, find the finite sum with `symsum()`, and finally, compute the limit as $n \to \infty$, which yields the infinite sum

$$S = \lim_{n\to\infty}\left[\left(1 + \frac{1}{n^2}\right)\sin\frac{\pi}{n^2} + \left(1 + \frac{2}{n^2}\right)\sin\frac{2\pi}{n^2} + \cdots + \left(1 + \frac{n-1}{n^2}\right)\sin\frac{(n-1)\pi}{n^2}\right].$$

It can be seen that the key point is to find the general term of the series. In each term, the quantity n^2 appears in the denominator in all the terms, while the numerator is changing. Thus denoting the numerator as k, the general term can be expressed as

$$a_k = \left(1 + \frac{k}{n^2}\right)\sin\frac{k\pi}{n^2}, \quad k = 1, 2, \ldots, n-1.$$

With the general term, the following statements can be used to compute the sum of the first n terms:

```
>> syms n k;
   s=symsum((1+k/n^2)*sin(k*pi/n^2),k,1,n-1);
   simplify(rewrite(s,'sin'))
```

If the function `rewrite()` is not used, the results are those with complex terms in exponential function, and the readability is not good. The simplified version of the sum is obtained below as

$$S = -\frac{(1 - n - n^2)\sin\pi/n + n^2\sin\pi/n^2 + (n + n^2)\sin(n-1)\pi/n^2}{2n^2(\cos\pi/n^2 - 1)}.$$

In fact, if you are not interested in finding "the sum of the first n terms", the following commands can be used directly, and the infinite sum can be obtained as $S = \pi/2$.

```
>> syms n k;
   S=simplify(limit(symsum((1+k/n^2)*sin(k*pi/n^2),k,1,n-1),n,inf))
```

6.1.3 Sum of functional series

The sum of a number series can be evaluated with the symbolic computation approach. Alternatively, pure numerical method can be used to evaluate the series. In real applications, if the series terms are functions, it is referred to as a functional series. In this section, the sum of functional series is considered.

Definition 6.2. The mathematical form of the sum of functional series is

$$S(x) = a_{k_0}(x) + a_{k_0+1}(x) + \cdots + a_{k_n}(x) = \sum_{k=k_0}^{k_n} a_k(x), \qquad (6.1.2)$$

where $a_k(x)$ is the general term, k is the independent variable, k_0 and k_n are the first and last terms, respectively.

In fact, the sum of functional series can also be computed with symsum() function. It should be noted that with symsum() function, the sum can be obtained, also the convergence condition can be returned.

Example 6.7. Compute the infinite sum of the functional series

$$J = 2\sum_{n=0}^{\infty} \frac{1}{(2n+1)(2x+1)^{2n+1}}.$$

Solutions. Since variable x appears in the general term, the sum cannot be obtained with numerical methods. Symbolic computation is the only choice. With the following statements, the simplified result is $2\operatorname{atanh}(1/(2x+1))$. Also, the convergence condition is obtained as $|2x+1| > 1$. In earlier versions, the result is $\ln[(x+1)/x]$, and it is hard to prove the two functions are equal. The function ezplot() can be used to draw the difference of the two functions, as shown in Figure 6.1. It can be seen that the two functions are almost identical.

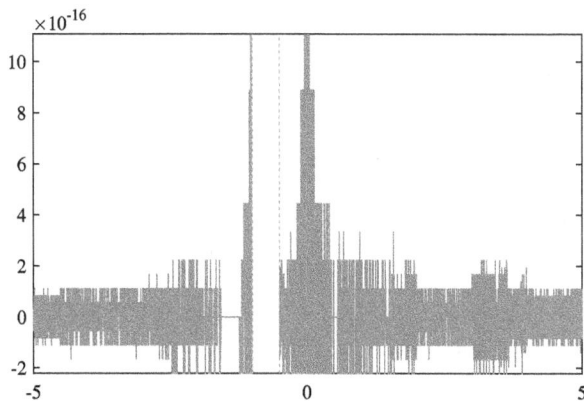

Figure 6.1: The curves of error between $\ln[(x+1)/x]$ and $2\operatorname{atanh}(1/(2x+1))$.

```
>> syms n x; s1=symsum(2/((2*n+1)*(2*x+1)^(2*n+1)),n,0,inf);
   simplify(s1), fplot(2*atanh(1/(2*x+1))-log((x+1)/x))
```

The function symsum() also gives the condition of convergence for the series. The form is different from those we are familiar with. Manual processing may sometimes be needed:

$$|2x + 1| > 1 \Longrightarrow 2x + 1 > 1 \text{ or } 2x + 1 < -1 \Longrightarrow x > 0 \text{ or } x < -1.$$

Example 6.8. Compute the sum of the infinite functional series

$$S = \frac{1}{x} + \frac{2}{x^2} + \frac{3}{x^3} + \cdots + \frac{n}{x^n} + \cdots.$$

Solutions. It can be seen that the general term of the infinite series is n/x^n, $n = 1, 2, \ldots$ The following MATLAB statements can be used directly:

```
>> syms n x; S=symsum(n/x^n,n,1,inf)
```

and the result is $x/(x - 1)^2$. Also the convergence condition is $(|x| - 1/2)^2 > 1/4$. It can be seen that with the use of symsum() function, the sum as well as its convergence conditions are returned.

The convergence conditions generated automatically by computers may not in the same form, as those we are used to. The conditions can further processed through manual deduction, as $|x| > 1$:

$$(|x| - 1/2)^2 > 1/4 \Rightarrow |x| - 1/2 > 1/2 \text{ or } |x| - 1/2 < -1/2 \text{ (negligible)} \Rightarrow |x| > 1.$$

6.1.4 Special infinite term problems

The sums of the series were studied in the previous sections, and ready-to-use functions are available to deal with the standard problems. In practical applications, some other infinite expressions may appear. For instance, consider the following infinite function expression:

$$\sqrt{\frac{1}{x} + \sqrt{\frac{1}{x} + \sqrt{\frac{1}{x} + \sqrt{\frac{1}{x} + \sqrt{\frac{1}{x} + \cdots}}}}}. \tag{6.1.3}$$

This kind of problem cannot be handled with symsum() or other functions. Alternative solution methodology should be presented.

Example 6.9. Compute the general term of the expression in (6.1.3), and compute the infinite expression.

Solutions. Since the expression is not a series, it is not possible to use function sym-
sum() or other functions to deal with this kind of problems. An alternative way should
be found. Assume that the expression of the nth term is expressed as s_n, then a recursive algorithm can be established

$$s_1 = 1/x, \quad s_{n+1} = \sqrt{1/x + s_n}, \quad n = 1, 2, 3, \ldots$$

It can be seen that the first ten terms can be obtained in LaTeX

```
>> syms x; s(1)=sqrt(1/x); n=10;
   for k=1:n-1, s(k+1)=sqrt(s(k)+1/x); end, latex(s(end))
```

The tenth term is obtained, in the automatically generated expression, as

$$\sqrt{\frac{1}{x} + \sqrt{\sqrt{\sqrt{\sqrt{\frac{1}{x} + \sqrt{\sqrt{\sqrt{\frac{1}{x} + \frac{1}{x} + \frac{1}{x} + \frac{1}{x} + \frac{1}{x} + \frac{1}{x} + \frac{1}{x} + \frac{1}{x}}}}}}}}.$$

It is the time now to compute the expression when $n \to \infty$. Assume that the expression settles down at S, then the equation $S = \sqrt{1/x + S}$ can be established. The
following statements can be used to solve the equation:

```
>> syms x S; A=solve(S==sqrt(1/x+S),S)
```

With the above commands, two solutions to the equation can be found

$$a_1 = 1/2 - \sqrt{(x+4)/x}/2, \quad a_2 = \sqrt{(x+4)/x}/2 + 1/2.$$

Which of the two should be retained? In fact, it is not difficult to find that when
$x > 0$, $a_1 < 0$. Of course, a_1 should be discarded. The expression will finally settles
down at $\sqrt{(x+4)/x}/2 + 1/2$.

If $x = 2$, the trends of the first 10 terms of the expression can be drawn as shown
on Figure 6.2. It can be seen that $a_1 = -0.366$, $a_2 = 1.366$. From the sixth term, the
expression is rather very close to a_2.

```
>> n=1:10; y=double(subs(s,x,2)); stem(n,y),
   A=double(subs(A,x,2)), hold on, plot([1,10],[A(2) A(2)],'--')
```

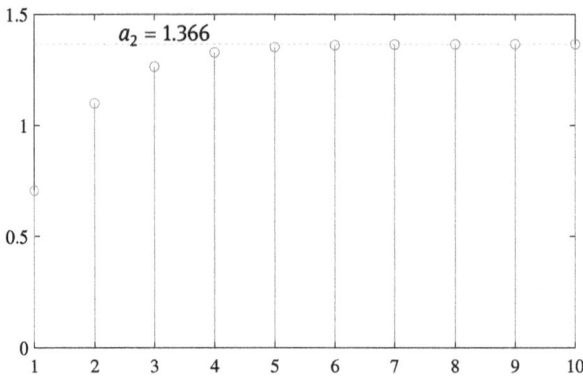

Figure 6.2: The trend of the expression when n increases.

Example 6.10. Compute the value of the infinite expression

$$S = \sqrt{2 + \sqrt{2 + \sqrt{2 + \sqrt{2 + \sqrt{2 + \cdots}}}}}.$$

Solutions. This is again not an infinite series problem. Similar to the example discussed earlier, the following recursive algorithm can be formulated

$$s_1 = \sqrt{2}, \; s_{n+1} = \sqrt{2 + s_n}, \quad n = 1, 2, 3, \ldots$$

The first 11 terms of the expression can be computed, and it can be seen that the sequence is quite close to 2. The difference between the 11th term and 2 is as small as 5.8827×10^{-7}. Further increasing the number of terms, the difference becomes smaller. If 10 in the loop is changed to 20, the difference is reduced to 10^{-13}.

```
>> s=sqrt(2); for i=1:10; s(i+1)=sqrt(2+s(i)); end % recursive evaluation
   s, 2-s(end)    % display the sequence, the distance of the last term to 2
```

Assuming the expression settles down at S, the following statements can be used. It is found that $S = 2$, which agrees well with that predicted earlier.

```
>> syms S; S=solve(S==sqrt(2+S))
```

Example 6.11. Complete $\lim\limits_{n \to \infty} \underbrace{\sin \sin \sin \cdots \sin x}_{n\text{-fold}}$.

Solutions. This is another infinite expression which is difficult to represent. Assume that the n-fold expression converges to a, then $a = \sin a$, whose unique solution is $a = 0$. Therefore, the function approaches zero, no matter what the value of x.

```
>> syms a; vpasolve(a==sin(a))
```

6.2 Convergence tests for infinite series

Different kinds of series may be encountered in real applications. Sometimes, even if the powerful function symsum() cannot find closed-form solutions of certain infinite series, convergence test may become an important issue. This kind of phenomenon was demonstrated in Example 6.4. Convergence tests of infinite series are presented in this section.

6.2.1 General description of a positive series

Definition 6.3. The general form of an infinite series is expressed as

$$S = a_1 + a_2 + \cdots + a_n + \cdots = \sum_{k=1}^{\infty} a_n. \tag{6.2.1}$$

Definition 6.4. If all the terms in the series are positive, the series is referred to as a positive series.

6.2.2 Convergence tests for positive series

Definition 6.5. If as $n \to \infty$, the limit of the sum S is finite, the series is referred to as convergent. If there is no limit for S, the series is divergent.

The following criteria can be used to test whether a given positive series is convergent or not.

Theorem 6.1. *The necessary condition for the convergency of a positive series is* $\lim_{n\to\infty} a_n = 0$. *If the condition is not satisfied, the infinite series is divergent.*

For a positive series, if the necessary condition is satisfied, the following criteria can be tried in turn to check its convergence.

Theorem 6.2. *Commonly used convergence tests are as follows:*
(1) *D'Alembert test. Compute* $L = \lim_{n\to\infty} a_{n+1}/a_n$. *If* $L < 1$, *the series is convergent; if* $L > 1$, *the series is divergent; if* $L = 1$, *other criteria should be tried.*
(2) *Raabe test. If Criterion (1) failed, compute* $R = \lim_{n\to\infty} n(a_n/a_{n+1} - 1)$. *If* $R > 1$, *the series is convergent; if* $R < 1$, *the series is divergent; if* $R = 1$, *further criteria should be tried.*
(3) *Bertrand test.[6] If Criterion (2) failed, compute*

$$L = \lim_{n\to\infty} \ln n\left(n\left(\frac{a_n}{a_{n+1}} - 1\right) - 1\right). \tag{6.2.2}$$

If $L > 1$, the series is convergent; if $L < 1$, the series is divergent; if $L = 1$, the test is inconclusive.

Example 6.12. Check whether the following infinite series is convergent:

$$S = \sum_{n=1}^{\infty} \frac{2^n}{1 \cdot 3 \cdot 5 \cdots (2n-1)} = \sum_{n=1}^{\infty} \frac{2^n}{\prod_{k=1}^{n}(2k-1)}.$$

Solutions. For positive series, the limit of a_{n+1}/a_n can easily be found

```
>> syms n k; a(n)=2^n/symprod(2*k-1,k,1,n)
   limit(a,n,inf), L=limit(a(n+1)/a(n),n,inf) % Criterion (1)
```

It can be seen that the limit is $L = 0$, satisfying Criterion (1), meaning the series in convergent. Although a_{n+1}/a_n can be derived with MATLAB, it is not in the simplest form. Manual derivation may be used to find the final simplified results:

$$\frac{a_{n+1}}{a_n} = \frac{4(2n)!(n+1)!}{(2n+2)!\,n!} \implies \frac{4(2n)!(n+1)n!}{(2n+2)(2n+1)(2n)!\,n!} = \frac{2}{2n+1}.$$

Example 6.13. Check the convergence of the series in Example 6.4.

$$S = \frac{1}{\sqrt{2}-1} - \frac{1}{\sqrt{2}+1} + \frac{1}{\sqrt{3}-1} - \frac{1}{\sqrt{3}+1} + \frac{1}{\sqrt{4}-1} - \frac{1}{\sqrt{4}+1} + \cdots.$$

Solutions. In Example 6.4, the general term of the series was given

$$a_n = \frac{1}{\sqrt{n+1}-1} - \frac{1}{\sqrt{n+1}+1}, \quad n = 1, 2, \ldots$$

With the general term of the positive series, since the necessary condition is satisfied, Criterion (1) can be tested. Unfortunately, it can be seen that the limit is $L = 1$, indicating that Criterion (1) cannot be used to test the convergence.

```
>> syms n; a(n)=1/(sqrt(n+1)-1)+1/(sqrt(n+1)+1);
   limit(a,n,inf), L=limit(a(n+1)/a(n),n,inf) % criterion (1)
```

Criterion (2) can be tried with the following statements:

```
>> R=limit(n*(a(n)/a(n+1)-1),n,inf)
```

and it can be seen that $R = 1/2 < 1$, meaning the series is divergent, which agrees well with the observed phenomenon in Example 6.4.

Example 6.14. Check whether the following positive series converges or not:

$$S = \sqrt{2} + \sqrt{2-\sqrt{2}} + \sqrt{2-\sqrt{2+\sqrt{2}}} + \sqrt{2-\sqrt{2+\sqrt{2+\sqrt{2}}}} + \cdots.$$

Solutions. It has been shown in Example 6.10 that the expression under the square root sign approaches 2. Therefore the limit of the general term is zero, satisfying the necessary condition in Theorem 6.1. Denoting the sum under the square root sign as s_n, the general term of the series can be expressed as $a_n = \sqrt{2 - s_n}$. Besides, it is known that $s_{n+1} = \sqrt{2 + s_n}$, so it follows that $a_{n+1} = \sqrt{2 - s_{n+1}} = \sqrt{2 - \sqrt{2 + s_n}}$, and $s_n \to 2$ as $n \to \infty$. If we use Criterion (1) then we need to find

$$\lim_{n \to \infty} \frac{a_{n+1}}{a_n} = \lim_{s_n \to 2} \frac{\sqrt{2 - \sqrt{2 + s_n}}}{\sqrt{2 - s_n}}.$$

With MATLAB, the following statements can be used to find the limit:

```
>> syms sn; L=limit(sqrt(2-sqrt(2+sn))/sqrt(2-sn),sn,2)
```

and the result is $1/2 < 1$, indicating that the series is convergent. With the numerical method, it can be seen that the sum approaches 2.962099833756542.

```
>> s=0; for i=1:2000, s(i+1)=sqrt(2+s(i)); end
   a=sqrt(2-s); for i=1:length(s), S(i)=sum(a(1:i)); end
```

6.2.3 Convergence test for alternating series

Definition 6.6. Alternating series can be expressed as follows:

$$S = b_1 - b_2 + b_3 - b_4 + \cdots + (-1)^{n-1}b_n + \cdots = \sum_{n=1}^{\infty} (-1)^{n-1}b_n, \tag{6.2.3}$$

where $b_n \geq 0$ is a positive sequence; we denote $a_n = (-1)^n b_n$.

Definition 6.7. If the series $\sum_{n=1}^{\infty} |a_n|$ is convergent, $\sum_{n=1}^{\infty} a_n$ is also convergent; we call such a series absolutely convergent.

Theorem 6.3. *Convergency tests for alternating series:*
(1) *Compute $L = \lim_{n \to \infty} b_{n+1}/b_n = L$. If $L < 1$, the series is absolutely convergent; if $L > 1$, the series is divergent; if $L = 1$, convergence cannot be decided by the method.*
(2) *Leibniz test. If $b_{n+1} \leq b_n$, and $\lim_{n \to \infty} b_n = 0$, the series is convergent.*
(3) *Compute $R = \lim_{n \to \infty} n(b_n/b_{n+1} - 1)$. If $R > 1$, the series is absolutely convergent; if $0 < R \leq 1$, the alternating series is conditionally convergent; otherwise, the series is divergent.*

Example 6.15. Check the convergence of the following infinite series:

$$\frac{1}{1} + \frac{1}{2} + \frac{1}{3} - \frac{1}{4} - \frac{1}{5} - \frac{1}{6} + \frac{1}{7} + \frac{1}{8} + \frac{1}{9} - \frac{1}{10} - \frac{1}{11} - \frac{1}{12} + \cdots.$$

Solutions. Since the well-known infinite series $\sum_{n=1}^{\infty} 1/n$ is divergent, the alternating series is not absolutely convergent. It is rather difficult to directly write the general term, an alternative method should be used. If the original series is grouped in triplets, an alternating series can be formulated, where

$$b_n = \left(\frac{1}{3n-2} + \frac{1}{3n-1} + \frac{1}{3n} \right), \quad n = 1, 2, 3, \ldots.$$

Also, $a_n = (-1)^{n-1} b_n$ is an alternating series. For the alternating series, from Criterion (1), it is found $L = 1$, so the convergence cannot be tested in this way. It can be seen that for any n, $b_{n+1} \leqslant b_n$, and $\lim_{n \to \infty} b_n = 0$, therefore, it is known from Criterion (2) that the alternating series is convergent. Criterion (3) can also be used to compute R.

```
>> syms n; b(n)=1/(3*n-2)+1/(3*n-1)+1/(3*n);
   limit(b,n,inf), L=limit(b(n+1)/b(n),n,inf) % necessary condition
   R=limit(n*(b(n)/b(n+1)-1),n,inf)
```

Since $R = 1$, the condition $0 < R \leqslant 1$ is satisfied. Thus, from Criterion (3), the alternating series in conditionally convergent.

6.2.4 Convergence interval of a functional series

The convergence of a number series can easily be investigated with the methods discussed earlier. If all the criteria fail, the numerical method can still be used to observe the tendency of the series. The convergence tests for a functional series are not so simple, since the numerical method cannot be used. Also the general term of the series is related to the variable x, and the convergence interval of x should be found. In this section, the convergence tests for functional series will be explored.

Example 6.16. Consider the functional series below. If p is real, find the interval of x, such that the infinite series is convergent.

$$\sum_{n=1}^{\infty} \left[\frac{1 \cdot 3 \cdot 5 \cdots (2n-1)}{2 \cdot 4 \cdot 6 \cdots (2n)} \right]^p \left(\frac{x-1}{2} \right)^n.$$

Solutions. Symbolic variables can be declared, and the general term a_n can be expressed. Taking the limit of a_{n+1}/a_n, it is found that the simplified value is $L = (x-1)/2$. In order to have the infinite series convergent, we need $|L| < 1$. Solving the inequality $|(x-1)/2| < 1$, it is found that the convergence interval is $x \in (-1, 3)$.

```
>> syms n k positive; syms p real; assume(n,'integer');
   a=(symprod(2*k-1,k,1,n)/symprod(2*k,k,1,n))^p*((x-1)/2)^n;
   F=simplify(subs(a,n,n+1)/a), L=simplify(limit(F,n,inf))
```

Letting $x = -1$, the original series becomes an alternating series. It can be seen from Criterion (3) that $L = p/2$, meaning that when $p > 0$, $x = -1$ is still admissible for convergence and $x = -1$ is the boundary of conditional convergence.

```
>> b=(symprod(2*k-1,k,1,n)/symprod(2*k,k,1,n))^p;  % general term
   L=limit(n*(b/subs(b,n,n+1)-1),n,inf)                % Raabe test
```

If $x = 3$, the series is a positive series. It can be seen from Criterion (2) that $L = p/2$, meaning the series is absolutely convergent when $p > 2$, otherwise, the series is divergent.

If $x = -1$, then, for $p > 2$, the series is absolutely convergent.

6.3 Products of sequences

6.3.1 Products of number sequences

Definition 6.8. A product of a sequence can be expressed as

$$P = a_{k_0} a_{k_0+1} \cdots a_{k_n} = \prod_{k=k_0}^{k_n} a_k. \tag{6.3.1}$$

A MATLAB function $P=\text{symprod}(a_k,k,k_0,k_n)$ is provided in the Symbolic Math Toolbox to handle directly the sequence product problems.

Example 6.17. Find the finite- and infinite-term product $P_n = \prod_{k=1}^{n}\left(1 + \dfrac{1}{k^3}\right)$.

Solutions. The following statements can be used to find the products:

```
>> syms k n; P1=symprod(1+1/k^3,k,1,n); P1=simplify(P1)   % finite-term
   P2=symprod(1+1/k^3,k,1,inf); P2=simplify(P2), vpa(P2) % infinite-term
```

and the results obtained are

$$P_1 = -\frac{(n+1)!\,\sin\left(\dfrac{-1+\sqrt{3}i}{2}\pi\right)\Gamma\left(n + \dfrac{1-\sqrt{3}i}{2}\right)\Gamma\left(n + \dfrac{1+\sqrt{3}i}{2}\right)}{\pi(n!)^3},$$

$$P_2 = \frac{\cos(\sqrt{3}i\pi/2)}{\pi} \approx 2.4281897920988703287360414361791.$$

Example 6.18. Find the sum of the infinite series

$$S = 1 - \frac{1}{2} + \frac{1 \cdot 3}{2 \cdot 4 \cdot 6} - \frac{1 \cdot 3 \cdot 5}{2 \cdot 4 \cdot 6} + \frac{1 \cdot 3 \cdot 5 \cdot 7}{2 \cdot 4 \cdot 6 \cdot 8} - \frac{1 \cdot 3 \cdot 5 \cdot 7 \cdot 9}{2 \cdot 4 \cdot 6 \cdot 8 \cdot 10} + \cdots.$$

Solutions. Just neglect the first term in S, the remaining part can be regarded as the sum of a series, whose general term is the finite product of a sequence given by

$$a_n = (-1)^n \prod_{k=1}^{n}(2k-1)/(2k), \quad n = 1, 2, \ldots$$

Since the general term a_n is the product, it can be computed by symprod(). Then, function symsum() can be used to find the sum. Finally, the 1 can be added back to the sum. The following statements can be used to find the sum $S = \sqrt{2}/2$:

```
>> syms k n,
   S=1+symsum((-1)^n*symprod((2*k-1)/(2*k),k,1,n),n,1,inf)
```

6.3.2 Products of functional sequences

Definition 6.9. If the general term of the sequence is a function of x, the sequence is referred to as a functional sequence.

Here, the product of a functional sequence is studied.

Example 6.19. Compute $P = \prod_{n=1}^{\infty}\left(1 + \frac{x}{n}\right)e^{-x/n}$.

Solutions. The original problem can be solved directly with

```
>> syms n x; P=symprod((1+x/n)*exp(-x/n),n,1,inf) % direct solution
```

Piecewise function can be obtained as

$$P = \begin{cases} 0, & x \text{ is a negative integer,} \\ e^{-\gamma x}/\Gamma(x+1), & \text{otherwise, where } \gamma \text{ is the Euler constant.} \end{cases}$$

In fact, it is known that when x is a negative integer, $\Gamma(x+1) = \pm\infty$. Therefore, the whole result can be simplified manually as $P = e^{-\gamma x}/\Gamma(x+1)$.

Example 6.20. Compute the limit of the infinite sequence product in Example 3.11.

$$\lim_{x \to 0} \frac{1 - \cos x \sqrt{\cos 2x} \sqrt[3]{\cos 3x} \cdots \sqrt[n]{\cos nx}}{x^2}$$

Solutions. With the updated knowledge, do you feel it is much easier to handle again the problem in Example 3.11? Since there is a product of a sequence in the numerator, with a general term of $\sqrt[k]{\cos kx}$, $k = 1, 2, \ldots, n$, the function `symprod()` can be used to find the product of the first n terms. The limit can then be found directly as $L = n(n + 1)/4$.

```
>> syms n x k;
   L=limit((1-symprod(cos(k*x)^(1/k),k,1,n))/x^2,x,0)
```

6.3.3 Convergence test for the products of positive sequences

A direct test for the convergence of the products of sequences may be rather difficult. Now, let us consider the property $\ln ab = \ln a + \ln b$ ($a > 0, b > 0$), then the convergence test for a sequence product can be converted into the test for an infinite series, which was well established and presented in Section 6.2.

Theorem 6.4. *For the positive sequence given in Definition 6.8,*

$$\prod_{k=k_0}^{k_n} a_k \text{ is convergent if and only if } \sum_{k=k_0}^{k_n} \ln a_k \text{ is convergent.} \quad (6.3.2)$$

It can be seen that the test for an infinite series of b_k, where $b_k = \ln a_k$ is simpler, and the methods are more mature.

Example 6.21. Check the convergence of the following infinite product:

$$P = \left(1 + \frac{1}{\sqrt{1}}\right)\left(1 - \frac{1}{\sqrt{3}}\right)\left(1 - \frac{1}{\sqrt{5}}\right)\left(1 + \frac{1}{\sqrt{2}}\right)\left(1 - \frac{1}{\sqrt{7}}\right)\left(1 - \frac{1}{\sqrt{9}}\right)\left(1 + \frac{1}{\sqrt{3}}\right)\cdots.$$

Solutions. In the analysis of series and products of sequences, the most important step is to find the general term. Careful introduction of the given expression is crucial. For the sequence given in this example, if the terms are grouped in triplets, the general term can be established as

$$a_k = \left(1 + \frac{1}{\sqrt{k}}\right)\left(1 - \frac{1}{\sqrt{4k - 1}}\right)\left(1 - \frac{1}{\sqrt{4k + 1}}\right), \quad k = 1, 2, \ldots$$

With the general term, the convergence of the product can be checked with the following statements:

```
>> syms k;
   a(k)=(1+1/sqrt(k))*(1-1/sqrt(4*k-1))*(1-1/sqrt(4*k+1));
   u(k)=log(a(k)); L0=limit(u(k),k,inf)
   L1=limit(u(k+1)/u(k),k,inf), L2=limit(k*(u(k)/u(k+1)-1),k,inf)
```

In the commands, a_k is the general term of the infinite product, while u_k is the general term after taking logarithms. The convergence of the series of u_k can be investigated instead. It can be seen that $L_0 = 0$, so the necessary condition is satisfied. Also it is found that $L_1 = L_2 = 1$, hence Criteria (1) and (2) fail to give an answer. Criterion (3), i. e., the Bertrand test, can be used

```
>> L3=limit(log(k*(k*(u(k)/u(k+1)-1)-1)),k,inf)
```

Since $L_3 \rightarrow +\infty > 1$, the series of u_k is convergent, which means that the infinite product a_k is also convergent.

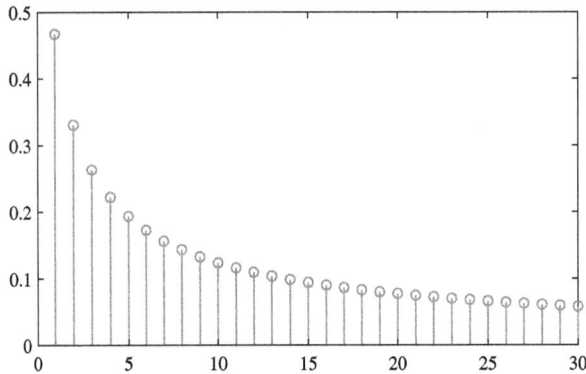

Figure 6.3: Evolution trends in the product of sequence.

The following statements can be used to draw the evolution trends for the original product of sequences, as shown in Figure 6.3. It can be seen that, with the increase of the number of terms, the sequence product may settle at a certain point.

```
>> k0=1:30; y0=1; y=[];
   for k1=k0, y0=y0*double(a(k1)); y=[y, y0]; end, stem(k0,y)
```

6.4 Taylor series

Power series is also known as polynomial series, and the structure is rather simple and easy to construct. In this section, the mathematical form of a power series is given, and Taylor series approximations to given functions are studied.

Definition 6.10. The general form of a power series is

$$P(x) = a_1 + a_2x + a_3x^2 + \cdots = \sum_{i=1}^{\infty} a_i x^{i-1}, \tag{6.4.1}$$

and its convergence region is $|x| < \lim_{i \to \infty} |a_i/a_{i+1}|$, when the latter limit exists.

6.4.1 Taylor series expansions for univariate functions

Taylor series was named after British mathematician Brook Taylor (1685–1731), who studied arbitrary function approximations by power series. An introduction to Taylor series expansion is presented first, and then we concentrate on finding Taylor series and assess their behavior.

Theorem 6.5. *The Taylor series expansion of $f(x)$ at $x = 0$ is given by*

$$f(x) = a_1 + a_2 x + a_3 x^2 + \cdots + a_k x^{k-1} + o(x^k), \tag{6.4.2}$$

where the coefficients a_i can be evaluated from

$$a_i = \frac{1}{(i-1)!} \lim_{x \to 0} \frac{d^{i-1}}{dx^{i-1}} f(x), \quad i = 1, 2, 3, \ldots \tag{6.4.3}$$

Definition 6.11. Taylor series about $x = 0$ is also known as Maclaurin series.

Theorem 6.6. *Taylor series about $x = a$ can be expressed as*

$$f(x) = b_1 + b_2(x-a) + b_3(x-a)^2 + \cdots + b_k(x-a)^{k-1} + o[(x-a)^k] \tag{6.4.4}$$

where the coefficients b_i can be evaluated from

$$b_i = \frac{1}{(i-1)!} \lim_{x \to a} \frac{d^{i-1}}{dx^{i-1}} f(x), \quad i = 1, 2, 3, \ldots \tag{6.4.5}$$

Symbolic Math Toolbox function `taylor()` can be used to construct Taylor series

`F=taylor(f,x,'Order',k)`, %kth order Taylor series about $x = 0$
`F=taylor(f,x,a,'Order',k)`, %kth order Taylor series about $x = a$

where f is the function expression, x is the independent variable, k is the order. If x is the only symbolic variable in f, it can be omitted. The default order k is 6. If a is not given, $a = 0$ is assumed. The following examples will be used to demonstrate Taylor series problems.

Example 6.22. Consider the function $f(x) = \sin x/(x^2 + 4x + 3)$ in Example 4.4. Find the first nine terms in the Taylor series and observe its fitting quality.

Solutions. The following statements can be used, and the function `taylor()` can be employed to find the first nine terms in Taylor (Maclaurin) series.

```
>> syms x; f(x)=sin(x)/(x^2+4*x+3);
   y=taylor(f,x,'Order',9) % Taylor series expansion
```

The Taylor (Maclaurin) series obtained is

$$y(x) \approx -\frac{386\,459}{918\,540}x^8 + \frac{515\,273}{1\,224\,720}x^7 - \frac{3\,067}{7\,290}x^6 + \frac{4\,087}{9\,720}x^5 - \frac{34}{81}x^4 + \frac{23}{54}x^3 - \frac{4}{9}x^2 + \frac{1}{3}x.$$

Infinite series is used in Taylor series approximation to given functions. So the immediate question is: If only a few terms are used, what is the fitting quality? Yet another question is: How many terms should one use in Taylor series approximation in a particular problem? In traditional calculus courses, since there was no support from powerful computer tools, these questions could not be answered. What was provided was merely a conservative mathematical notation $o(x^n)$.

With MATLAB, these questions can be answered immediately with ease. In Figure 6.4, the fitting over the interval $(-1, 1)$ with nine terms is shown, and it is seen that when x is large, the fitting is not good.

```
>> fplot([f,y],[-1,1]), ylim([-2.4 0.3])   % comparisons
```

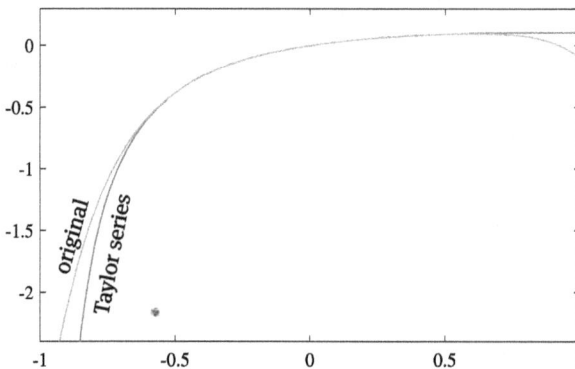

Figure 6.4: Finite Taylor series approximation.

Zooming facilities in MATLAB graphics window can be tried, and the interval with satisfactory fitting quality can be found. It can be observed that for the Taylor series expansion, the fitting in the interval $[-0.6, 0.6]$ is satisfactory.

Example 6.23. Consider the function in Example 6.22. If we want to have good fitting in the interval $(-0.7, 0.7)$, how many terms of Taylor series should be retained?

Solutions. In mathematics it can be shown that the error of Taylor series approximation is an infinitesimal of x^k, however, this conservative mathematical notation may not be useful in this particular example. With the powerful tools provided in MATLAB, this question can be answered immediately. We can try different orders and visualize the results. In this way, the number of terms can be shown.

```
>> syms x; f(x)=sin(x)/(x^2+4*x+3);
   n=15, f1=taylor(f,'Order',n); fplot([f1 f],[0.52 0.7])
```

After some trials, it is found that the 14th order Taylor series yields the best fitting. Further order increase may not improve the fitting quality. The zoomed comparison is shown in Figure 6.5, which yields the best quality.

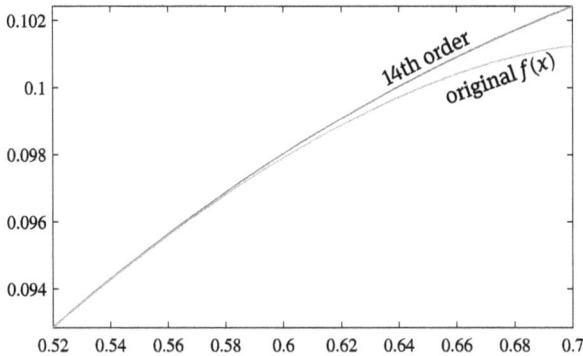

Figure 6.5: High-order fitting quality (zoomed plot).

Example 6.24. Consider again the function in Example 6.22. Find the Taylor series expansion about $x = 2$ and $x = a$.

Solutions. The following statements can be used to find the Taylor series expansion about $x = 2$.

```
>> F=taylor(f,x,2,'Order',9)   % the result is lengthy
```

the first three terms are

$$f(x) \approx \frac{\sin 2}{15} + \left(\frac{\cos 2}{15} - \frac{8\sin 2}{225}\right)(x-2) - \left(\frac{127\sin 2}{6750} + \frac{8\cos 2}{225}\right)(x-2)^2 + \cdots.$$

If one wants to expand about $x = a$, the following commands are needed:

```
>> syms a; taylor(f,x,a,'Order',5)  % axpand about x = a
```

Again the first three terms are displayed here

$$f(x) \approx \frac{\sin a}{a^2 + 3 + 4a} + \left[\frac{\cos a}{a^2 + 3 + 4a} - \frac{(4+2a)\sin a}{(a^2+3+4a)^2}\right](x-a)$$
$$+ \left[-\frac{\sin a}{(a^2+3+4a)^2} - \frac{\sin a}{2(a^2+3+4a)}\right.$$
$$\left.- \frac{(a^2\cos a + 3\cos a + 4a\cos a - 4\sin a - 2a\sin a)(4+2a)}{(a^2+3+4a)^3}\right](x-a)^2.$$

Example 6.25. Approximate sinusoidal function $y = \sin x$ with Taylor series, and observe the fitting quality for different orders.

Solutions. The following MATLAB statements can be tried in a loop structure. Taylor series of different orders can be obtained and compared in Figure 6.6. It can be seen that when the order is increased, the fitting quality gets better. For this example, a good choice is $n = 16$, and the fitting is satisfactory in the interval $(-2\pi, 2\pi)$.

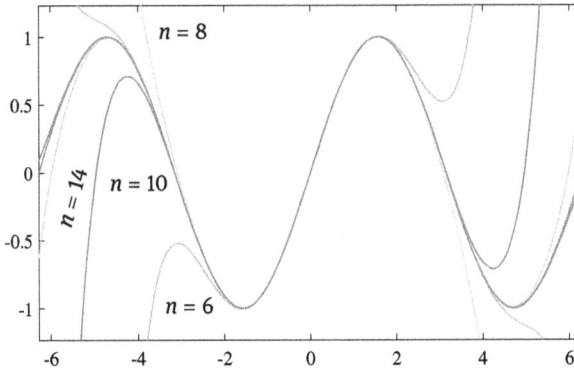

Figure 6.6: Taylor series approximation to a sinusoidal function.

```
>> syms x; y=sin(x); fplot(y), hold on % hold the coordinates
   for n=[6:2:16], p=taylor(y,x,'Order',n),
   fplot(p), end, hold off % try Taylor approximations under different orders
```

where the 16th order Taylor series expansion is

$$\sin x \approx x - \frac{x^3}{6} + \frac{x^5}{120} - \frac{x^7}{5\,040} + \frac{x^9}{362\,880} - \frac{x^{11}}{39\,916\,800} + \frac{x^{13}}{6\,227\,020\,800} - \frac{x^{15}}{1\,307\,674\,368\,000}.$$

6.4.2 Taylor series for multivariate functions

Definition 6.12. The Taylor series expansion of a multivariate function $f(\boldsymbol{x}) = f(x_1, x_2, \ldots, x_n)$ is

$$
\begin{aligned}
f(\boldsymbol{x}) = f(\boldsymbol{a}) + & \left[(x_1 - a_1)\frac{\partial}{\partial x_1} + \cdots + (x_n - a_n)\frac{\partial}{\partial x_n} \right] f(\boldsymbol{x}) \bigg|_{\boldsymbol{x}=\boldsymbol{a}} \\
& + \frac{1}{2!} \left[(x_1 - a_1)\frac{\partial}{\partial x_1} + \cdots + (x_n - a_n)\frac{\partial}{\partial x_n} \right]^2 f(\boldsymbol{x}) \bigg|_{\boldsymbol{x}=\boldsymbol{a}} + \cdots \qquad (6.4.6) \\
& + \frac{1}{k!} \left[(x_1 - a_1)\frac{\partial}{\partial x_1} + \cdots + (x_n - a_n)\frac{\partial}{\partial x_n} \right]^k f(\boldsymbol{x}) \bigg|_{\boldsymbol{x}=\boldsymbol{a}} + \cdots,
\end{aligned}
$$

where $\boldsymbol{a} = [a_1, a_2, \ldots, a_n]$ is the center point is the Taylor series.

With MATLAB function `taylor()`, the Taylor series for multivariate functions can be obtained directly, with the syntax

F=taylor$(f, [x_1, x_2, \ldots, x_n], [a_1, a_2, \ldots, a_n],$'Order'$, k)$

where $k - 1$ is the highest order and f is the expression of the multivariate function. The center point can also be specified in the function call.

Example 6.26. For the function $f(x, y) = (x^2 - 2x)e^{-x^2-y^2-xy}$ studied in Example 4.15, find the Taylor series expansion.

Solutions. The Taylor series about the origin can be found directly

```
>> syms x y;
   f(x,y)=(x^2-2*x)*exp(-x^2-y^2-x*y);  % the original function
   F=taylor(f,[x,y],[0,0],'Order',8);   % Taylor series expansion
   collect(F,x),  xx=[-0.6 0.45];       % collect terms
   fsurf(f,xx), hold on; fsurf(F,xx), hold off  % observe results
```

Different fitting regions around the origin can be tried, and it can be found that in the interval $(-0.6, 0.45)$ the fitting is satisfactory. The surfaces of the original function and the Taylor series can be drawn together, as shown in Figure 6.7. It can be seen that the two surfaces are close, and the maximum error is about 10^{-3}.

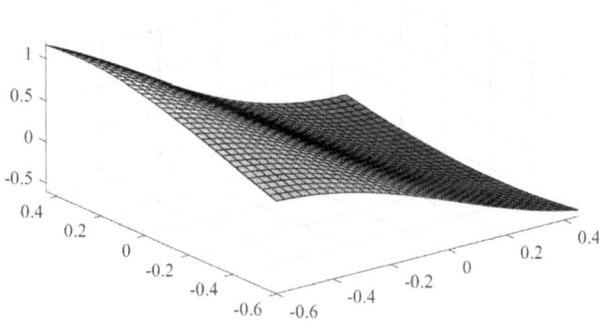

Figure 6.7: Taylor series fitting of a 2D function.

The 2D Taylor series expansion can also be found as

$$F(x,y) \approx \frac{x^7}{3} + \left(y + \frac{1}{2}\right)x^6 + (2y^2 + y - 1)x^5 + \left(\frac{7y^3}{3} + \frac{3y^2}{2} - 2y - 1\right)x^4$$

$$+ (2y^4 + y^3 - 3y^2 - y + 2)x^3 + \left(y^5 + \frac{y^4}{2} - 2y^3 - y^2 + 2y + 1\right)x^2$$

$$+ \left(\frac{y^6}{3} - y^4 + 2y^2 - 2\right)x.$$

Selecting the center point at $x = 1$, $y = a$, the Taylor series expansion can also be found.

```
>> syms a; F=taylor(f,[x,y],[1,a],'Order',3),
   F1(x)=simplify(F)
```

The simplified result is

$$F(x,y) \approx -e^{-a^2-a-1}[(a/2+1)(a+2)-2](x-1)^2 - e^{-a^2-a-1}(2a+1)(a-y)$$
$$-e^{-a^2-a-1}(a-y)^2[(2a+1)(a+1/2)-1] + e^{-a^2-a-1}(a+2)(x-1) - e^{-a^2-a-1}$$
$$+e^{-a^2-a-1}(a-y)(x-1)[(2a+1)(a/2+1)+(a+2)(a+1/2)-1],$$

$$F_1(x) = -\frac{1}{2}e^{-a^2-a-1}(4a^4 - 4a^3x - 8a^3y + 8a^3 + a^2x^2 + 4a^2xy - 12a^2x + 4a^2y^2 - 12a^2y$$
$$+14a^2 + 4ax^2 + 10axy - 12ax + 4ay^2 - 12ay + 10a + 2xy - 4x - y^2 - 4y + 6).$$

6.5 Fourier series expansions

While French mathematician Jean-Baptiste Joseph Fourier (1768–1830) was studying the heat transfer and vibration problems, he introduced the method to approximate periodic signals with trigonometric functions. Such approximations are now called Fourier series.

In this section, Fourier series approximation is presented, followed by a universal MATLAB function, with which Fourier series and fitting quality assessment are made possible with MATLAB.

6.5.1 Mathematical description of Fourier series

Before analyzing further Fourier series expansions of ordinary functions, the function defined in the symmetric interval $(-L, L)$ is presented.

Definition 6.13. Consider a given periodic function $f(x)$, where $x \in [-L, L]$, and the period is $T = 2L$. The function values in other intervals can be artificially extended such that $f(x) = f(kT + x)$, where k is an integer. The Fourier series of such periodic function can be written as

$$f(x) = \frac{a_0}{2} + \sum_{n=1}^{\infty} \left(a_n \cos \frac{n\pi}{L} x + b_n \sin \frac{n\pi}{L} x \right), \tag{6.5.1}$$

where

$$
\begin{cases}
a_n = \dfrac{1}{L} \displaystyle\int_{-L}^{L} f(x) \cos \dfrac{n\pi x}{L}\, dx, & n = 0, 1, 2, \ldots, \\[6mm]
b_n = \dfrac{1}{L} \displaystyle\int_{-L}^{L} f(x) \sin \dfrac{n\pi x}{L}\, dx, & n = 1, 2, 3, \ldots
\end{cases}
\tag{6.5.2}
$$

The series is referred to as Fourier series, while a_n, b_n are known as Fourier coefficients.

Similar to Taylor series expansion, finite-term Fourier series can also be used to approximate periodic functions, for instance, for the first p pairs of terms, (6.5.1) can be rewritten as

$$
f(x) \approx \frac{a_0}{2} + \sum_{n=1}^{p} \left(a_n \cos \frac{n\pi}{L} x + b_n \sin \frac{n\pi}{L} x \right).
\tag{6.5.3}
$$

Definition 6.14. If $f(x)$ is defined in an arbitrary interval $x \in (a, b)$, the period $L = (b - a)/2$ can be found. Introducing a new variable $\hat{x}$, and using variable substitution $x = \hat{x} + L + a$, function $f(\hat{x})$ is then mapped to a function in $(-L, L)$. Performing Fourier series expansion to the result, and then mapping back with $\hat{x} = x - L - a$, the Fourier series of $f(x)$ can be obtained.

6.5.2 MATLAB implementation of Fourier series

There is no existing MATLAB function which can be used in dealing with Fourier series and coefficient problems. Based on the above mathematical presentation, analytical solutions of Fourier series expansions can be obtained using

```
function [F,A,B]=fseries(f,x,varargin)
[p,a,b]=default_vals({6,-pi,pi},varargin{:}); L=(b-a)/2; % defaults
if a+b, f=subs(f,x,x+L+a); end, A=int(f,x,-L,L)/L; % variable substitution
B=[]; F=A/2;
for n=1:p % finding Fourier series in loop structure
    an=int(f*cos(n*pi*x/L),x,-L,L)/L; bn=int(f*sin(n*pi*x/L),x,-L,L)/L;
    A=[A,an]; B=[B,bn]; F=F+an*cos(n*pi*x/L)+bn*sin(n*pi*x/L);
end
if a+b, F=subs(F,x,x-L-a); end % if asymmetrical, mapping back
```

A low-level function `default_vals()` can be written to accept default values. This function is also used later. The listing of the function is

```
function varargout=default_vals(vals,varargin) % subfunction
if nargout=length(vals), error('number of arguments mismatch');
else, nn=length(varargin)+1; % assign default values
    varargout=varargin; for i=nn:nargout, varargout{i}=vals{i};
end, end, end
```

The syntax of fseries() function is [F,A,B]=fseries(f,x,p,a,b), where f is the given function, x is the independent variable, p is the number of terms in expansion, with a default value of 6; a and b are the bounds of the interval of x, with the default being $[-\pi, \pi]$. The returned vectors A and B are the Fourier coefficients, while F is the Fourier series expression.

Example 6.27. Find Fourier series for the function $y = x(x - \pi)(x - 2\pi)$, $x \in (0, 2\pi)$.

Solutions. The expected Fourier series can be obtained immediately with

```
>> syms x; f=x*(x-pi)*(x-2*pi);
   [F,A,B]=fseries(f,x,12,0,2*pi); F % Fourier series
```

The Fourier series with 12 terms can be expressed as

$$f(x) = 12\sin x + \frac{3}{2}\sin 2x + \frac{4}{9}\sin 3x + \frac{3}{16}\sin 4x + \frac{12}{125}\sin 5x + \frac{1}{18}\sin 6x + \frac{12}{343}\sin 7x$$
$$+ \frac{3}{128}\sin 8x + \frac{4}{243}\sin 9x + \frac{3}{250}\sin 10x + \frac{12}{1331}\sin 11x + \frac{1}{144}\sin 12x.$$

In fact, closely observing the results, it is easy to summarize:

$$f(x) \approx \sum_{n=1}^{\infty} \frac{12}{n^3}\sin nx.$$

With the following commands, the 12th order Fourier series can be obtained, and the fitting is shown in Figure 6.8. It can be seen that the fitting quality is very good, and it is almost not possible to find the error between the curves.

```
>> fplot([f,F],[0,2*pi]) % comparison
```

It is natural to compare the fitting quality over a larger interval, e. g., $x \in (-\pi, 3\pi)$. The following commands can be used:

```
>> fplot([f,F],[-pi,3*pi]) % larger interval
```

The fitting comparison is shown in Figure 6.9. It can be seen that in the interval $(0, 2\pi)$, the fitting is perfect, while outside this region, it is completely different from the original function. Since the "original function" was redefined as a periodic function, it is not the function $y = x(x - \pi)(x - 2\pi)$ there.

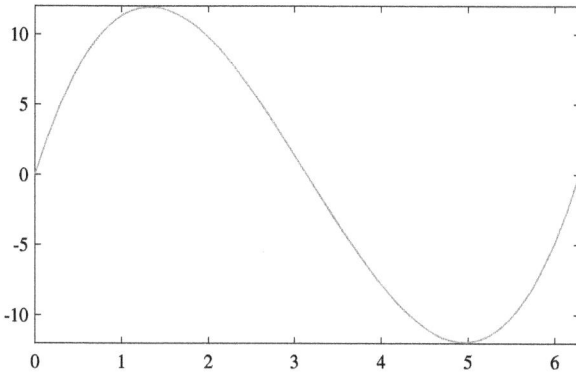

Figure 6.8: Fourier series fitting with finitely many terms.

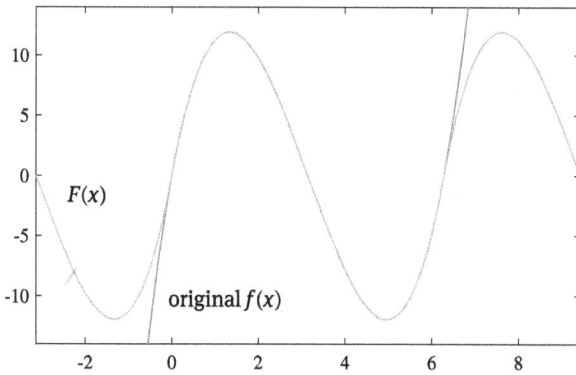

Figure 6.9: Fourier series fitting over a larger interval.

Example 6.28. Consider the square-wave signal defined in $(-\pi, \pi)$. When $x \geqslant 0$, $y = 1$, otherwise, $y = -1$. Construct the Fourier series to approximate the square-wave, and find out how many terms are needed to approximate the signal.

Solutions. The square-wave can be mathematically expressed by $f(x) = |x|/x$. Some samples on the x axis can be generated. The point $x = 0$ can be removed, in its place two more points $-\epsilon$, ϵ can be added, and the sample vector can be sorted.

Fourier series of different orders can be tried, and compared as shown in Figure 6.10.

```
>> syms x; f(x)=abs(x)/x;   % square-wave signal
   xx=[-pi:pi/200:pi]; xx=xx(xx~=0); xx=sort([xx,-eps,eps]);
   yy=double(f(xx)); plot(xx,yy),   % draw the function
   for n=1:20,      % try different orders and compare
       f1=fseries(f,x,n); y1=subs(f1,x,xx); line(xx,y1);
   end
```

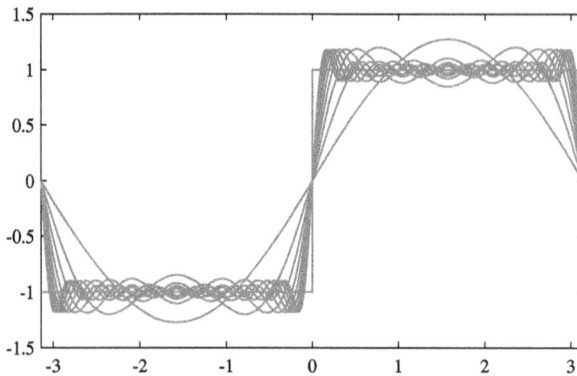

Figure 6.10: Approximation of the square-wave signal.

It can be seen that when the order is selected around 10, the fitting is good. Even though the order is increased, the fitting may not be improved significantly.

Letting $n = 14$, the Fourier series can be constructed with

```
>> f1=fseries(f,x,14) % 14th order Fourier series
```

It is found that

$$f(x) \approx 4\frac{\sin x}{\pi} + \frac{4\sin 3x}{3\pi} + \frac{4\sin 5x}{5\pi} + \frac{4\sin 7x}{7\pi} + \frac{4\sin 9x}{9\pi} + \frac{4\sin 11x}{11\pi} + \frac{4\sin 13x}{13\pi}.$$

The general form of the series can be established as

$$f(x) \approx \frac{4}{\pi} \sum_{k=1}^{\infty} \frac{\sin(2k-1)x}{2k-1}.$$

6.6 Rational function approximation of univariate functions

Apart from the Taylor and Fourier series studied earlier, continued fractions and rational functions can also be used to approximate given functions. In this section, these methods will be explored.

6.6.1 Continued fraction expansions

The continued fraction expansion is a very effective form in approximating functions or even numbers. Some examples are given here to demonstrate the use of continued fractions.

Definition 6.15. The general form of a continued fraction can be expressed as

$$f(x) = b_1 + \cfrac{(x-a)^{c_1}}{b_2 + \cfrac{(x-a)^{c_2}}{b_3 + \cfrac{(x-a)^{c_3}}{b_4 + \cfrac{(x-a)^{c_4}}{b_5 + \cfrac{(x-a)^{c_5}}{\cdots}}}}}, \tag{6.6.1}$$

where b_i are constants, c_i are rational numbers, and a is the center point.

There is no function provided in MATLAB for finding a continued fraction approximation, while a low-level function contfrac() is provided in MuPAD, and an interface can be designed so that contfrac() can be called from MATLAB. The syntax of the function is

cf=contfrac(f,n) or [cf,r]=contfrac(f,n,a)

where f is the symbolic expression of the original function, or a constant, x is the independent variable, a is the center point, with a default value 0, n is the number of expected stages in the expansion. The returned variable cf is the MuPAD expression of the continued fraction, r is the rational function expression.
    The listing of the interface function is

```
function [cf,r]=contfrac(f,varargin)
[n,a]=default_vals({6,0},varargin{:});   % defaults
if isanumber(f),  cf=feval(symengine,'contfrac',f,n);  % for a number
     p1=char(cf);  k=strfind(p1,',');  k1=strfind(p1,'/');
     if nargout>1,  r=sym(p1(k(end)+1:k1-1))/sym(p1(k1+1:end-1));  end
else, if isfinite(a),  str=num2str(a); else, str='infinity'; end
     cf=feval(symengine,'contfrac',f,['x=' str],n);
     if nargout>1,  r=feval(symengine,'contfrac::rational',cf);
end, end
%   , a low-level function to check a number of function
function key=isanumber(a)   % subfunction
key=0; if length(a)~=1,  return;  end  % if not scalar, returns 0
switch class(a)   % check data type
   case 'double',  key=1;     % if double precision, returns 1
   case 'sym',  try, double(a); key=1; catch   % a symbolic number?
end, end  % if input is number, returns 1, otherwise 0
```

Example 6.29. Find the continued fraction of a rational approximation of π.

Solutions. Assuming the number of stages is 15, the following statements can be used in LaTeX.

Table 6.2: Rational approximation to π.

stage	rational number	digits	stage	rational number	digits
11	312 689/99 532	11	16	165 707 065/52 746 197	16
12	1 146 408/364 913	12	17	411 557 987/131 002 976	18
13	5 419 351/1 725 033	14	18	1 068 966 896/340 262 731	18
14	5 419 351/1 725 033	15	19	6 167 950 454/1 963 319 607	20
15	80 143 857/25 510 582	16	20	14 885 392 687/4 738 167 652	20

```
>> [cf r]=contfrac(pi,15), latex(cf) % approximate π
```

The continued fraction expression of π can be expressed as

$$\pi \approx 3 + \cfrac{1}{7 + \cfrac{1}{15 + \cfrac{1}{1 + \cfrac{1}{292 + \cfrac{1}{1 + \cfrac{1}{1 + \cfrac{1}{1 + \cfrac{1}{2 + \cfrac{1}{1 + \cfrac{1}{3 + \cfrac{1}{1 + \cfrac{1}{14 + \cdots}}}}}}}}}}}.$$

A rational approximation of π can also be obtained, with 16 digits of accurate results, as $r = 80\,143\,857/25\,510\,582$.

It can be seen from the continued fraction that the term 292 in the denominator is relatively large, compared with others. Truncation can be made after this term, and the rational approximation obtained is $103\,993/33\,102 \approx 3.141592653012$, which is already very close to π. If in the command, 15 is replaced by other numbers the approximations are listed in Table 6.2. It can be seen that a continued fraction is a very effective tool in number and function approximation.

If the command [cf,r]=contfrac(pi,120) is used, the approximated π is

$$\pi \approx \frac{1\,244\,969\,988\,745\,789\,040\,106\,366\,155\,256\,015\,114\,976\,454\,553\,337\,906\,033\,890\,313}{396\,286\,255\,419\,907\,262\,612\,262\,286\,579\,004\,636\,343\,912\,658\,949\,984\,351\,074\,158}.$$

The error of the rational approximation to π can be measured with the following statement, and the error is 2.975×10^{-120}:

```
>> 10^log10(abs(vpa(sym(['1244969988745789040106366155256015601511497',...
   '6454553337906033890313/3962862554199072626122262286657900463',...
   '6343912658949984351074158-pi']),200))))
```

Example 6.30. Find the continued fraction for $f(x) = e^{-x} \sin x/(x+1)^3$, and also find the rational function approximation.

Solutions. The original function should be entered first, then the continued fraction with 10 stages can be found with the following statements:

```
>> syms x; f(x)=sin(x)*exp(-x)/(x+1)^3;
   [cf,y]=contfrac(f,10) % continued fraction
```

and the result obtained is

$$f(x) \approx \cfrac{x}{1 + \cfrac{x}{\cfrac{1}{4} + \cfrac{x}{-\cfrac{12}{5} + \cfrac{x}{\cfrac{25}{43} + \cfrac{x}{-\cfrac{7396}{1685} + \cfrac{x}{\cfrac{2839\,225}{4\,863\,128} + \cfrac{x}{-\cfrac{44\,767\,468\,256}{2\,592\,461\,805} + \cdots}}}}}}}.$$

Also, from the 8- and 10-stage continued fraction, rational function approximations can be obtained via

```
>> [cf1,f8]=contfrac(f,8), [cf2,f10]=contfrac(f,10) % rational function
```

The two rational functions can be obtained as

$$f_8(x) \approx \frac{-8\,457\,130x^4 + 49\,735\,600x^3 - 118\,414\,380x^2 + 107\,698\,710x}{-5\,864\,273x^4 + 83\,147\,900x^3 + 294\,069\,480x^2 + 312\,380\,460x + 107\,698\,710},$$

$$f_{10}(x) \approx \frac{\begin{aligned}-170\,455\,846\,739x^5 + 472\,453\,225\,650x^4 + 3\,615\,529\,382\,220x^3\\ -\,20\,275\,122\,684\,600x^2 + 28\,175\,852\,788\,020x\end{aligned}}{\begin{aligned}2\,071\,713\,977\,216x^5 + 14\,187\,032\,489\,655x^4 + 58\,214\,153\,847\,990x^3\\ +\,110\,354\,057\,230\,620x^2 + 92\,428\,288\,467\,480x + 28\,175\,852\,788\,020\end{aligned}}.$$

The curves obtained by the fitting functions and the original function over the interval $(0,2)$ can be seen in Figure 6.11 using

```
>> fplot([f,f8,f10],[0,2])
```

It can be observed that the fitting is satisfactory, and the fitting when $n = 10$ is better, since the curves cannot be distinguished from the figure. If a larger interval $(0,5)$ is compared, the curves in Figure 6.12 can be obtained. It can be seen that when the in-

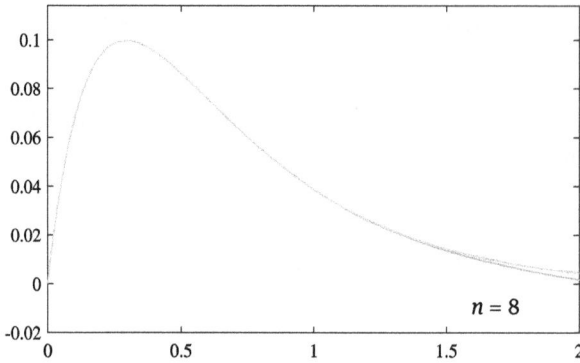

Figure 6.11: Fitting quality over the interval $(0, 2)$.

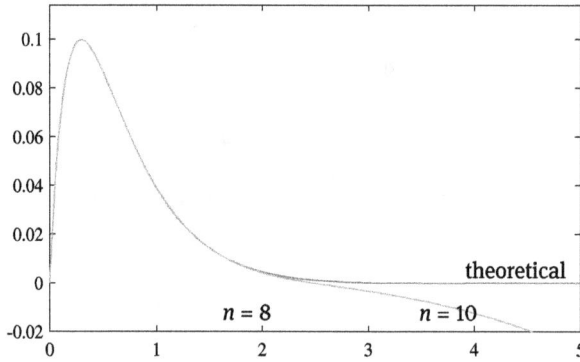

Figure 6.12: Fitting over a larger interval.

terval is increased, the fitting becomes poorer. Further increasing the number of stages in the continued fraction, the fitting quality may be improved.

```
>> fplot([f,f8,f10],[0,5])
```

An alternative solution is to select another center point, e. g., $x = 0.5$, and the following statements can be used:

```
>> [cf1,f8]=contfrac(f,8,0.5), fplot([f,f8],[0,5])
```

The rational function obtained is

$$f_8(x) \approx -\frac{5.79734386x^4 - 41.483839x^3 + 111.860576x^2 - 110.51792x + 0.004}{116.95748x^3 + 329.5053x^2 + 330.558218x + 110.44266}.$$

The corresponding continued fraction expansion is

$$f_1(x) \approx 0.0862 + \cfrac{x - 0.5}{-9.9242 + \cfrac{x - 0.5}{0.1429 + \cfrac{x - 0.5}{1.3 + \cfrac{x - 0.5}{-0.2942 + \cfrac{x - 0.5}{22.139 + \cfrac{x - 0.5}{0.1588 + \cfrac{x - 0.5}{-33.69 + \cdots}}}}}}}.$$

6.6.2 Padé approximations

Padé approximation uses rational functions to fit any function. This technique was proposed by a French mathematician Henri Eugène Padé (1863–1953) in 1890.

Definition 6.16. For a given function $f(x)$, and the expected orders of numerator and denominator being $m \geqslant 0$, $n \geqslant 1$, the rational function can be written as

$$R(x) = \frac{b_0 + b_1 x + b_2 x^2 + b_3 x^3 + \cdots + b_m x^m}{1 + a_1 x + a_2 x^2 + a_3 x^3 + \cdots + a_n x^n}. \tag{6.6.2}$$

If such rational function is used to approximate a given function $f(x)$, it is referred to as a Padé approximant.

A low-level MuPAD function pade() is provided, and an interface padefrac() can be written as

```
function p=padefrac(f,varargin)
[x,n,m]=default_vals(symvar(f),2,2,varargin:);
orders=['[' int2str(n) ',' int2str(m) ']'];
p=feval(symengine,'pade',f,x,orders);
```

The syntax of the function is p=padefrac (f, x, m, n), where f is the original function, x is the independent variable, m and n are respectively the expected orders of the numerator and denominator, with the default being 2 for each. The returned p is the Padé approximation.

Example 6.31. Find the Padé approximation to the sinusoidal signal.

Solutions. The sine function can be entered in MATLAB, and the 3rd to 6th order Padé approximants can be obtained and compared in Figure 6.13. It can be seen that the 6th order Padé approximant is already very close to the sine function.

```
>> syms x; f(x)=sin(x); xx=[-2*pi,2*pi]; fplot(f,xx), hold on
   for n=3:7, P=padefrac(f,x,n,n), fplot(P,xx); end
   ylim([-1.5 1.5])
```

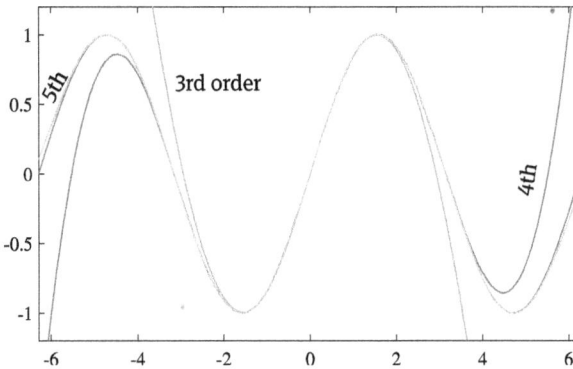

Figure 6.13: Fitting quality of Padé approximants.

It can be seen that the following Padé approximants are found:

$$P_3(x) = \frac{-7x^3 + 60x}{3(x^2 + 20)}, \quad P_4(x) = \frac{x(551x^4 - 22\,260x^2 + 166\,320)}{15(5x^4 + 364x^2 + 11\,088)},$$

$$P_5(x) = \frac{551x^5 - 22\,260x^3 + 166\,320x}{15(5x^4 + 364x^2 + 11\,088)},$$

$$P_6(x) = -\frac{x(479\,249x^6 - 52\,785\,432x^4 + 1\,640\,635\,920x^2 - 11\,511\,339\,840)}{7(2\,623x^6 + 453\,960x^4 + 39\,702\,960x^2 + 1\,644\,477\,120)}.$$

Example 6.32. Consider Example 6.30 again with the function

$$f(x) = \frac{e^{-x}\sin x}{(x + 1)^3}.$$

Find the 4th order Padé approximant and compare it with continued fraction approximation.

Solutions. The original function can be entered into MATLAB environment and the following statements can be used to approximate the function. It can be seen that the Padé approximant and the continued fraction model are the same.

```
>> syms x; f=exp(-x)*sin(x)/(x+1)^3;
   [cf1,y1]=contfrac(f,8), y2=padefrac(f,x,3,4), simplify(y1-y2)
```

6.7 Laurent series

6.7.1 Laurent series expansion

It has been shown that for a given function $f(t)$, Taylor series can approximate this function by polynomials of $(x - x_0)$ with positive powers. If the powers are extended such that both positive and negative integers are allowed, the series is a Laurent series. Laurent series was proposed by French mathematician Pierre Alphonse Laurent (1813–1854) in 1832.

Definition 6.17. If a function $f(z)$ is analytic in the disc $\mathscr{D}$: $R_1 < |z - z_0| < R_2$, and $0 \leqslant R_1 < R_2 < +\infty$, the Laurent series can be written as

$$f(z) = \sum_{k=-\infty}^{\infty} c_k (z - z_0)^k, \tag{6.7.1}$$

where the coefficients can be evaluated from

$$c_k = \frac{1}{2\pi j} \int_{|z-z_0|=\rho} \frac{f(\zeta)}{(\zeta - z_0)^{k+1}} \, d\zeta, \tag{6.7.2}$$

where $|z - z_0| = \rho$ is any circle satisfying $R_1 < \rho < R_2$. If $f(z)$ is analytic in $\mathscr{D}$, Laurent series is unique.

The computation of the coefficients from (6.7.2) is very complicated, so alternative approaches to construct Laurent series should be introduced.

If $f(z)$ can be decomposed as a product of two subfunctions, denoted as $f(z) = f_1(z)f_2(z)$, where $f_2(z)$ is suitable for ordinary Taylor series expansion, while the other, $f_1(z)$, can be expanded as a series of $(z - z_0)^k$, with k being a negative integer, variable substitution $x = 1/(z - z_0)$ can be performed first to transform the function into a function of x, i.e., using substitution $z = (1 + xz_0)/x$. Taylor series about x can be obtained, denoted as $F_1(x)$. Variable substitution $x = 1/(z - z_0)$ should be performed again, converting back to the function of z.

An example is given below to find Laurent series.

Example 6.33. Find the Laurent series for $f(z) = z^2 e^{1/z}$.

Solutions. Due to the existence of the term $e^{1/z}$, there might be odd behavior of this function at $z = 0$. However, $z = 0$ is not a pole, but an essential singularity. It can be seen that $f(z)$ can be factorized into a product of two functions, $f_1(z) = e^{1/z}$ and $f_2(z) = z^2$. Letting $z = 1/x$, the subfunction $f_1(x)$ can be expanded as a Taylor series or x. Then letting $x = 1/z$, we convert it back to a function of z. Multiplying by $f_2(z) = z^2$, Laurent series is obtained.

```
>> syms x z;  f1(z)=exp(1/z);  f2(z)=z^2;  % factorization
   f1a=f1(1/x);  F1a(x)=taylor(f1a,x,'Order',7);
   F=simplify(f2*F1a(1/z))        %    Laurent series
```

Laurent series can be obtained as

$$F(z) = z^2 + z + \frac{1}{2} + \frac{z^{-1}}{6} + \frac{z^{-2}}{24} + \frac{z^{-3}}{120} + \frac{z^{-4}}{720} + \frac{z^{-5}}{5\,040} + \cdots.$$

In a relatively large interval $z \in (-20, 20)$, the curves of the original function and its Laurent series can be drawn as shown in Figure 6.14. It can be seen that the two curves are almost identical, apart from the tiny neighborhood of the essential singularity $z = 0$.

```
>> fplot([F,f1*f2],[-20,20]),  hold on;  plot(0,1/6,'o')  % comparison
```

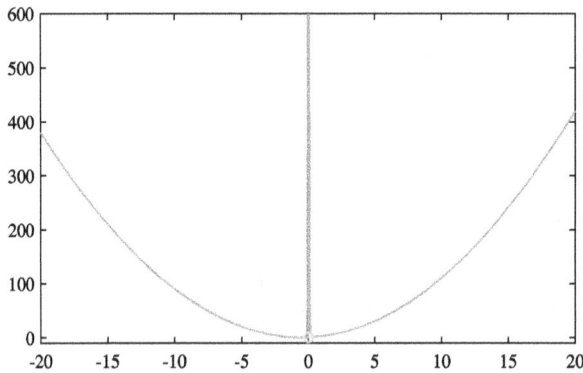

Figure 6.14: The curves of the original function and its Laurent series.

For the Laurent series $F(z)$, due to the existence of the essential singularity, there exist z^{-k} terms, and the residue is defined as c_{-1}, the coefficient of the term z^{-1}.

Example 6.34. Now consider the rational function $f(z) = 1/(z - 1) + 1/(z - 2j)$. How to find its Laurent series?

Solutions. It can be seen that function $f(z)$ is not analytic at the two poles, $z = 1$ and $z = 2j$. The z region can be partitioned into three regions: (1) the disc $|z| < 1$; (2) the ring $1 < |z| < 2$, and (3) and $\infty > |z| > 2$. Three cases should be considered for the Laurent series expansion:

(1) If $|z| < 1$, it means that $|z| < 2$ or $|z/2| < 1$ is satisfied. Taylor series is sufficient. The Taylor series formula can be used

$$\frac{1}{1 - u} = \sum_{k=0}^{\infty} u^k, \quad \text{if } |u| < 1, \text{ the series is convergent.}$$

The Laurent series, i. e., Taylor series in this case, can be written as

$$F_1(z) = \frac{-1}{1-z} + \frac{-1/2j}{1-z/(2j)} = -\sum_{k=0}^{\infty}\left(1 + \frac{1}{(2j)^{k+1}}\right)z^k.$$

(2) If $1 < |z| < 2$, function $f(z)$ is analytic. Since the Taylor series in z of $1/(z-1)$ is not convergent, this function should be rewritten as $(1/z)/(1-1/z)$. Now Taylor series expansion with respect to $1/z$ can be found. For the second term, Laurent series can be found

```
>> syms z x; f1(z)=1/(z-1); f1a=f1(1/x);      % factorization
   F2a(x)=taylor(f1a,'Order',6); F2=F2a(1/z) % Laurent series
```
Laurent series can be obtained as

$$F_2(z) = \sum_{k=-\infty}^{-1} z^k - \sum_{k=0}^{\infty} \frac{1}{(2j)^{k+1}}z^k.$$

(3) If $|z| > 2$, the two Taylor series in z are both divergent, so Taylor series expansion with respect to $1/z$ should be obtained

```
>> f3(z)=1/(z-1)+1/(z-2i); f3a=f3(1/x);
   F3a(x)=taylor(f3a,'Order',6); F3=F3a(1/z)
```
Laurent series in this case is

$$F_3(z) = \sum_{k=-\infty}^{-1}\left(1 + \frac{1}{(2j)^{k+1}}\right)z^k.$$

It can be seen that, in order to get $f(z)$ analytic in a domain, the z plane should be partitioned into three regions. Laurent series for each region can be found separately. The actual Laurent series can be expressed as a piecewise function.

6.7.2 Laurent series of rational functions

According to the above idea, the region can be partitioned automatically, based on the poles of the partial fraction expansion. Laurent series expansion for each partition can be obtained accordingly. A MATLAB function `laurent_series()` can be written as follows to compute piecewise Laurent series expansion for a given rational function:

```
function [F0,p,m,F]=laurent_series(f,n), [p,m]=poles(f);
STR=''; if nargin==1, n=6; end
syms z x; assume(z~=0); assume(x~=0); F2=0;
if length(p)==0, error('The poles cannot be found, failed.'); end
v=sort(unique([sym(0); abs(p)])); v0=[v; inf];
Fx=feval(symengine,'partfrac',f,'List'); % partial fraction expansion
```

```
nv=Fx(1); dv=Fx(2); f=feval(symengine,'partfrac',f);
for i=1:length(v), F1=f-F2; % process each pole in loop structure
    f1=taylor(F1,'Order',n); f2=subs(F2,z,1/x);
    f2=taylor(f2,'Order',n); f2=subs(f2,x,1/z); F(i)=f1+f2;
    v1=[char(v(i)) '<abs(z)']; F2=0; % generate the ring expressions
    if i==length(v), str1=v1;
    else, str1=[v1 ' and abs(z)<' char(v(i+1))]; end
    str2=char(F(i)); STR=[STR,  str1 ',' str2 ','];
    for j=1:length(nv), x0=solve(dv(j)); x0=x0(1);
        if abs(x0)<v0(i+1)+eps, F2=F2+nv(j)/dv(j); end
end, end
F0=eval(['piecewise(' STR(1:end-1) ');']) % piecewise Laurent series
```

Example 6.35. For the rational function below, find the Laurent series expansion about $z = 0$,

$$G(z) = \frac{2z^7 + 2z^3 + 8}{z^8 + 30z^7 + 386z^6 + 2772z^5 + 12093z^4 + 32598z^3 + 52520z^2 + 45600z + 16000}.$$

Solutions. With the new function `laurent_series()`, Laurent series of the function about $z = 0$ point can be obtained using

```
>> syms z;
   G=(2*z^7+2*z^3+8)/(z^8+30*z^7+386*z^6+...
     2772*z^5+12093*z^4+32598*z^3+52520*z^2+45600*z+16000);
   F=laurent_series(G) % piecewise Laurent series
```

Therefore, Laurent series can be obtained as a piecewise function:

(1) If $|z| < 1$,

$$F_1(z) = -\frac{22\,818v679z^5}{6\,400\,000\,000} + \frac{221\,063z^4}{64\,000\,000} - \frac{4\,981z^3}{1\,600\,000} + \frac{121z^2}{50\,000} - \frac{57z}{40\,000} + \frac{1}{2\,000}.$$

(2) If $1 < |z| < 2$,

$$F_2(z) = -\frac{216\,104\,333z^5}{172\,800\,000\,000} + \frac{1\,968\,701z^4}{1\,728\,000\,000} - \frac{34\,487z^3}{43\,200\,000} + \frac{71z^2}{675\,000} + \frac{961z}{1\,080\,000}$$
$$-\frac{49}{27\,000} + \frac{1}{432z} - \frac{1}{432z^2} + \frac{1}{432z^3} - \frac{1}{432z^4} + \frac{1}{432z^5}.$$

(3) If $2 < |z| < 4$,

$$F_3(z) = \frac{3\,083\,895\,667z^5}{172\,800\,000\,000} - \frac{64\,031\,299z^4}{1\,728\,000\,000} + \frac{3\,265\,513z^3}{43\,200\,000} - \frac{51\,527z^2}{337\,500} + \frac{330\,961z}{1\,080\,000}$$
$$-\frac{16\,549}{27\,000} + \frac{529}{432z} - \frac{1\,057}{432z^2} + \frac{2\,113}{432z^3} - \frac{4\,225}{432z^4} + \frac{8\,449}{432z^5}.$$

(4) If $4 < |z| < 5$,

$$F_4(z) = -\frac{342\,479\,989z^5}{56\,250\,000} + \frac{62\,140\,157z^4}{2\,250\,000} - \frac{56\,159\,897z^3}{450\,000} + \frac{252\,776\,089z^2}{450\,000}$$
$$- \frac{226\,630\,597z}{90\,000} + \frac{202\,363\,009}{18\,000} + \frac{31\,883\,669}{144z^2} - \frac{7\,198\,645}{144z} - \frac{140\,679\,637}{144z^3}$$
$$+ \frac{618\,454\,229}{144z^4} - \frac{2\,709\,385\,813}{144z^5}.$$

(5) If $|z| > 5$,

$$F_5(z) = \frac{2}{z} - \frac{60}{z^2} + \frac{1\,028}{z^3} - \frac{13\,224}{z^4} + \frac{142\,048}{z^5} - \frac{1\,346\,208}{z^6} + \frac{11\,631\,876}{z^7}.$$

6.8 Exercises

6.1 Compute the first n term and infinite term sums:

(1) $\dfrac{1}{1 \cdot 6} + \dfrac{1}{6 \cdot 11} + \cdots + \dfrac{1}{(5n-4)(5n+1)} + \cdots$,

(2) $\left(\dfrac{1}{2} + \dfrac{1}{3}\right) + \left(\dfrac{1}{2^2} + \dfrac{1}{3^2}\right) + \cdots + \left(\dfrac{1}{2^n} + \dfrac{1}{3^n}\right) + \cdots$,

(3) $\dfrac{1}{3}\left(\dfrac{x}{2}\right) + \dfrac{1 \cdot 4}{3 \cdot 6}\left(\dfrac{x}{2}\right)^2 + \dfrac{1 \cdot 4 \cdot 7}{3 \cdot 6 \cdot 9}\left(\dfrac{x}{2}\right)^3 + \dfrac{1 \cdot 4 \cdot 7 \cdot 10}{3 \cdot 6 \cdot 9 \cdot 12}\left(\dfrac{x}{2}\right)^4 + \cdots$.

6.2 Compute the infinite sums:

(1) $\displaystyle\sum_{n=1}^{\infty} \dfrac{\sin^2 n\alpha \sin nx}{n}, \left(0 < \alpha < \dfrac{\pi}{2}\right)$, (2) $\displaystyle\sum_{n=0}^{\infty} \dfrac{(-1)^n n^3}{(n+1)!}x^n$, (3) $\displaystyle\sum_{n=0}^{\infty} \dfrac{x^{4n+1}}{4n+1}$,

(4) $\dfrac{1}{3}\dfrac{x}{2} + \dfrac{1 \cdot 4}{3 \cdot 6}\left(\dfrac{x}{2}\right)^2 + \dfrac{1 \cdot 4 \cdot 7}{3 \cdot 6 \cdot 9}\left(\dfrac{x}{2}\right)^3 + \dfrac{1 \cdot 4 \cdot 7 \cdot 10}{3 \cdot 6 \cdot 9 \cdot 12}\left(\dfrac{x}{2}\right)^4 + \cdots$.

6.3 Compute the sums of the first n term and infinite term series:

(1) $\sqrt[3]{x} + \left(\sqrt[5]{x} - \sqrt[3]{x}\right) + \left(\sqrt[7]{x} - \sqrt[5]{x}\right) + \cdots + \left(\sqrt[2k+1]{x} - \sqrt[2k-1]{x}\right) + \cdots$,

(2) $1 + \dfrac{m}{1!}x + \dfrac{m(m-1)}{2!}x^2 + \cdots + \dfrac{m(m-1)\cdots(m-n+1)}{n!}x^n + \cdots$.

6.4 For the general terms a_n, compute the sums of the infinite series:

(1) $a_n = (\sqrt{1+n} - \sqrt{n})^p \ln\dfrac{n-1}{n+1}$, (2) $a_n = \dfrac{1}{n^{1+k/\ln n}}$.

6.5 Compute the sums of the following series:

(1) $\displaystyle\sum_{n=1}^{\infty} \dfrac{x^n}{(1+x)(1+x^2)\cdots(1+x^n)}$,

(2) $\displaystyle\sum_{n=2}^{\infty} \dfrac{(-1)^n}{n^2 + n - 2}$, (3) $\displaystyle\sum_{n=2}^{\infty} \dfrac{1}{n^2(n+1)^2(n+2)^2}$.

6.6 Compute the following limits:

(1) $\lim\limits_{n\to\infty}\left(\dfrac{1}{2^2-1}+\dfrac{1}{4^2-1}+\dfrac{1}{6^2-1}+\cdots+\dfrac{1}{(2n)^2-1}\right)$,

(2) $\lim\limits_{n\to\infty}n\left(\dfrac{1}{n^2+\pi}+\dfrac{1}{n^2+2\pi}+\dfrac{1}{n^2+3\pi}+\cdots+\dfrac{1}{n^2+n\pi}\right)$.

6.7 Show that $\cos\theta+\cos 2\theta+\cdots+\cos n\theta = \dfrac{\sin(n\theta/2)\cos[(n+1)\theta/2]}{\sin\theta/2}$.

6.8 Compute the products of the infinite sequences:

(1) $\displaystyle\prod_{n=1}^{\infty}\dfrac{(2n+1)(2n+7)}{(2n+3)(2n+5)}$, (2) $\displaystyle\prod_{n=1}^{\infty}\dfrac{9n^2}{(3n-1)(3n+1)}$, (3) $\displaystyle\prod_{n=1}^{\infty}a^{(-1)^n/n}$, $a>0$.

6.9 If the general term of a series is $a_n = \displaystyle\int_0^{\pi/4}\tan^n x\,dx$, compute $S = \displaystyle\sum_{n=1}^{\infty}\dfrac{1}{n}(a_n + a_{n+2})$.

6.10 Check the convergence of the following infinite series:

(1) $\displaystyle\sum_{n=2}^{\infty}\left(\dfrac{n}{1+n^2}\right)^n$, (2) $\displaystyle\sum_{n=10}^{\infty}\dfrac{1}{\ln n\ln(\ln x)}$, (3) $\displaystyle\sum_{n=1}^{\infty}(-1)^n\dfrac{n+1}{(n+1)\sqrt{n+1}-1}$,

(5) $\dfrac{3}{2}-\dfrac{3\cdot 5}{2\cdot 5}+\dfrac{3\cdot 5\cdot 7}{2\cdot 5\cdot 8}+\cdots+(-1)^{n-1}\dfrac{3\cdot 5\cdot 7\cdots(2n+1)}{2\cdot 5\cdot 8\cdots(3n-1)}+\cdots$.

6.11 Find the intervals of x such that the infinite series are convergent:

(1) $\displaystyle\sum_{n=1}^{\infty}(-1)^n\left(\dfrac{2^n(n!)^2}{(2n+1)!}\right)^p x^n$, (2) $\displaystyle\sum_{n=1}^{\infty}\dfrac{3^{2n}n}{2^n}x^n(1-x)^n$, (3) $\displaystyle\sum_{n=1}^{\infty}\dfrac{1}{x^n}\sin\dfrac{\pi}{2^n}$.

6.12 Compute the limit $\lim\limits_{x\to 1}\dfrac{(1-\sqrt{x})(1-\sqrt[3]{x})\cdots(1-\sqrt[n]{x})}{(1-x)^{n-1}}$.

6.13 Evaluate the infinite expression $S = \sqrt{x+\sqrt{x+\sqrt{x+\sqrt{x+\sqrt{x+\cdots}}}}}$.

6.14 The convergence of a positive number series can be assessed in turn by the following criteria: (1) necessary condition; (2) D'Alembert test; (3) Raabe test; (4) Bertrand test. Please write out a universal MATLAB function to test convergence of a given series.

6.15 Assess the fitting quality for the solutions in Example 6.26. Find high-order Taylor series expansions for large fitting regions.

6.16 Compute the Taylor series of the following functions, and also assess the fitting quality and orders:

(1) $\displaystyle\int_0^x\dfrac{\sin t}{t}\,dt$, (2) $\ln\left(\dfrac{1+x}{1-x}\right)$, (3) $\ln(x+\sqrt{1+x^2})$, (4) $(1+4.2x^2)^{0.2}$,

(5) $e^{-5x}\sin(3x+\pi/3)$ about $x=0$ and $x=a$.

6.17 Compute the first ten terms of Taylor series of a given function $f(t) = e^t$, and find the range of good fitting.

6.18 Find the Taylor series for the 2D functions
(1) $f(x, y) = e^x \cos y$ about $x = 0, y = 0$ and $x = a, y = b$,
(2) $f(x, y) = \ln(1 + x)\ln(1 + y)$ about $x = 0, y = 0$ and $x = a, y = b$.

6.19 For $f(x, y) = \dfrac{1 - \cos(x^2 + y^2)}{(x^2 + y^2)e^{x^2 + y^2}}$, find the 2D Taylor series expansion about $x = 1$, $y = 0$, and assess its behavior.

6.20 Find the Fourier series expansion for the following functions:
(1) $f(x) = (\pi - |x|) \sin x, -\pi \leqslant x < \pi$, (2) $f(x) = e^{|x|}, -\pi \leqslant x < \pi$
(3) $f(x) = \begin{cases} 2x/l, & 0 < x < l/2, \\ 2(l - x)/l, & l/2 < x < l, \end{cases}$ and $l = \pi$.

6.21 Find the continued fraction expansion for the following constants: e, $\sqrt{19}$, lg2, sin 1°, and Euler constant γ. Also find accurate rational approximations to them.

6.22 Find the rational function approximations for the following functions with respectively continued fraction expansion and Padé approximations. Observe the fitting behaviors and appropriate orders.
(1) $f(x) = e^{-2x} \sin 5x$, (2) $f(x) = \dfrac{x^3 + 7x^2 + 24x + 24}{x^4 + 10x^3 + 35x^2 + 50x + 24} e^{-3x}$.

6.23 Find the Padé approximations to $\cos x$ and e^{-x}, and decide the orders such that good approximation can be obtained.

6.24 Write Laurent series for the given transcendental functions, and compute the residues
(1) $f(z) = ze^{-1/z^2}\left[\sin \dfrac{1}{z} - \cos \dfrac{1}{z}\right]$, (2) $f(z) = z^5 \cos \dfrac{1}{z^2}$.

6.25 Find Laurent series for the following rational function:
$$f(z) = \frac{3}{z - 1} + \frac{1}{(z - 1)^2} + \frac{1}{z - 2} + \frac{1}{(z - 2)^2} + \frac{5}{z + i} + \frac{5}{z - i}.$$

7 Numerical derivatives and differentials

In the previous chapters, we have shown that when the mathematical form of a function is given, analytical expressions of its derivatives of different orders can be obtained directly with the function `diff()`. Is was also shown that even if the order is as high as 100, the derivatives can be obtained in MATLAB within a few seconds. It should be emphasized here that to find the derivatives in this way, the original function must be known.

If the mathematical form of a function is not known, with only some measured data obtained in some way, e. g., from experiments, the derivatives of the function may also be needed in practical applications. It is obvious that the analytical solution approaches discussed earlier cannot be used. Numerical approaches should be introduced for solving these problems. Since there is almost no existing function for finding numerically the derivatives in MATLAB, further explorations will be carried out in this book for finding numerical derivatives and differentials.

In Section 7.1, some of the numerical algorithms will be presented for finding numerical derivatives for univariate functions. A simple approximation derived from the concept of first-order derivative will be given first, then several other advanced algorithms with high precision will be presented. The algorithms for numerical derivatives of any order, precision, and selection of sample will be presented. In Section 7.2, some of the algorithms will be implemented in MATLAB, and there behaviors can be demonstrated through examples. In Section 7.3, some proposals will be implemented in a universal MATLAB function for forward difference based numerical derivative algorithm of any order and precision. In Section 7.4, numerical partial derivatives for 2D functions will be given, and a general-purpose MATLAB implementation will be provided. In Section 7.5, two kinds of commonly used spline – cubic and B-spline – will be presented, and based on them, numerical derivatives will be evaluated.

7.1 Numerical derivative algorithms

Assume that a set of data (t_i, y_i) are obtained, where they are equally spaced according to t, with a step-size of h. The mathematical form of the original function $y(t)$ is not known. How can we find the derivative $y'(t)$?

Let us review Definition 4.1, where the first-order derivative is defined as

$$y'(t) = \frac{dy(t)}{dt} = \lim_{h \to 0} \frac{y(t+h) - y(t)}{h}. \tag{7.1.1}$$

It can be seen that, if $h \to 0$, the difference of two adjacent points can be measured, and, divided by the step-size h, used as the first-order derivative of the function.

https://doi.org/10.1515/9783110666977-007

However, in real applications, the condition $h \to 0$ is too stringent, and normally cannot be achieved. Approximate solutions are needed to compute numerical derivatives. Simple numerical derivative formulas will be introduced in this section, followed by high order, high precision algorithms.

7.1.1 Forward and backward difference algorithms

Assume that a set of samples $y(k)$, briefly denoted as y_k, $k = 1, 2, \ldots, m$, is provided. How can we find the first-order derivative of function y from the samples? One may drop the limit sign in (7.1.1) and use the formula to compute the derivative with difference algorithms. For instance, the forward and backward difference summarized below can be introduced.

Definition 7.1. The forward difference formula for numerical derivative is

$$y'_k \approx \frac{\Delta y_k}{h} = \frac{y_{k+1} - y_k}{h}, \quad k = 1, 2, \ldots, m - 1. \tag{7.1.2}$$

Definition 7.2. The backward difference numerical derivative is defined as

$$y'_k \approx \frac{\Delta y_k}{h} = \frac{y_k - y_{k-1}}{h}, \quad k = 2, 3, \ldots, m. \tag{7.1.3}$$

If a data vector y based on equally spaced time instances is provided, with step-size of h, the MATLAB function y_1=diff(y)/h can be used directly to find the first-order derivative of function $y(t)$.

The function diff() was used previously for finding analytical solutions of given functions, however, if the input argument is a numerical vector, the difference, i. e., each term is subtracted from a previous one, can be obtained instead. The function call above implements the forward difference formula, and the length of y_1 is one shorter than that of y.

The precision of the two algorithms is $o(h)$ in both cases. When h is relatively large, the error may be large. Imprecisely speaking, $o(h)$ means that, if h is selected as 0.1, the error in derivatives is also on the same level as 0.1. Therefore, in real applications, high precision algorithms are needed, and universal MATLAB solvers are expected.

7.1.2 Central difference algorithms with $o(h^2)$ precision

Since the precisions of the forward and backward algorithms are very low, high precision algorithms are needed. For instance, central difference algorithms can be used, and will be summarized in the subsequent presentations.

Definition 7.3. The central difference first-order derivative is defined as

$$y_k' \approx \frac{\Delta y_k}{h} = \frac{y_{k+1} - y_{k-1}}{2h}. \tag{7.1.4}$$

Denote

$$\tilde{f}'(x) = \frac{f(x+h) - f(x-h)}{2h}. \tag{7.1.5}$$

Using the Taylor expansion technique, the above equation can further be written as

$$\tilde{f}'(x) = \frac{f(x) + hf'(x) + h^2 f''(x)/2! + h^3 f'''(\xi)/3! + o(h^4)}{2h}$$
$$- \frac{f(x) - hf'(x) + h^2 f''(x)/2! - h^3 f'''(\xi)/3! + o(h^4)}{2\Delta t} \tag{7.1.6}$$
$$\approx f'(x) + \frac{h^3}{3!} f''(\xi).$$

Therefore, mathematically the precision of the algorithm is $o(h^2)$.

Theorem 7.1. *Similarly, second- up to fourth-order derivatives can be evaluated from*

$$y_k'' \approx \frac{y_{k+1} - 2y_k + y_{k-1}}{h^2},$$
$$y_k''' \approx \frac{y_{k+2} - 2y_{k+1} + 2y_{k-1} - y_{k-2}}{2h^3}, \tag{7.1.7}$$
$$y_k^{(4)} \approx \frac{y_{k+2} - 4y_{k+1} + 6y_k - 4y_{k-1} + y_{k-2}}{h^4}.$$

It can be shown that the central difference algorithms have precision $o(h^2)$.

In the first two formulas, the so-called "three-point" approximations are used in y_k' and y_k'', i. e., the information at the current instance k and at two other adjacent points, $k-1$ and $k+1$, is used in evaluating the numerical derivatives. For third- and fourth-order derivatives with precision $o(h^2)$, the three-point algorithm is not adequate. Two extra points are needed to build the five-point algorithms. The two extra points are the points at $k-2$ and $k+2$.

Compared with the $o(h)$ algorithms, $o(h^2)$ algorithms have some advantages. If $h = 0.1$, the error may reach the level of 0.01.

7.1.3 Central difference algorithm with $o(h^4)$ precision

If high precision algorithms are applied, the following central difference algorithms can be adopted. In the formulas, the first- and second-order derivatives need the five-point central difference approximation, while for third- and fourth-order derivatives, seven-point central difference approximations are needed.

Theorem 7.2. *Another set of high precision central difference algorithms are*

$$y_k' \approx \frac{-y_{k+2} + 8y_{k+1} - 8y_{k-1} + y_{k-2}}{12h},$$

$$y_k'' \approx \frac{-y_{k+2} + 16y_{k+1} - 30y_k + 16y_{k-1} - y_{k-2}}{12h^2},$$

$$y_k''' \approx \frac{-y_{k+3} + 8y_{k+2} - 13y_{k+1} + 13y_{k-1} - 8y_{k-2} + y_{k-3}}{8h^3},$$

$$y_k^{(4)} \approx \frac{-y_{k+3} + 12y_{k+2} - 39y_{k+1} + 56y_k - 39y_{k-1} + 12y_{k-2} - y_{k-3}}{6h^4},$$

(7.1.8)

and it can be shown that these algorithms have precision $o(h^4)$. Imprecisely speaking, if $h = 0.1$, the $o(h^4)$ algorithm may lead to an error of 0.0001.

7.1.4 Central difference algorithms of higher precision

The coefficients for high-order derivatives with higher precision are summarized in [7]. An algorithm is proposed to build the finite difference formulas of any order and points, together with pseudocode. The coefficient table in Table 7.1 is extended and unified from the existing ones. It can be seen that the formulas in (7.1.7) and (7.1.8) are only a small subset. If needed, MATLAB functions can be written.

If the order selected is 1, and the precision is $o(h^2)$, i. e., the first row of the table, the formula is exactly the same as that in (7.1.4). For the $o(h^4)$ precision, i. e., the second row of the table, the coefficients are $1/12$, $-2/3$, $2/3$, and $-1/12$, which are rewritten

Table 7.1: Coefficients in high-precision central difference algorithms.

order	precision	$k-4$	$k-3$	$k-2$	$k-1$	k	$k+1$	$k+2$	$k+3$	$k+4$
1	$o(h^2)$				$-1/2$	0	$1/2$			
	$o(h^4)$			$1/12$	$-2/3$	0	$2/3$	$-1/12$		
	$o(h^6)$		$-1/60$	$3/20$	$-3/4$	0	$3/4$	$-3/20$	$1/60$	
	$o(h^8)$	$1/280$	$-4/105$	$1/5$	$-4/5$	0	$4/5$	$-1/5$	$4/105$	$-1/280$
2	$o(h^2)$				1	-2	1			
	$o(h^4)$			$-1/12$	$4/3$	$-5/2$	$4/3$	$-1/12$		
	$o(h^6)$		$1/90$	$-3/20$	$3/2$	$-49/18$	$3/2$	$-3/20$	$1/90$	
	$o(h^8)$	$-1/560$	$8/315$	$-1/5$	$8/5$	$-205/72$	$8/5$	$-1/5$	$8/315$	$-1/560$
3	$o(h^2)$			$-1/2$	1	0	-1	$1/2$		
	$o(h^4)$		$1/8$	-1	$13/8$	0	$-13/8$	1	$-1/8$	
	$o(h^6)$	$-7/240$	$3/10$	$-169/120$	$61/30$	0	$-61/30$	$169/120$	$-3/10$	$7/240$
4	$o(h^2)$			1	-4	6	-4	1		
	$o(h^4)$		$-1/6$	2	$-13/2$	$28/3$	$-13/2$	2	$-1/6$	
	$o(h^6)$	$7/240$	$-2/5$	$169/60$	$-122/15$	$91/8$	$-122/15$	$169/60$	$-2/5$	$7/240$
5	$o(h^2)$		$-1/2$	2	$-5/2$	0	$5/2$	-2	$1/2$	
6	$o(h^2)$		1	-6	15	-20	15	-6	1	

here as $[1, -8, 8, -1]/12$, in the same format as in (7.1.8). Therefore, the $o(h^2)$ and $o(h^4)$ formulas are special examples in Table 7.1.

7.1.5 Deriving high-order high precision algorithms

A generating algorithm is proposed in [7] for finding the numerical derivatives of any order, precision, and sample arrangements. A pseudocode for computing the coefficients is also presented in the reference.

Theorem 7.3. *If $(N + 1)$-point interpolation algorithm is used, the mth order derivative of $f(t)$ with respect to t can be computed from*

$$\left.\frac{d^m f(t)}{dt^m}\right|_{t=t_0} \approx \sum_{v=0}^{n} \delta_{m,v}^n f(t - \alpha_v),\tag{7.1.9}$$

where, $m = 0, 1, \ldots, M$, $n = m, m+1, \ldots, N$. The values $\alpha_0, \alpha_1, \ldots, \alpha_N$ are the samples of t. Assigning the initial value $\delta_{0,0}^0 = 1$, the other coefficients can be evaluated recursively from

$$\delta_{n,v}^m = \frac{1}{\alpha_n - \alpha_v}(\alpha_n \delta_{n-1,v}^m - m\delta_{n-1,v-1}^m), \quad v = 0, 1, \ldots, n - 1.\tag{7.1.10}$$

Constructing the function

$$w_m(x) = \prod_{m=0}^{m}(t - \alpha_m),\tag{7.1.11}$$

the boundaries $\delta_{n,m}^m$ can be computed directly from

$$\delta_{n,m}^m = \frac{w_{n-2}(\alpha_{n-1})}{w_{n-1}(\alpha_n)}(m\delta_{n-1,n-1}^{m-1} - \alpha_{n-1}\delta_{n-1,n-1}^m).\tag{7.1.12}$$

Based on the theorem and the pseudocode in [7], and with reference to the Fortran source code in [8], the following MATLAB can be written for computing the coefficients:

```
function [coefs,delta]=fdcoef(M,N,alpha,t0)
if nargin==3, t0=0; end, M=M+1;
delta=sym(zeros(M,N,N)); delta(1,1,1)=sym(1); c1=1; nm=min(N,M);
for n=2:N, c2=1;
   for nu=1:n-1
      c3=alpha(n)-alpha(nu); c2=c2*c3; c4=1/c3;
      a0=alpha(n)-t0; delta(1,n,nu)=c4*(a0*delta(1,n-1,nu));
      for m=2:nm
```

```
            delta(m,n,nu)=c4*(a0*delta(m,n-1,nu)...
                -(m-1)*delta(m-1,n-1,nu));
        end
    end
    a0=(alpha(n-1)-t0);
    delta(1,n,n)=c1/c2*(-a0*delta(1,n-1,n-1)); c4=c1/c2;
    for m=2:nm
        delta(m,n,n)=c4*((m-1)*delta(m-1,n-1,n-1)-a0*delta(m,n-1,n-1));
    end, c1=c2;
end
coefs=delta(M,N,:); coefs=coefs(:).';
```

The syntax of the function is $[c, \delta]$=fdcoef(m, n, α, t_0), with m being the derivative order and n the number of points. When a central difference algorithm is used, n should be an odd number; t_0 is the reference, normally $t_0 = 0$. The vector α is defined as $\alpha = [\alpha_0, \alpha_1, \ldots, \alpha_n]$, where in a central difference algorithm, $\alpha = [-(n-1)/2, -(n-1)/2+1, \ldots, (n-1)/2]$, while in a forward algorithm, $\alpha = [0, 1, \ldots, n]$. The returned c is the coefficient vector, and δ is a 3D array, returning the intermediate results of $\delta^m_{n,v}$.

Example 7.1. Considering the central difference formulas in Table 7.1, derive the fifth- and sixth-order central difference algorithms. Then derive the fifth- and sixth-order derivatives with nine- and eleven-point central difference algorithms. Derive also the seventh-order derivatives.

Solutions. Consider the seven-point formulas in Table 7.1. The parameters $m = 5$ and $m = 6$, with $n = 7$ can be selected, and the central difference sample of $\alpha = [-3, -2, -1, 0, 1, 2, 3]$ can be selected. The following statements can be used:

```
>> m=6; n=7; alpha=[-3:3];
   c5=fdcoef(m,n,alpha), m=6; c6=fdcoef(m,n,alpha)
```

The coefficient vectors obtained are $c_5 = [-1/2, 2, -5/2, 0, 5/2, -2, 1/2]$ and $c_6 = [1, -6, 15, -20, 15, -6, 1]$. They agree well with those in Table 7.1.

Now let us consider the nine-point formula, i. e., $n = 9$, and also let $m = 5$ and $m = 6$, respectively. Updating the α vector, the following MATLAB statements can be used:

```
>> n=9; alpha=-4:4;
   c52=fdcoef(5,n,alpha), c62=fdcoef(6,n,alpha)
```

and it is found that nine-point central difference algorithms can be

$$c_{52} = [1/6, -3/2, 13/3, -29/6, 0, 29/6, -13/3, 3/2, -1/6],$$
$$c_{62} = [-1/4, 3, -13, 29, -75/2, 29, -13, 3, -1/4],$$

with the corresponding mathematical formulas given by

$$y^{(5)}(k) \approx y(k-4)/6 - 3y(k-3)/2 + 13y(k-2)/3 - 29y(k-1)/6$$
$$+ 29y(k+1)/6 - 13y(k+2)/3 + 3y(k+3)/2 - y(k+4)/6,$$
$$y^{(6)}(k) \approx -y(k-4)/4 + 3y(k-3) - 13y(k-2) + 29y(k-1) - 75y(k)/2$$
$$+ 29y(k+1) - 13y(k+2) + 3y(k+3) - y(k+4)/4,$$

with precision $o(h^4)$.

If eleven-point algorithms are needed, setting $n = 11$, and selecting $m = 5$ and $m = 6$, respectively, the corresponding central difference formulas can be obtained via

```
>> n=11; alpha=-5:5;
   c52=fdcoef(5,n,alpha), c62=fdcoef(6,n,alpha)
```

with precision $o(h^6)$. The coefficient vectors are

$$c_{53} = \left[-\frac{13}{288}, \frac{19}{36}, -\frac{87}{32}, \frac{13}{2}, -\frac{323}{48}, 0, \frac{323}{48}, -\frac{13}{2}, \frac{87}{32}, -\frac{19}{36}, \frac{13}{28} \right],$$

$$c_{63} = \left[\frac{13}{240}, -\frac{19}{24}, \frac{87}{16}, -\frac{39}{2}, \frac{323}{8}, -\frac{1023}{20}, \frac{323}{8}, -\frac{39}{2}, \frac{87}{16}, -\frac{19}{24}, \frac{13}{240} \right].$$

Further, seventh-order derivative can also be obtained from

```
>> c7=fdcoef(7,n,alpha)
```

with precision $o(h^4)$, and the coefficients are

$$c_7 = \left[\frac{5}{24}, -\frac{13}{6}, \frac{69}{8}, -17, \frac{63}{4}, 0, -\frac{63}{4}, 17, -\frac{69}{8}, \frac{13}{6}, -\frac{5}{24} \right].$$

7.1.6 High precision forward and backward difference algorithms

Apart from the central difference algorithms presented earlier, forward difference formulas are also presented in [7]. An extended version of the table is shown in Table 7.2. The high-order numerical derivatives can easily be obtained with given function samples.

Table 7.2: Coefficients in high precision forward difference algorithms.

order	precision	y_k	y_{k+1}	y_{k+2}	y_{k+3}	y_{k+4}	y_{k+5}	y_{k+6}	y_{k+7}	y_{k+8}
	$o(h)$	-1	1							
	$o(h^2)$	$-3/2$	2	$-1/2$						
	$o(h^3)$	$-11/6$	3	$-3/2$	$1/3$					
1	$o(h^4)$	$-25/12$	4	-3	$4/3$	$-1/4$				
	$o(h^5)$	$-137/60$	5	-5	$10/3$	$-5/4$	$1/5$			
	$o(h^6)$	$-49/20$	6	$-15/2$	$20/3$	$-15/4$	$6/5$	$-1/6$		
	$o(h^7)$	$-363/140$	7	$-21/2$	$35/3$	$-35/4$	$21/5$	$-7/6$	$1/7$	
	$o(h^8)$	$-761/280$	8	-14	$56/3$	$-35/2$	$56/5$	$-14/3$	$8/7$	$-1/8$
	$o(h)$	1	-2	1						
	$o(h^2)$	2	-5	4	-1					
	$o(h^3)$	$35/12$	$-26/3$	$19/2$	$-14/3$	$11/12$				
2	$o(h^4)$	$15/4$	$-77/6$	$107/6$	-13	$61/12$	$-5/6$			
	$o(h^5)$	$203/45$	$-87/5$	$117/4$	$-254/9$	$33/2$	$-27/5$	$137/180$		
	$o(h^6)$	$469/90$	$-223/10$	$879/20$	$-949/18$	41	$-201/10$	$1\,019/180$	$-7/10$	
	$o(h^7)$	$29\,531/5\,040$	$-962/35$	$621/10$	$-4\,006/45$	$691/8$	$-282/5$	$2\,143/90$	$-206/35$	$363/560$
	$o(h)$	-1	3	-3	1					
	$o(h^2)$	$-5/2$	9	-12	7	$-3/2$				
	$o(h^3)$	$-17/4$	$71/4$	$-59/2$	$49/2$	$-41/4$	$7/4$			
3	$o(h^4)$	$-49/8$	29	$-461/8$	62	$-307/8$	13	$-15/8$		
	$o(h^5)$	$-967/120$	$638/15$	$-3\,929/40$	$389/3$	$-2\,545/24$	$268/5$	$-1\,849/120$	$29/15$	
	$o(h^6)$	$-801/80$	$349/6$	$-18\,353/120$	$2\,391/10$	$-1\,457/6$	$4\,891/30$	$-561/8$	$527/30$	$-469/240$
	$o(h)$	1	-4	6	-4	1				
	$o(h^2)$	3	-14	26	-24	11	-2			
4	$o(h^3)$	$35/6$	-31	$137/2$	$-242/3$	$107/2$	-19	$17/6$		
	$o(h^4)$	$28/3$	$-111/2$	142	$-1\,219/6$	176	$-185/2$	$82/3$	$-7/2$	
	$o(h^5)$	$1\,069/80$	$-1\,316/15$	$15\,289/60$	$-2\,144/5$	$10\,993/24$	$-4\,772/15$	$2\,803/20$	$-536/15$	$967/240$
5	$o(h^4)$	$-27/2$	$575/6$	$-895/3$	$1\,065/2$	$-1\,790/3$	$2\,581/6$	-195	$305/6$	$-35/6$
6	$o(h^3)$	$39/4$	-73	239	-447	$1\,045/2$	-391	183	-49	$23/4$

The forward difference formulas can also be derived with the function `fdcoef()`, where n is not necessarily an odd number, and vector α should be selected as $\alpha = [0, 1, \ldots, n]$.

Example 7.2. Derive the nine-point forward difference algorithm for the fourth-order derivatives. Compare the results with those in Table 7.2. Also derive a nine-point forward difference algorithm for the fifth- and sixth-order derivatives.

Solutions. Selecting the parameters below, the fourth-order derivative can be obtained with the following MATLAB statements:

```
>> m=4; n=9; alpha=0:8; c41=fdcoef(m,n,alpha)
```

and the results are as follows, which are exactly the same as those in Table 7.2:

$$c_4 = \left[\frac{1069}{80}, -\frac{1316}{15}, \frac{15289}{60}, -\frac{2144}{5}, \frac{10993}{24}, -\frac{4772}{15}, \frac{2803}{20}, -\frac{536}{15}, \frac{967}{240}\right].$$

The corresponding mathematical representation of the algorithm is

$$y^{(4)}(k) \approx \frac{1069}{80}y(k) - \frac{1316}{15}y(k+1) + \frac{15289}{60}y(k+2) - \frac{2144}{5}y(k+3) + \frac{10993}{24}y(k+4)$$
$$- \frac{4772}{15}y(k+5) + \frac{2803}{20}y(k+6) - \frac{536}{15}y(k+7) + \frac{967}{240}y(k+8).$$

If fifth- and sixth-order derivatives with nine-point forward difference are expected, the following statements can be entered:

```
>> c5=fdcoef(5,n,alpha), c6=fdcoef(6,n,alpha)
```

The coefficients of fifth- and sixth-order derivatives are

$$c_5 = [-27/2, 575/6, -895/3, 1065/2, -1790/3, 2581/6, -195, 305/6, -35/6],$$
$$c_6 = [39/4, -73, 239, -447, 1045/2, -391, 183, -49, 23/4].$$

Example 7.3. Derive the fifth-order derivative with a nine-point backward difference formula.

Solutions. If the mth order derivative formula with n-segment backward difference is expected, the vector $\alpha = [-(n-1), -(n-2), \ldots, 0]$ should be assigned. The nine-point backward formula can be derived from

```
>> m=5; n=9; alpha=-8:0; c5=fdcoef(m,n,alpha)
```

with the results

$$c_5 = [35/6, -305/6, 195, -2581/6, 1790/3, -1065/2, 895/3, -575/6, 27/2].$$

The mathematical form of the formula is

$$y^{(5)}(k) \approx 35y(k-8)/6 - 305y(k-7)/6 + 195y(k-6)$$
$$- 2581y(k-5)/6 + 1790y(k-4)/3 - 1065y(k-3)/2$$
$$+ 895y(k-2)/3 - 575y(k-1)/6 + 27y(k)/2,$$

with precision $o(h^4)$.

7.2 MATLAB implementations of the numerical derivative algorithms

Various numerical derivative algorithms are discussed earlier. One may assign the number of points, orders, and arrangement of the samples. Relevant formulas can be derived. In particular, with Theorem 7.1, the derivatives with $o(h^2)$ precision can be obtained. When the formulas in Table 7.1 are considered, the derivatives of first- up to sixth-order can be derived with seven-point differences. Finally, a general-purpose forward difference based MATLAB solver for numerical derivatives of any order and precision will be written and presented.

7.2.1 Implementation of the $o(h^2)$ algorithms

If the practical precision required is not high, the $o(h^2)$ algorithms in Theorem 7.1 can be used to evaluate numerical derivatives. Based on the theorem, the following MATLAB function can be written:

```
function [f,t]=diff_ctr2(y,h,n)
y1=[y 0 0 0 0]; y2=[0 y 0 0 0]; y3=[0 0 y 0 0];
y4=[0 0 0 y 0]; y5=[0 0 0 0 y];
switch n % select derivative formulas according to order
    case 1, f=(y1-y3)/2/h;
    case 2, f=(y1-2*y2+y3)/h^2;
    case 3, f=(y1-2*y2+2*y4-y5)/2/h^3;
    case 4, f=(y1-4*y2+6*y3-4*y4+y5)/h^4;
end
k=3+2*(n>2); f=f(k:end-k-1); t=([1:length(f)]+(n>2))*h;
```

The syntax of the function is $[f,t]$=diff_ctr2(y,h,n)], where y is the vector of the samples, h is the step-size, n is the expected order of derivative, where 1 to 4 can be selected, such that f returns the nth order numerical derivative. Vector t is the offset of the time vector, and the actual time vector should be t_0+t, where t_0 is the first instance in t. The length of the vector f is shorter than that in y.

Example 7.4. Consider the function $f(x) = \sin x/(x^2 + 4x + 3)$ studied in Example 4.4. In real applications, if the mathematical expression of the function is known, there is no point of using numerical algorithms to compute the derivatives. Here, the precision of the algorithms is to be assessed, therefore, a known function is considered.

In this example, a set of samples can be generated from the given function. The objective is to compute numerically from the samples the first- up to fourth-order derivatives, and assess the precision of the algorithms, by comparing the results with the known analytical solutions.

Solutions. The analytical solutions of the required derivatives can be evaluated directly with function diff(). Also, a set of samples can be generated, and their theoretical values can be created.

```
>> syms t; y(t)=sin(t)/(t^2+4*t+3); h=0.05; x=0:h:pi;% theoretical
   yy1=diff(y); f1=double(yy1(x)); yy2=diff(yy1); f2=double(yy2(x));
   yy3=diff(yy2); f3=double(yy3(x)); yy4=diff(yy3); f4=double(yy4(x));
```

Assume that the prototype function can be used to generate the samples y_i. The first- to fourth-order numerical derivatives can be obtained with the following statements, and their curves, together with those from the theoretical values, are as shown in Figure 7.1. It can be seen that the two sets of curves are almost identical, and almost cannot be distinguished from the picture.

```
>> y0=double(y(x)); % Comparison of numerical and analytical solutions
   subplot(221), [y1,dx1]=diff_ctr2(y0,h,1); plot(x,f1,dx1,y1,':');
   subplot(222), [y2,dx2]=diff_ctr2(y0,h,2); plot(x,f2,dx2,y2,':');
   subplot(223), [y3,dx3]=diff_ctr2(y0,h,3); plot(x,f3,dx3,y3,':');
   subplot(224), [y4,dx4]=diff_ctr2(y0,h,4); plot(x,f4,dx4,y4,':')
```

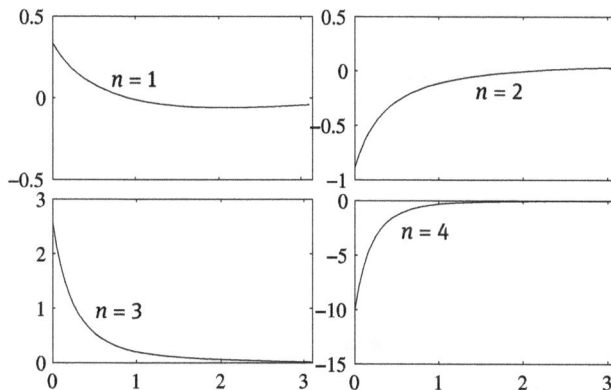

Figure 7.1: Comparisons of derivatives of different orders.

To quantitatively address the error, let us consider only the fourth-order derivative obtained. Since vector y_4 is shorter than vector f_4, the vectors at the same time instances are extracted and compared. The maximum error of 0.0653 is found for this example. It can be seen that the error is rather small.

```
>> max(abs(y4-f4(3:end-4))) %maximum error
```

If the error cannot be accepted in real applications, algorithms with higher precision should be introduced.

7.2.2 Implementation of the seven-point central difference algorithms

Considering the algorithms in Table 7.1, it is found that the seven-point algorithms are common, and while for the first- and second-order derivatives, the precision is $o(h^6)$, for the third- and fourth-order derivatives, the precision is $o(h^4)$, whereas for the fifth- and sixth-order derivatives, the expected precision is $o(h^2)$, which is the highest possible with seven-point algorithms.

Based on all the seven-point central difference algorithms, a general purpose MATLAB function can be written as follows:

```
function [f,t]=diff_ctr7(y,h,n)
y1=[y 0 0 0 0 0 0]; y2=[0 y 0 0 0 0 0]; y3=[0 0 y 0 0 0 0];
y4=[0 0 0 y 0 0 0]; y5=[0 0 0 0 y 0 0];
y6=[0 0 0 0 0 y 0]; y7=[0 0 0 0 0 0 y];
switch n
    case 1, f=(y1-9*y2+45*y3-45*y5+9*y6-y7)/60/h;
    case 2, f=(2*y1-27*y2+270*y3-490*y4+270*y5-27*y6+2*y7)/180/h^2;
    case 3, f=(-y1+8*y2-13*y3+13*y5-8*y6+y7)/8/h^3;
    case 4, f=(-y1+12*y2-39*y3+56*y4-39*y5+12*y6-y7)/6/h^4;
    case 5, f=(-y1+4*y2-5*y3+5*y5-4*y6+y7)/28/h^5;
    case 6, f=(y1-6*y2+15*y3-20*y4+15*y5-6*y6+y7)/h^6;
end
k=7; f=f(k:end-k); t=([3:length(f)+2])*h;
```

The syntax of the function is $[f,t]$=diff_ctr7(y,h,n), and it is quite similar to that presented in function diff_ctr2().

Example 7.5. Solve the problem in Example 7.4 again with the new function.

Solutions. The following MATLAB lines can be entered and the numerical derivatives can be computed again with the new function:

```
>> syms t; y(t)=sin(t)/(t^2+4*t+3); h=0.05; x=0:h:pi;% theoretical
   yy1=diff(y); f1=double(yy1(x)); yy2=diff(yy1); f2=double(yy2(x));
   yy3=diff(yy2); f3=double(yy3(x)); yy4=diff(yy3); f4=double(yy4(x));
   y0=double(y(x)); % comparison of the curves
   [y1,dx1]=diff_ctr7(y0,h,1); [y2,dx2]=diff_ctr7(y0,h,2);
   [y3,dx3]=diff_ctr7(y0,h,3); [y4,dx4]=diff_ctr7(y0,h,4);
   max(abs(y4-f4(4:end-4))) % maximum error
```

The maximum error found this time is 9.0152×10^{-4}, and the precision here is much higher than that provided in Example 7.4.

Example 7.6. Consider $x \in (1.5, 3.5)$, select a step-size of $h = 0.02$, and generate samples from

$$f(x) = \frac{1}{2}\ln(x+1) - \frac{1}{4}\ln(x^2 - x + 1) + \frac{1}{\sqrt{3}}\arctan\frac{2x-1}{\sqrt{3}},$$

first, then, based on the samples, compute the sixth-order derivative and compare the results with the analytical function, by finding the maximum error.

Solutions. The following statements can be used to generate directly the samples. Based on them, the sixth-order numerical derivatives can be found. It is also known from the results that the vector x_6 is, in fact, the offset in time, and the initial time of 1.5 should be added back to form the time vector. The numerical and analytical solutions of the sixth-order derivatives can also be found, as illustrated in Figure 7.2. There is almost no error found between the two curves, and numerically the maximum error is 0.0311.

```
>> syms x; h=0.02; x0=1.5:h:3.5;
   f(x)=log(1+x)/2-log(x^2-x+1)/4+atan((2*x-1)/sqrt(3))/sqrt(3);
```

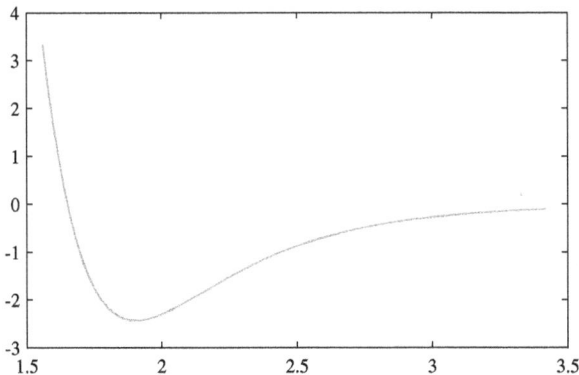

Figure 7.2: Comparisons of the sixth-order derivatives.

```
y0=double(f(x0)); [y6,x6]=diff_ctr7(y0,h,6); x6=x6+1.5;
f6=diff(f,6); y60=double(f6(x6));
plot(x6,y6,x6,y60), max(abs(y60-y6))
```

7.2.3 Implementation of forward difference numerical derivative algorithms

According to Table 7.2, several forward difference algorithms are presented. Selecting the nine-point algorithms, a general purpose MATLAB function is written as follows:

```
function [f,t]=diff_forward(y,h,n)
y0=[0 0 0 0 0 0 0 0 y];y1=[0 0 0 0 0 0 0 y 0];y2=[0 0 0 0 0 0 y 0 0];
y3=[0 0 0 0 0 y 0 0 0];y4=[0 0 0 0 y 0 0 0 0];y5=[0 0 0 y 0 0 0 0 0];
y6=[0 0 y 0 0 0 0 0 0];y7=[0 y 0 0 0 0 0 0 0];y8=[y 0 0 0 0 0 0 0 0];
switch n            % evaluate different order numerical derivatives
   case 1, f=(-761/280*y0+8*y1-14*y2+56/3*y3-35/2*y4+...
               56/5*y5-14/3*y6+8/7*y7-1/8*y8)/h;
   case 2, f=(29531/5040*y0-962/35*y1+621/10*y2-4006/45*y3+...
               691/8*y4-282/5*y5+2143/90*y6-206/35*y7+363/560*y8)/h^2;
   case 3, f=(-801/80*y0+349/6*y1-18353/120*y2+2391/10*y3-...
               1457/6*y4+4891/30*y5-561/8*y6+527/30*y7-469/240*y8)/h^3;
   case 4, f=(1069/80*y0-1316/15*y1+15289/60*y2-2144/5*y3+...
               10993/24*y4-4772/15*y5+2803/20*y6-536/15*y7+967/240*y8)/h^4;
   case 5, f=(-27/2*y0+575/6*y1-895/3*y2+1065/2*y3-1790/3*y4+...
               2581/6*y5-195*y6+305/6*y7-35/6*y8)/h^5;
   case 6, f=(39/4*y0-73*y1+239*y2-447*y3+1045/2*y4-391*y5+183*y6-...
               49*y7+23/4*y8)/h^6;
end
f=f(9:end-8); t=(0:length(f)-1)*h;
```

The syntax of the function is $[f,t]$=diff_forward(y,h,n), and it is the same as that explained for diff_ctr2(). The order n can be selected 1~6. The precision of the function is $o(h^{9-n})$.

Example 7.7. Solve the problem in Example 7.4 with forward difference algorithms.

Solutions. The following MATLAB lines can be used:

```
>> syms t; y(t)=sin(t)/(t^2+4*t+3); h=0.05; x=0:h:pi;
   yy4=diff(y,4); f4=double(yy4(x)); y0=double(y(x));
   [y4,dx4]=diff_forward(y0,h,4); max(abs(y4-f4(1:length(y4))))
```

The starting time instance of forward difference algorithms is at $t = 0$, which is different from those in central difference algorithms. This can be regarded as one of the advantages in numerical computation.

Example 7.8. Solve the problem in Example 7.6 again with forward difference algorithms.

Solutions. The function `diff_ctr7()` can be replaced by function `diff_forward()` in Example 7.6. The following statements can be employed, and the maximum error is 0.0953:

```
>> syms x; h=0.02; x0=1.5:h:3.5;
   f(x)=log(1+x)/2-log(x^2-x+1)/4+atan((2*x-1)/sqrt(3))/sqrt(3);
   y0=double(f(x0)); [y6,x6]=diff_forward(y0,h,6); x6=x6+1.5;
   f6=diff(f,6); y60=double(f6(x6));
   plot(x6,y6,x6,y60), max(abs(y60-y6))
```

It can be seen that all these MATLAB functions can be used in finding the numerical derivatives for univariate samples.

7.3 Numerical derivatives of any orders

Comparing the three types of difference manipulations, it is found that although the central difference algorithms are slightly more accurate, the initial points may be missing in the interpolation process, while the forward difference algorithms start directly with an offset of zero, which may be preferred in real applications. Here, only the forward difference algorithms are considered, with MATLAB implementations.

Analyzing the forward difference algorithms discussed earlier, if the derivative order is selected as n, and the expected precision is $o(h^p)$, the number of points required is $m = n + p$. Based on this fact, the following general purpose MATLAB function can be written:

```
function [f,t]=num_diff(y,h,n,p)
m=n+p; alpha=0:m; c=double(fdcoef(n,m,alpha)); t=[0:length(y)-m]*h;
for i=1:length(y)-m+1, yy=y(i:i+m-1); f(i)=c*yy(:)/h^n; end
```

The syntax of the function is $[f,t]$=num_diff(y,h,n,p), and it is quite similar to those discussed earlier. The precision p is used explicitly in the function call.

Example 7.9. Solve the problem in Example 7.6 using the forward difference algorithms.

Solutions. With the following statements, the results can be obtained, which are exactly the same as those obtained in Example 7.8.

```
>> syms x; h=0.02; x0=1.5:h:3.5;
   f(x)=log(1+x)/2-log(x^2-x+1)/4+atan((2*x-1)/sqrt(3))/sqrt(3);
   y0=double(f(x0)); [y6,x6]=num_diff(y0,h,6,3); x6=x6+1.5;
   f6=diff(f,6); y60=double(f6(x6));
   plot(x6,y6,x6,y60), max(abs(y60-y6))
```

Now let us consider increasing the precision p. For example, setting $p = 7$, the actual maximum error obtained is 0.1330, which is higher than those obtained earlier. The derivative curves are obtained as shown in Figure 7.3. It can be seen that for this particular example, the value of p is selected too high, which leads to errors in the results.

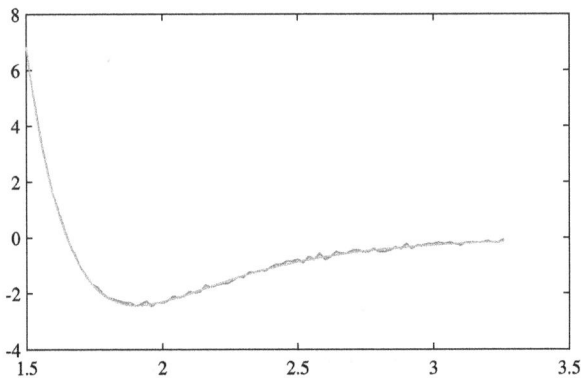

Figure 7.3: The sixth-order derivative with larger values of p.

```
>> [y6,x6]=num_diff(y0,h,6,7); x6=x6+1.5;
   f6=diff(f,6); y60=double(f6(x6));
   plot(x6,y6,x6,y60), max(abs(y60-y6))
```

The seventh- and eighth-order derivatives can also be found with similar MATLAB commands, as shown in Figure 7.4. It can be seen that the eighth-order derivatives obtained are oscillating, and lead to large errors. In practice, this function may sometimes be unsuitable for high-order derivatives.

```
>> [y7,x7]=num_diff(y0,h,7,3); x7=x7+1.5;
   f7=diff(f,7); y70=double(f7(x7));
   [y8,x8]=num_diff(y0,h,8,3); x8=x8+1.5;
   f8=diff(f,8); y80=double(f8(x8));
   plot(x7,y7,x7,y70,x8,y8,x8,y80)
```

It can be seen that the general purpose function given here is quite effective, however, it is a pity that the samples must be selected equally spaced. Unequally spaced

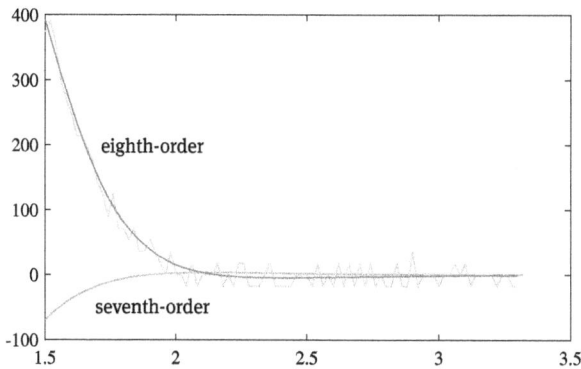

Figure 7.4: Curves of even higher order derivatives.

samples cannot be handled. Another serious problem is that, if the samples are not quite adequate since there are some points missing, the original shape of the derivative may not be retained.

7.4 Numerical partial derivatives of 2D functions

In the previous sections, the solution strategies for univariate functions are fully addressed. Here, we will concentrate on the computation and MATLAB implementations of partial derivatives of 2D functions.

7.4.1 Gradient computation

Definition 7.4. Consider Definition 4.4. If the limits are removed, the approximate partial derivatives of the 2D function $z = f(x,y)$, with respect to x and y, can be written as

$$\frac{\partial f(x,y)}{\partial x} \approx \frac{f(x + \Delta x, y) - f(x,y)}{\Delta x}, \tag{7.4.1}$$

$$\frac{\partial f(x,y)}{\partial y} \approx \frac{f(x,y + \Delta x) - f(x,y)}{\Delta y}, \tag{7.4.2}$$

with precision of $o(\Delta x)$ and $o(\Delta y)$, respectively. If the step-sizes are small enough, these equations can be used to approximate partial derivatives.

For the given data matrix z of 2D function samples, where z is a mesh grid matrix, the function `gradient()` can be used to compute the gradient of the function, with the syntax $[f_x, f_y]$=`gradient(z)`. Thus obtained results f_x and f_y are not really gradients, since the scales of x and y were not considered at all. The actual gradients can be evaluated from $f_x = f_x / \Delta x$ and $f_y = f_y / \Delta y$, where Δx and Δy are respectively the step-sizes in the x and y axes.

Example 7.10. Consider the problem in Example 4.15. If the mesh grid samples are generated, compute the numerical gradients from the data.

Solutions. A set of mesh grid data can be generated with the following statements, and the numerical gradients can be obtained. The contour plots can be obtained, and the results are exactly the same as those in Figure 4.4.

```
>> syms x y; z(x,y)=(x^2-2*x)*exp(-x^2-y^2-x*y);
   [x0,y0]=meshgrid(-3:0.1:3,-2:0.1:2); z0=double(z(x0,y0));
   [fx,fy]=gradient(z0); fx=fx/0.1; fy=fy/0.1; % gradients
   contour(x0,y0,z0,30); hold on;
   quiver(x0,y0,-fx,-fy) % negative gradients
```

It can be found from the following statements that the maximum error is $e_1 = 0.0238$, and the error surface can be drawn as shown in Figure 7.5. It can be seen that in most of the regions the error is small, while in certain areas the error is large, due the large spacing of the mesh grid. Accurate solutions cannot be obtained using simple gradient algorithms.

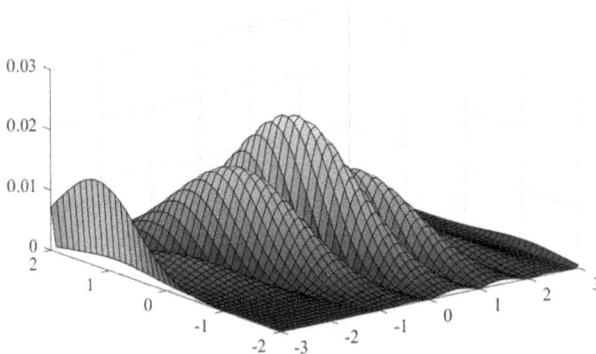

Figure 7.5: Error surface of numerical $\partial z / \partial x$.

```
>> zx=diff(z,x); zx0=double(zx(x0,y0)); % theoretical values
   err=abs(fx-zx0); e1=max(err(:))        % maximum error
   surf(x0,y0,err); axis([-3 3 -2 2 0,0.03]) % error surface
```

The error surface of $\partial z / \partial y$ can also be obtained, as shown in Figure 7.6, with the maximum error of $e_2 = 0.0557$.

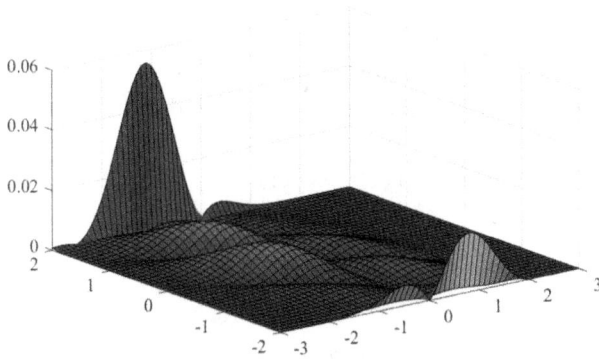

Figure 7.6: Error surface of $\partial z/\partial y$.

```
>> zy=diff(z,y); zy0=double(zy(x0,y0)); err=abs(fy-zy0);
   e2=max(err(:)), surf(x0,y0,err); axis([-3 3 -2 2 0,0.06])
```

It can be seen that the errors are inevitable if simple gradient algorithms are used. How to reduce them? An obvious approach is to have more densely distributed samples. However, this is not what we are expecting. An attractive idea is to use accurate numerical algorithms to get high precision results with the same samples.

7.4.2 High precision algorithms for derivatives with respect to a single variable

As in the case of univariate functions, if a simple forward difference algorithm is used, the error is large, while if the three-point central difference algorithm is used, the error is reduced. In the computation of partial derivatives with respect to one variable, the other variable can be regarded as a constant, so that higher precision algorithms can be proposed.

Theorem 7.4. *Numerical formulas for first-order partial derivatives are given by*

$$\frac{\partial}{\partial x}f(x,y) \approx \frac{f(x+h,y)-f(x-h,y)}{2h},$$
$$\frac{\partial}{\partial y}f(x,y) \approx \frac{f(x,y+k)-f(x,y-k)}{2k},$$

(7.4.3)

with their respective precisions being $o(h^2)$ and $o(k^2)$.

Since the partial derivatives with respect to x or to y are considered separately, the high precision forward difference algorithms in Section 7.3 can be employed, and the numerical partial derivatives can obtained by the following MATLAB function:

```
function [f,t]=part_diff(z,h,n,p,key)
if key==2, z=z.'; end, f=[]; [n1,m1]=size(z);
m=n+p; alpha=0:m; c=double(fdcoef(n,m,alpha)); t=[0:m1-m]*h;
for k=1:n1,
    for i=1:m1-m+1, yy=z(k,i:i+m-1); f1(i)=c*yy(:)/h^n; end
    f=[f; f1];
end
if key==2, f=f.'; end
```

The syntax of the function is $[f,t]$=part_diff(z,h,n,p,key), where z is the sample matrix, h is the step-size, n is the expected order, p is the precision, and key is the flag. If key is 1, the derivative with respect to x is evaluated, otherwise, we compute it with respect to y. The returned variable f is the numerical partial derivative matrix, while t is the offset vector of the independent variable, x or y.

Example 7.11. Consider again the first-order derivative problems in Example 7.10. Compute the numerical first-order derivatives with precision $o(h^5)$ or $o(h^6)$.

Solutions. Selecting $p = 5$ and $p = 6$, respectively, the following MATLAB commands can be used, and more accurate derivatives can be obtained. It is found for $p = 5$ that the maximum errors are $e_1 = 8.6625 \times 10^{-4}$, $e_2 = 2.7507 \times 10^{-4}$, while for $p = 6$, the maximum errors are $e_1 = 3.128 \times 10^{-4}$ and $e_2 = 7.8462 \times 10^{-5}$, which are several orders of magnitude higher than those obtained in Example 7.10, with function gradient().

```
>> syms x y; z(x,y)=(x^2-2*x)*exp(-x^2-y^2-x*y); h=0.1; p=5;
   [x0,y0]=meshgrid(-3:h:3,-2:h:2); z0=double(z(x0,y0));
   [fx,t1]=part_diff(z0,h,1,p,1); [x1,y1]=meshgrid(-3+t1,-2:h:2);
   zx=diff(z,x); zx0=double(zx(x1,y1)); err=abs(fx-zx0); e1=max(err(:))
   [fy,t1]=part_diff(z0,h,1,p,2); [x1,y1]=meshgrid(-3:h:3,-2+t1);
   zy=diff(z,y); zy0=double(zy(x1,y1)); err=abs(fy-zy0); e2=max(err(:))
```

Now consider the computation of $\partial^4 z/\partial x^4$. Of course, the functions like gradient() cannot be used for finding high-order numerical partial derivatives. Using again the universal function, with $p = 10$, the following commands can be applied, and the surface of the fourth-order derivative can be obtained as shown in Figure 7.7.

```
>> [fx,t1]=part_diff(z0,h,4,10,1); [x1,y1]=meshgrid(-3+t1,-2:h:2);
   zx=diff(z,x,4); zx0=double(zx(x1,y1));
   err=abs(fx-zx0); e1=max(err(:)), surf(x1,y1,fx)
```

The maximum error is $e = 0.0689$. Considering the magnitudes of the values of the derivatives in Figure 7.7, the error here is acceptable.

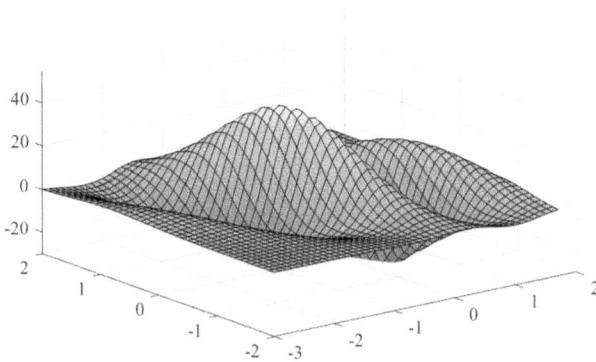

Figure 7.7: Surface of $\partial^4 z/\partial x^4$.

7.4.3 Numerical computation of mixed partial derivatives

It can be seen that for 2D functions, the numerical partial derivatives with respect to one of the independent variable can easily be obtained with the universal function of `part_diff()`, with high precision. How can we find the derivatives for mixed partial derivatives? Here the problem is explored, and two different conventional algorithms are presented first.

Theorem 7.5. *The numerical formula for the mixed partial derivative,*

$$\frac{\partial^2 f(x,y)}{\partial x \partial y} \approx \frac{1}{hk}[f(x+h,y+k)-f(x+h,y-k)-f(x-h,y+k)+f(x-h,y-k)], \quad (7.4.4)$$

has precision $o(\max(h,k))$.

Theorem 7.6. *A mixed partial derivative formula,*

$$\frac{\partial^2 f(x,y)}{\partial x \partial y} \approx \frac{1}{2hk}[f(x+h,y+k)-f(x+h,k)-f(x,y+k)+2f(x,y)$$

$$-f(x,y-k)-f(x-h,y)+f(x-h,y-k)], \quad (7.4.5)$$

has a higher precision of $o(\max(h,k)^2)$.

It is obvious that Theorem 7.6 yields higher-precision results. A MATLAB implementation of the algorithm is given below, where a double-loop structure can be introduced to compute the mixed partial derivatives.

```
function [fxy,x,y]=pdiffxy(z,h,k);
[n,m]=size(z);
for i=1:n-2, for j=1:m-2
    fxy(i,j)=z(i+2,j+2)-z(i+2,j+1)-z(i+1,j+2)+2*z(i+1,j+1) ...
              -z(i+1,j)-z(i,j+1)+z(i,j);
```

```
end, end
x=(1:n-2)*h; y=(1:m-2)*k;
```

Example 7.12. For the 2D function in Example 7.10, compute the mixed second-order partial derivative.

Solutions. The following statements can be used to compute the second-order partial derivative of z with respect to x and y. Compared with theoretical results, it is found that the maximum error is $e_1 = 0.0709$, which is large in certain applications.

```
>> syms x y; z(x,y)=(x^2-2*x)*exp(-x^2-y^2-x*y); h=0.1;
   [x0,y0]=meshgrid(-3:h:3,-2:h:2); z0=double(z(x0,y0));
   [fxy,x10,y10]=pdiffxy(z0,h,h);
   x10=-3+x10; y10=-2+y10; [x1,y1]=meshgrid(x10,y10);
   zxy=diff(z,x,y); zxy0=double(zxy(x1,y1));
   err=abs(fxy-zxy0); e1=max(err(:))
```

7.4.4 Numerical computation of high-order mixed partial derivatives

The mixed partial derivative algorithms studied earlier have certain limitations, since only $\partial^2 z / \partial x \partial y$ can be computed, while other high-order partial derivatives cannot be handled.

Considering the property of high-order partial derivatives,

$$\frac{\partial^{m+n} z(x,y)}{\partial x^m \partial y^n} = \frac{\partial^n}{\partial y^n}\left(\frac{\partial^m z(x,y)}{\partial x^m}\right) = \frac{\partial^m}{\partial x^m}\left(\frac{\partial^n z(x,y)}{\partial y^n}\right), \tag{7.4.6}$$

the nested call of function part_diff() can be applied for computing high-order partial derivatives of the 2D functions. This method is referred to as the sequential derivative method here.

Example 7.13. Solve again the problem in Example 7.6 again, with the sequential derivative method, and assess it precision.

Solutions. Letting $p = 6$, the partial derivatives can be computed for two different orders, and the maximum error between the two results is as small as $e_1 = 4.5741 \times 10^{-12}$, indicating that the order in the sequential call may not affect the final precision. The maximum error from the theoretical result is $e_2 = 6.8172 \times 10^{-4}$, and it can be seen that the precision is by far better than for the algorithms studied earlier.

```
>> syms x y; z(x,y)=(x^2-2*x)*exp(-x^2-y^2-x*y); h=0.1;
   [x0,y0]=meshgrid(-3:h:3,-2:h:2); z0=double(z(x0,y0)); p=6;
   [fx,t11]=part_diff(z0,h,1,p,1); [fxy,t21]=part_diff(fx,h,1,p,2);
   [fy,t22]=part_diff(z0,h,1,p,2); [fyx,t12]=part_diff(fy,h,1,p,1);
```

```
[x1,y1]=meshgrid(-3+t11,-2+t21); err=abs(fxy-fyx); e1=max(err(:))
zxy=diff(z,x,y); zxy0=double(zxy(x1,y1));
err=abs(fxy-zxy0); e2=max(err(:))
```

Example 7.14. Find $\partial^5 z(x,y)/(\partial^3 x\partial^2 y)$ for the 2D function in Example 7.10.

Solutions. Selecting $p = 10$, the maximum error of $e_1 = 0.0347$ can be obtained with the sequential derivative method, and the surface of the partial derivative is as shown in Figure 7.8.

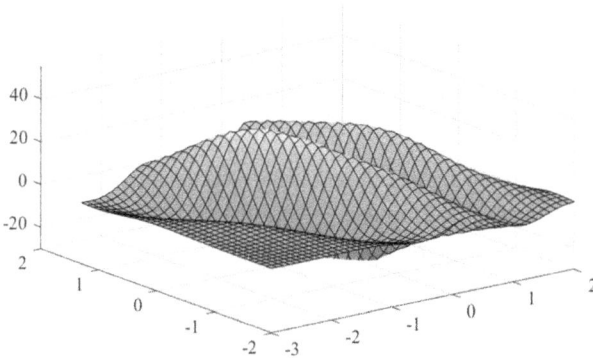

Figure 7.8: Surface of $\partial^5 z(x,y)/(\partial x^3\partial y^2)$.

```
>> syms x y; z(x,y)=(x^2-2*x)*exp(-x^2-y^2-x*y); h=0.1;
   [x0,y0]=meshgrid(-3:h:3,-2:h:2); z0=double(z(x0,y0)); p=6;
   [fx,t1]=part_diff(z0,h,3,p,1); [fxy,t2]=part_diff(fx,h,2,p,2);
   [x1,y1]=meshgrid(-3+t1,-2+t2);
   zxy=diff(z,x,x,x,y,y); zxy0=double(zxy(x1,y1));
   err=abs(fxy-zxy0); e1=max(err(:)), surf(x1,y1,fxy)
```

Numerical partial derivatives for 2D functions are presented in this section, and the sequential derivative method is recommended for high precision computation. The accuracy presented here is much higher than for the conventional algorithms. There is also a deadly defect in the algorithm, since the samples in a mesh grid should be provided, and in some applications, this requirement cannot be satisfied. Other effective approaches will be explored in the next section.

7.5 Spline interpolation and numerical derivatives

In real scientific and engineering applications, normally a set of experimental or simulation data can be obtained. The data can be used as samples. They can be equally-spaced or scattered data. The spline interpolation technique is a good way to restore unknown information from samples.

In this section, the concepts and MATLAB solutions of spline problems are presented, and spline-based numerical derivatives are addressed.

7.5.1 Cubic spline

Definition 7.5. Assume that the set of n samples is rearranged as (x_i, y_i), $i = 1, 2, \ldots, n$, where $x_1 < x_2 < \cdots < x_n$.

Cubic spline is the most widely used spline in real applications, since the mathematical form is simple and easy to implement. The mathematical form of the cubic spline is presented here first, followed by its use in MATLAB.

Definition 7.6. If a function $S(x)$ satisfies the following three conditions, $S(x)$ is referred to as a cubic spline function of the n samples:
(1) $S(x_i) = y_i$ ($i = 1, 2, \ldots, n$), i. e., all the samples are on $S(x)$;
(2) $S(x)$ is a cubic function in each subinterval $[x_i, x_{i+1}]$,

$$S(x) = c_{i1}(x - x_i)^3 + c_{i2}(x - x_i)^2 + c_{i3}(x - x_i) + c_{i4}. \tag{7.5.1}$$

(3) $S(x)$ has continuous first- and second-order derivatives in the whole interval $[x_1, x_n]$.

Theorem 7.7. *The mathematical form of a cubic function in the kth subinterval $[x_k, x_{k+1}]$ can be expressed as*[9]

$$s(x) = y_k + b_k(x - x_k) + c_k(x - x_k)^2 + d_k(x - x_k)^3, \tag{7.5.2}$$

with $k = 1, 2, \ldots, n$, and step-sizes $h_k = x_{k+1} - x_k$, where

$$b_k = \frac{y_{k+1} - y_k}{h_k} - h_k(\sigma_{k+1} + 2\sigma_k), \quad c_k = 3\sigma_k, \quad d_k = \frac{\sigma_{k+1} - \sigma_k}{h_k}. \tag{7.5.3}$$

The coefficients σ_k can be found recursively from

$$\sigma_n = \frac{\beta_n}{\alpha_n}, \quad \sigma_i = \frac{\beta_i - h_i\sigma_{i+1}}{\alpha_i}, \quad i = n - 1, n - 2, \ldots, 1, \tag{7.5.4}$$

with

$$\alpha_1 = -h_1, \quad \alpha_i = 2(h_{i+1} + h_i) - \frac{h_{i-1}^2}{\alpha_{i-1}}, \quad \alpha_n = -h_{n-1} - \frac{h_{n-1}^2}{\alpha_{n-1}}, \tag{7.5.5}$$

$$\beta_1 = h_1^2\sigma_1^{(3)}, \quad \beta_i = (\delta_i - \delta_{i-1}) - \frac{h_{i-1}\beta_{i-1}}{\alpha_{i-1}}, \quad \beta_n = -h_{n-1}\sigma_{n-3}^{(3)} - \frac{h_{n-1}\beta_{n-1}}{\alpha_{n-1}}. \tag{7.5.6}$$

where i = 2, 3, ..., n − 1, and

$$\delta_i = \frac{y_{i+1} - y_i}{x_{i+1} - x_i}, \quad \delta_i^{(2)} = \frac{\delta_{i+1} - \delta_i}{x_{i+2} - x_i}, \quad \delta_i^{(3)} = \frac{\delta_{i+1}^{(2)} - \delta_i^{(2)}}{x_{i+3} - x_i}. \quad\quad (7.5.7)$$

It might be rather trivial to implement the recursive algorithm. The MATLAB function `csapi()` can be used directly to create a cubic spline object, with the syntax `S=csapi(x,y)`, where vectors $x = [x_1, x_2, \ldots, x_n]$ and $y = [y_1, y_2, \ldots, y_n]$ are the samples. The fields, or membership variables, in object S include:

(1) `breaks` which stores the subinterval boundaries in a row vector $[x_1, x_2, \ldots, x_n]$;
(2) `coefs` which is a matrix, whose kth row stores the kth polynomial coefficients;
(3) Other fields, including `form` with content `'pp'`; `pieces` being the number of samples n; `order` which is the order of 4; and `dim` which stores the dimension number.

The spline object S can be shown graphically with function `fnplt(S)`. For a given interpolation vector x_p, function `fnval()` can be used to evaluate interpolation, with the syntax y_p=`fnval(S,`x_p`)`, where y_p contains the interpolation results.

Example 7.15. For a set of sparsely selected samples $x = [0, 0.4, 1, 2, \pi]$ and their sinusoidal values $y = \sin x$, restore the sine curve using a cubic spline.

Solutions. This example is an extreme and challenging one. The number of samples is very small, and one may suspect with the original function can hardly be restored with so few samples. The following commands can be used to establish a cubic sample. The restored curve and the sine curve can both be obtained, as shown in Figure 7.9.

```
>> x0=[0,0.4,1 2,pi]; y0=sin(x0);
   sp=csapi(x0,y0), sp.breaks, sp.coefs, fnplt(sp,':'); % cubic
   hold on, ezplot('sin(t)',[0,pi]); plot(x0,y0,'o')    % theoretical
```

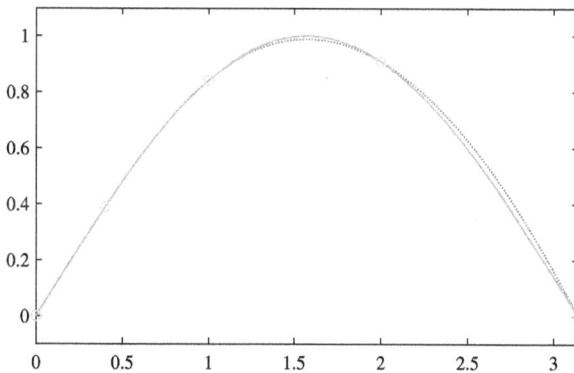

Figure 7.9: Cubic spline interpolation results.

where sp.breaks stores the boundaries, $[0, 0.4, 1, 2, \pi]$. These five points divide the whole interval into four subintervals $(0, 0.4)$, $(0.4, 1)$, $(1, 2)$, $(2, \pi)$. With command sp.coefs, the coefficients can be displayed as shown in Table 7.3, whose kth row stores the polynomial coefficients in the kth subinterval. For instance, in the subinterval $(0.4, 1)$, the interpolation polynomial can be expressed as

$$S_2(x) = -0.1627(x - 0.4)^3 - 0.1876(x - 0.4)^2 + 0.9245(x - 0.4) + 0.3894.$$

Table 7.3: Coefficients of the cubic spline.

interval	c_1	c_2	c_3	c_4
$(0, 0.4)$	−0.16265031	0.007585654	0.99653564	0
$(0.4, 1)$	−0.16265031	−0.18759472	0.92453202	0.38941834
$(1, 2)$	0.024435717	−0.48036529	0.52375601	0.84147098
$(2, \pi)$	0.024435717	−0.40705814	−0.36366741	0.90929743

Function csapi() can also be used to handle multivariate samples, though mesh grid data are needed. The syntax of the function is

S=csapi({$x_1, x_2, \ldots, x_n$},z)

where x_i is the vector of the ith independent variable and z is the mesh grid data of the samples. The cubic spline object S can be established.

Example 7.16. Consider the 2D function $z = f(x, y) = (x^2 - 2x)e^{-x^2 - y^2 - xy}$. Generate a set of sparsely distributed mesh grid samples, and based on the them, draw the interpolated surface.

Solutions. The spline object sp can easily be generated, and the surface can be obtained directly as shown in Figure 7.10.

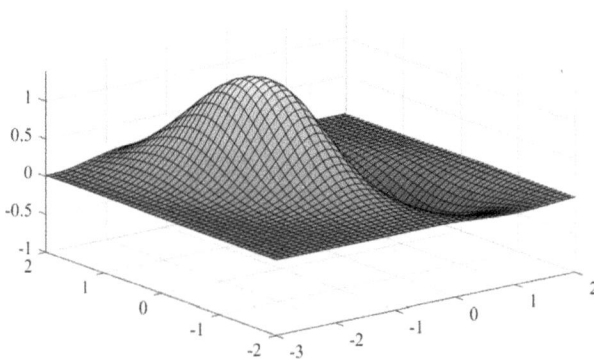

Figure 7.10: Interpolation result of a 2D function.

```
>> x0=-3:.6:3; y0=-2:.4:2; [x,y]=ndgrid(x0,y0); % mesh grid
   z=(x.^2-2*x).*exp(-x.^2-y.^2-x.*y);          % samples
   sp=csapi({x0,y0},z); fnplt(sp);              % cubic spline
```

It should be noted that the matrix z should be established upon the matrices x and y, based on the format of ndgrid() function. The function meshgrid() cannot be used to generate the mesh grid samples. Besides, scattered data cannot be used in the interpolation process.

7.5.2 B-splines

B-spline is another commonly used spline, whose mathematical form is more complicated to understand and implement than cubic spline. The quality of B-spline interpolation is superior to that of the cubic spline, and will be demonstrated later. The mathematical form is briefly presented, and we concentrate again on how to use B-splines in MATLAB.

Definition 7.7. The interval of interest (a, b) can be divided into several subintervals $a = t_0 < t_1 < t_2 < \cdots < t_m = b$, where t_i are referred to as knots. An approximate piecewise function can be written as

$$F(t) = \sum_{i=0}^{m} p_i B_{i,k}(t),\tag{7.5.8}$$

where p_i are coefficients, and k is the order, with $k \leqslant m$. Functions $B_{i,k}(x)$ are referred to as the kth B-spline basis, whose initial values are selected as

$$B_{i,0}(t) = \begin{cases} 1, & \text{if } t_i < t < t_{i+1}, \\ 0, & \text{otherwise.} \end{cases}\tag{7.5.9}$$

The B-spline basis can be recursively computed as

$$B_{i,j}(t) = \frac{t - t_i}{t_{i+j} - t_i} B_{i,j-1}(t) + \frac{t_{i+j+1} - t}{t_{i+j+1} - t_{i+1}} B_{i+1,j-1}(t),\tag{7.5.10}$$

with $j = 1, 2, \ldots, k$, $i = 0, 1, 2, \ldots, m$.

A MATLAB function spapi() can be used to create a B-spline object S for samples x and y, with the syntax S=spapi(k,x,y), where k is the selected order. Normally, one may select $k = 4$ or 5, for good interpolation results. For some specific problems, a higher order k may lead to better performance.

The B-spline object S has also its own fields, such as form which is set to B-; knots is the vector of knots; coefs stores the coefficients in B-splines; number is the number of samples; order is the order of the spline; and dim stores the dimension of the samples.

Example 7.17. Explore the fifth-order B-spline to interpolate the samples in Example 7.15.

Solutions. Consider the samples in Example 7.15. With the following statements, the interpolation results can be obtained, as shown in Figure 7.11. The difference between the B-spline interpolation curve and the original sine function cannot be distinguished from the curves.

```
>> x0=[0,0.4,1 2,pi]; y0=sin(x0);
   ezplot('sin(t)',[0,pi]); hold on
   sp2=spapi(5,x0,y0); fnplt(sp2,':'), v=sp.coefs  % fifth-order B-spline
```

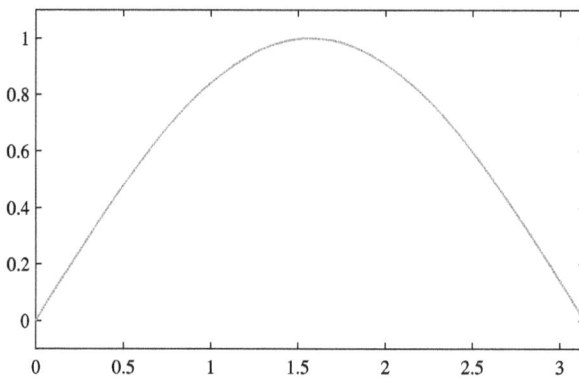

Figure 7.11: B-spline interpolation of the results.

It can be seen that the quality of the fifth-order B-spline is much better than that of the piecewise cubic polynomials. The coefficient vector of the B-spline is $v = [1.8076 \times 10^{-16}, 0.7784, 1.6244, 0.7802, 1.2246 \times 10^{-16}]$.

7.5.3 Numerical derivatives with splines

Theorem 7.8. *From (7.5.1), it can be seen that the first-order derivative of a cubic spline is a spline with degree one less*

$$S(x) = 3c_{i1}(x - x_i)^2 + 2c_{i2}(x - x_i) + c_{i3}. \tag{7.5.11}$$

Theorem 7.9. *The derivative of a kth order B-spline of is a function of $(k - 1)$th-order B-splines*

$$\frac{d}{dt}B_{i,k}(t) = k\left(\frac{B_{i,k-1}(t)}{t_{i+k} - t_i} - \frac{B_{i+1,k-1}(t)}{t_{i+k+1} - t_{i+1}}\right), \tag{7.5.12}$$

which implies[4]

$$\frac{d}{dt}\sum_i p_i B_{i,k}(t) = \sum_{r=i-k+2}^{s-1} k\frac{\alpha_i - \alpha_{i-1}}{t_{i+k} - t_i}B_{i,k-1}(t) \quad on\ t \in [t_s, t_r].\qquad (7.5.13)$$

A MATLAB function `fnder()` is provided to compute derivatives from the splines, with the following syntaxes:

S_d=fnder(S,k), % kth-order derivative
S_d=fnder($S, [k_1, \cdots, k_n]$), % partial derivatives multivariate spline

The former computes the kth-order derivatives from a univariate spline object S, and the obtained S_d is also an object. Using the latter, numerical partial derivatives of multivariate splines are evaluated.

Example 7.18. Consider the samples in Example 7.17. Find the second-order derivatives with cubic and B-splines.

Solutions. Since only five samples are known, the algorithms in Section 7.3 cannot be used at all. Also, the step-sizes vary. Splines can be tried to find numerical derivatives.

Samples can be generated with the following statements. Cubic and B-splines can be established. Function `fnder()` can be used to compute the second-order derivatives, and also the theoretical one, as shown in Figure 7.12.

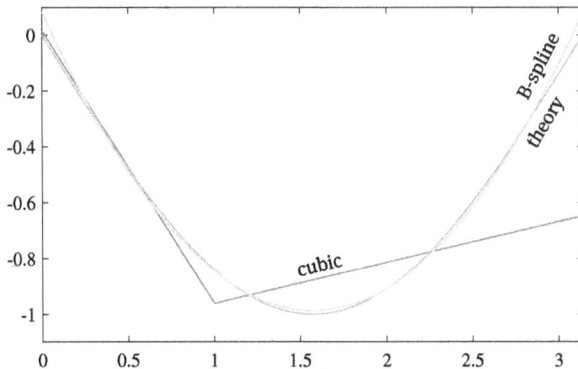

Figure 7.12: Numerical derivative results with splines.

```
>> x0=[0,0.4,1 2,pi]; y0=sin(x0); S=csapi(x0,y0);
   S1=fnder(S,2), fnplt(S1,[0,pi]), hold on, S=spapi(6,x0,y0);
   ezplot('-sin(t)',[0,pi]), S2=fnder(S,2), fnplt(S2,[0,pi]),
```

Obviously, since a cubic spline uses piecewise cubic functions as it kernel, second-order derivatives may lead to straight lines. Therefore, cubic splines cannot be used in numerical derivatives, while with B-spline the result is satisfactory, even though, only five samples are merely known.

It can be seen that for the problems with unevenly distributed samples and very few samples, the numerical derivatives cannot be obtained with the methods in the previous sections, while with B-splines, the problem can be solved easily, witḧ satisfactory results.

Example 7.19. Try to find the numerical partial derivative surface of $\dfrac{\partial^5 z(x,y)}{\partial x^3 \partial y^2}$ for Example 7.14, and compare it with the results in the example.

Solutions. Before computing high-order derivatives, the derivative $\dfrac{\partial^2 z(x,y)}{\partial x \partial y}$ can be evaluated first, with the B-spline technique. It can be found that the maximum error is 4.4003×10^{-4}.

```
>> h=0.1; x1=-3:h:2; y1=-2:h:2; [x0,y0]=ndgrid(x1,y1); % samples
   syms x y; z(x,y)=(x^2-2*x)*exp(-x^2-y^2-x*y);
   z0=double(z(x0,y0)); sp=spapi({5,5},{x1,y1},z0);
   pxy=fnder(sp,[1,1]); fxy=fnval(pxy,{x1,y1});
   dxy=diff(z,x,y); err=fxy-double(dxy(x0,y0)); max(abs(err(:)))
```

Now let us try to compute $\dfrac{\partial^5 z(x,y)}{\partial x^3 \partial y^2}$. The error surface can be obtained as shown in Figure 7.13. It can be seen that the error is large, with a maximum error of 1.0142.

```
>> pxy=fnder(sp,[3,2]); fxy=fnval(pxy,{x1,y1});
   dxy=diff(z,x,x,x,y,y); err=fxy-double(dxy(x0,y0));
   max(abs(err(:))), surf(x0,y0,err)
```

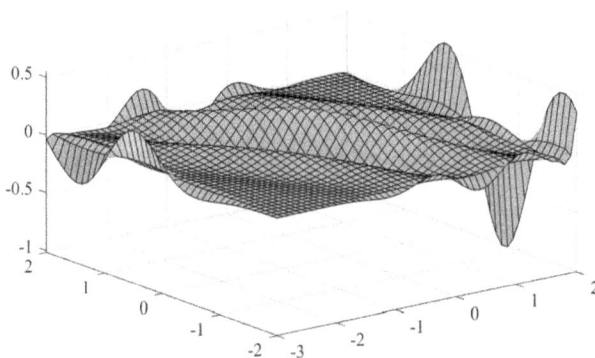

Figure 7.13: Error surface of the high-order partial derivatives.

7.5.4 Unequally spaced and scattered samples

For numerical partial derivatives of multivariate functions, the sequential derivative method in Section 7.4 was recommended. However, for unevenly distributed samples, the method cannot be used at all. For scatter samples, how can we accurately obtain numerical partial derivatives?

Two steps can be tried to compute the partial derivatives:
(1) A MATLAB function `griddata()` can be used to interpolate scattered data. The syntax of the function is z=`griddata`$(x_0,y_0,z_0,x,y,$'v4'$)$, where x_0, y_0, and z_0 are vectors storing the samples, and they can be of any format. The arguments x and y are expected interpolation points, which can be a point, a vector, or a matrix. The size of the returned variable z is exactly the same as of x and y. Argument 'v4' refers to the interpolation algorithm presented in MATLAB 4.0. For computing numerical partial derivatives, x and y should be selected as evenly spaced mesh grids.
(2) Compute partial derivatives with the method studied in Section 7.4.

Example 7.20. Consider the function in Example 7.10. Generate a set of 800 scattered samples, and see whether the partial derivative $\dfrac{\partial^2 z(x,y)}{\partial x \partial y}$ can be restored with the samples.

Solutions. Input first the original symbolic expression into MATLAB, for later comparisons. Use `rand()` function to generate randomly a set of 800 pairs of x and y, as scattered samples, with the distribution shown in Figure 7.14. It can been that the distribution is good.

```
>> syms x y; z(x,y)=(x^2-2*x)*exp(-x^2-y^2-x*y); h=0.1;
   x1=-3+5*rand(800,1); y1=-2+4*rand(800,1); plot(x1,y1,'o')
```

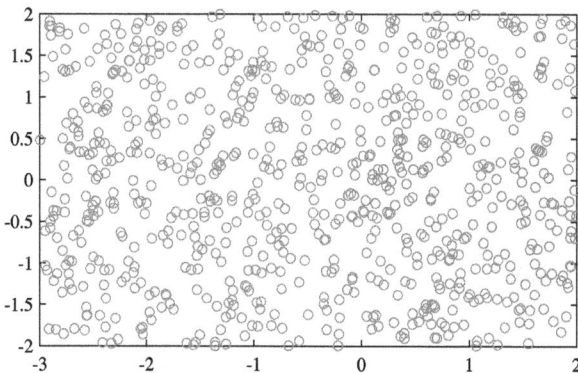

Figure 7.14: Distribution of the samples in the xy plane.

Computing mesh grid interpolation points, the original function surface can be restored, and the error surface is shown in Figure 7.15. It can be seen that the errors are small, while at certain points or on the boundaries, oscillation can be witnessed. These may lead to errors in numerical derivatives.

```
>> z1=double(z(x1,y1)); [x0,y0]=meshgrid(-3:h:2,-2:h:2);
   z0=griddata(x1,y1,z1,x0,y0,'v4');
   zx=double(z(x0,y0)); surf(x0,y0,zx-z0)
```

Due to the existence of unexpected oscillations, the precision p should not be set to a large number. One may select $p = 2$, and the error surface of the numerical mixed partial derivatives can be obtained, as shown in Figure 7.16. The maximum error is as high as 0.5071. Also, since this problem cannot be solved by other methods, this method might be the only choice.

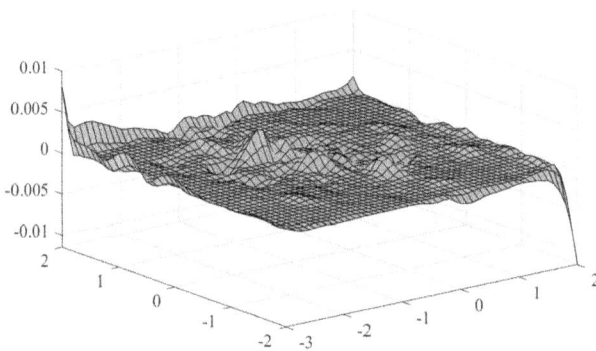

Figure 7.15: Error surface between the original and interpolated functions.

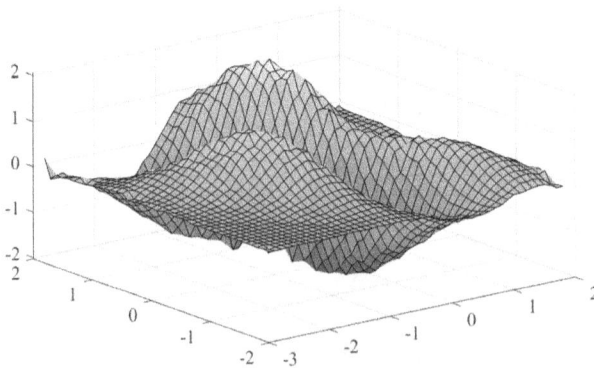

Figure 7.16: The surface of numerical partial derivative.

```
>> p=2;  [fx,t1]=part_diff(z0,h,1,p,1);
   [fxy,t2]=part_diff(fx,h,1,p,2);  [x1,y1]=meshgrid(-3+t1,-2+t2);
   zxy=diff(z,x,y);  zxy0=double(zxy(x1,y1));
   err=abs(fxy-zxy0);  e1=max(err(:)),  surf(x1,y1,fxy)
```

It can also be seen that for this example, the high-order partial derivatives cannot be obtained from the scattered samples.

7.6 Exercises

7.1   Using forward difference algorithm, find the numerical derivative for the data provided in Table 7.4.

Table 7.4: Samples in Problem 7.1.

x_i 0	0.1	0.2	0.3	0.4	0.5	0.6	0.7	0.8	0.9	1	1.1	1.2
y_i 0	2.2077	3.2058	3.4435	3.241	2.8164	2.311	1.8101	1.3602	0.98172	0.67907	0.4473	0.27684

7.2   Derive forward and central difference algorithms for the tenth-order derivative with 15 samples.

7.3   Generate a set of samples for the following functions, and then compute the numerical first-order derivatives. Compare the results with theoretical ones, and assess their precision.

(1) $y(t) = \arccos^2 x + (\ln^2 \arccos x - \ln \arccos x + 1/2)$,

(2) $y(t) = \dfrac{1}{2} \arctan \sqrt[4]{1 + x^4} + \dfrac{1}{4} \ln \dfrac{\sqrt[4]{1 + x^4} + 1}{\sqrt[4]{1 + x^4} - 1}$,

(3) $y(x) = \dfrac{e^{-x^2} \arcsin e^{-x^2}}{\sqrt{1 - e^{-2x^2}}} + \dfrac{1}{2} \ln(1 - e^{-2x^2})$.

7.4   Generate samples from the 2D function

$$z(x,y) = \frac{1}{3x^2 + y^2} e^{-x^2 - y^4} \sin(xy^2 + x^2 y).$$

The data can be in mesh grid format, or as scattered data. Fit the surface of the function and also find the numerical high-order partial derivatives of the function $z(x,y)$, finally, assess the accuracy.

7.5   Compute numerical gradients from the data given in Table 7.5. If the prototype function is $f(x,y) = 4 - x^2 - y^2$, assess the accuracy.

7.6   Generate a set of samples for the 2D function

$$u(x,y) = x - y + x^2 + 2xy + y^2 + x^3 - 3x^2 y - y^3 + x^4 - 4x^2 y^2 + y^4,$$

Table 7.5: Samples in Problem 7.5.

t	0	0.2	0.4	0.6	0.8	1	1.2	1.4	1.6	1.8	2
0	4	3.96	3.84	3.64	3.36	3	2.56	2.04	1.44	0.76	0
0.2	3.96	3.92	3.8	3.6	3.32	2.96	2.52	2	1.4	0.72	−0.04
0.4	3.84	3.8	3.68	3.48	3.2	2.84	2.4	1.88	1.28	0.6	−0.16
0.6	3.64	3.6	3.48	3.28	3	2.64	2.2	1.68	1.08	0.4	−0.36
0.8	3.36	3.32	3.2	3	2.72	2.36	1.92	1.4	0.8	0.12	−0.64
1	3	2.96	2.84	2.64	2.36	2	1.56	1.04	0.44	−0.24	−1
1.2	2.56	2.52	2.4	2.2	1.92	1.56	1.12	0.6	0	−0.68	−1.44
1.4	2.04	2	1.88	1.68	1.4	1.04	0.6	0.08	−0.52	−1.2	−1.96
1.6	1.44	1.4	1.28	1.08	0.8	0.44	0	−0.52	−1.12	−1.8	−2.56
1.8	0.76	0.72	0.6	0.4	0.12	−0.24	−0.68	−1.2	−1.8	−2.48	−3.24
2	0	−0.04	−0.16	−0.36	−0.64	−1	−1.44	−1.96	−2.56	−3.24	−4

and find the following partial derivatives with numerical methods. Assess the accuracy of the results.

(1) $\dfrac{\partial^4 u(x,y)}{\partial x^4}$, (2) $\dfrac{\partial^4 u(x,y)}{\partial x^3 \partial y}$, (3) $\dfrac{\partial^4 u(x,y)}{\partial x^2 \partial y^2}$.

8 Numerical integrals

Numerical integrals, or quadratures, are an important topic in traditional numerical analysis courses. There are typically two cases where numerical integrals must be used. In the first case, if the integrand is not known, and only a set of measured samples are known, trapezoidal or some other algorithms should be used to compute the numerical integrals of the given functions. In the second case, if the integrands are known, while analytical solutions for the definite integrals of univariate functions, improper integrals, or multiple integrals do not exist, numerical approaches must be employed. With the analytical approach discussed earlier, with the support of function vpa(), high precision integrals of integrals of any complexity can almost be obtained. Pure numerical algorithms may not be badly needed, however, for multiple integrals, with inner integrals which do not have analytical expressions, numerical algorithms are musts.

In Section 8.1, numerical integrals are discussed for functions with given samples. Trapezoidal algorithm is introduced first for solving the problems. High precision algorithms for numerical integrals from samples are then presented. Subsequently numerical integration algorithms and implementations when the integrands are known are given. In Section 8.2, numerical integral problems for univariate functions are explored first, and the simple Simpson integration algorithms are presented. Integral functions can also be evaluated. In Section 8.3, MATLAB implementations of numerical integrals of double integrands are proposed, and the integral surface can also be reconstructed numerically. In Section 8.4, numerical solutions to triple or even multiple integral problems are formulated and implemented. For triple integrals, functional boundaries can be processed, while for multiple integrals, hyper-rectangular integral boundaries can be handled. In Section 8.5, some special algorithms such as Monte Carlo approximation and spline interpolation-based algorithms are presented and demonstrated.

8.1 Numerical integrals from samples

8.1.1 Direct computation of numerical definite integrals

In practical situations, sometimes, samples of a function can be measured through experiment, while the mathematical expression of the function cannot be reconstructed. In this case, how can we find the definite integral of the unknown function numerically? In this section, numerical integral is proposed.

Definition 8.1. The definite integral is mathematically described as

$$I = \int_a^b f(x)\mathrm{d}x. \tag{8.1.1}$$

https://doi.org/10.1515/9783110666977-008

198

198

198198198198198198

198198198198

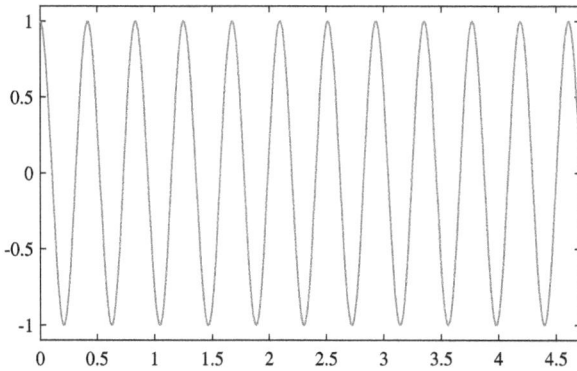

Figure 8.1: Curve of integrand function $f(x) = \cos 15x$.

```
>> x=linspace(0,3*pi/2,500); y=cos(15*x);
   plot(x,y) % oscillatory integrand
```

For different step-sizes, h = 0.1, 0.01, 0.001, 0.0001, 0.00001, 0.000001, the following statements can be executed, and the results, as well as time needed, are obtained as shown in Table 8.1.

```
>> syms x, A=int(cos(15*x),0,3*pi/2); h0=10.^[-1:-1:-6]; v=[];
   for h=h0, % trapezoidal evaluation for different step-sizes
       x=[0:h:3*pi/2, 3*pi/2]; y=cos(15*x);
       tic, I=trapz(x,y); v=[v; h,I,1/15-I]; toc,
   end
```

With the decrease of step-size h, the accuracy increases. For instance, when $h = 10^{-6}$, 11 digits are accurate, and the time required is large, 0.079 seconds. If higher accuracy is expected, the step-size h should further be reduced, until out-of-memory error occurs.

Table 8.1: Step-size selection and computation results.

h	integral	error	time
0.1	0.05389175150	0.01277491	0.0083
0.01	0.06654169547	0.000125	0.0018
0.001	0.06666541668	1.25×10^{-6}	0.0031
0.0001	0.066666654167	1.25×10^{-8}	0.021
10^{-5}	0.066666666542	1.25×10^{-10}	0.014
10^{-6}	0.066666666665	1.250×10^{-12}	0.079

If the integral interval is changed from $(0, 3\pi/2)$ to a larger one of $(0, 1\,000)$, to ensure accuracy, the step-size should be very small. If we still have $h = 10^{-6}$, the "out of memory" error may occur. Therefore, effective algorithms are needed for numerical integration tasks.

```
>> h=1e-6; x=[0:h:1000];
   tic, y=cos(15*x); I=trapz(x,y); toc % cannot be used
```

8.1.2 Reconstruction of integral function

Assume again that a set of samples is measured and given by

$$(x_1, y_1), (x_2, y_2), \dots, (x_{n+1}, y_{n+1}),$$

and x_i is a strictly increasing sequence, so that the reconstruction of integral functions can be made.

Definition 8.3. Assuming the samples are known, the integral function can be reconstructed recursively from the samples

$$F(x_i) = \int_a^{x_i} f(x)dx \approx \sum_{i=1}^{x_i} s_i = F(x_{i-1}) + \frac{1}{2}(y_i + y_{i+1})(x_{i+1} - x_i), \qquad (8.1.4)$$

where $x_1 = a$, $F(x_1) = 0$, $i = 2, 3, \dots, n$.

Based on the above recursive algorithm, the following universal MATLAB function can be written:

```
function F=trapz_fun(x,y)
n=length(x)-1; F=0;
for i=2:n, F(i)=F(i-1)+0.5*(y(i)+y(i-1))*(x(i+1)-x(i)); end
```

The syntax of the function is $F=\texttt{trapz\_fun}(x, y)$, where x and y are the sample vectors, and vector F is the approximation of the integral function, the length of F is one less than that of x.

Example 8.3. Reconstruct the integral function from the samples in Example 8.2 with the trapezoidal method.

Solutions. A set of samples can be generated from the original function, and the primitive function can be reconstructed. The analytical result of the primitive function is $\sin 15x/15$, therefore, the integral function can be reconstructed as shown in Figure 8.2. It can be seen that the result reconstructed is exactly the same as the theoretical one.

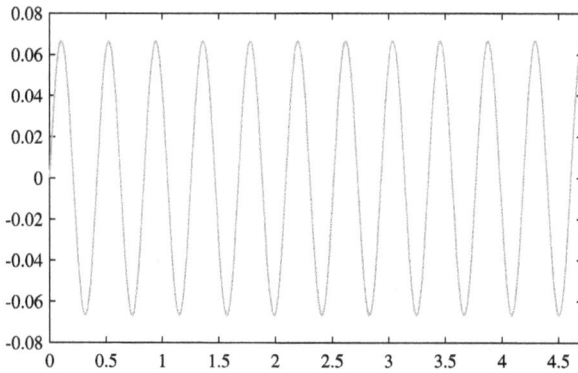

Figure 8.2: Integral function and theoretically precise sin 15x/15.

```
>> x=linspace(0,3*pi/2,500); y=cos(15*x); x1=x(1:end-1);
   y1=trapz_fun(x,y); plot(x1,y1,x1,sin(15*x1)/15)
```

8.1.3 High precision numerical integration algorithm for equally spaced samples

To evaluate numerically the integral $\int_a^b f(x)\mathrm{d}x$, Newton–Cotes formulas are a set of important and practical algorithms. These formulas are classified into open and closed. The values at the terminal point of the integration interval are not used in the open formulas, while the terminal points are used in the closed formulas.

From the practical application viewpoint, closed formulas are more suitable to evaluate numerical integrals, while open formulas are better for improper integrals, since the terminal function values may not exist. Open formulas, on the other hand, avoid the terminal function values.

In real applications, the functions are often defined at one of the terminal points, a or b, while undefined at the other, for improper integral problems. In this case, semi-open Newton–Cotes formulas can be used instead. If the interval (a, b) is too large, it should be divided into smaller subintervals, which can be evaluated independently.

Definition 8.4. If a set samples $f_0, f_1, \ldots, f_n$ are known, and $n \leqslant 6$, with a step-size of h, the integral in the interval $(x_0, x_0 + (n-1)h)$ can be evaluated from Table 8.2. The algorithms are referred to as closed Newton–Cotes formulas, and in [13], the coefficients for up to $o(h^{22})$ are listed.

Although any precision can be obtained in theory, extra high-order algorithm may lead to the Runge phenomenon in numerical stability, which may be catastrophic in polynomial fitting,[28] and the fitting may well be beyond the data range. In real ap-

Table 8.2: High precision formula.

algorithm	formula	error
trapezoidal	$h(f_0 + f_1)/2$	$o(h^2)$
Simpson 1/3 formula	$h(f_1 + 4f_1 + f_2)/3$	$o(h^4)$
Simpson 3/8 formula	$3h(f_0 + 3f_1 + 3f_2 + f_3)/8$	$o(h^4)$
Boole formula	$2h(7f_0 + 32f_1 + 12f_2 + 32f_3 + 7f_4)/45$	$o(h^6)$
six-point	$5h(19f_0 + 75f_1 + 50f_2 + 50f_3 + 75f_4 + 19f_5)/288$	$o(h^6)$
seven-point	$h(41f_0 + 216f_1 + 27f_2 + 272f_3 + 27f_4 + 216f_5 + 41f_6)/140$	$o(h^8)$

plications, moderate precision requirement should be assigned, for instance, seven-point algorithm or Boole formula in Table 8.2.

It can be seen from the table that the larger the order n, the higher precision can be expected. If equally spaced samples y_i, $i = 1, 2, \ldots, n + 1$ are obtained, they can be divided into subgroups, with seven samples each, as f_i, the $o(h^8)$ formula can then be used to compute the definite integral

$$(y_1, \ldots, y_7), (y_7, \ldots, y_{14}), \ldots, (y_{6i+1}, \ldots, y_{6i+7}), \ldots, (y_{6m+1}, \ldots, y_{n+1}), \tag{8.1.5}$$

where $m = \lfloor n/6 \rfloor$, $i = 1, 2, \ldots, m$.

In the final group, the length is quite likely to be less than seven. The integral in the final group can be solved directly with the remaining samples, and the integral in $[x_1, x_{n+1}]$ can be evaluated. The precision depends entirely on the length of the final group. Therefore, to ensure the overall precision, the total numbers of points should be multiples of 6 plus one, such that there are seven samples in the final group.

Based on the above ideas, the following MATLAB function can be written:

```
function S=num_integral(y,h)
y=y(:); S=0; m=floor((length(y)-1)/6)+1;
A(1,1:2)=[1 1]/2; A(2,1:3)=[1 4 1]/3; A(3,1:4)=3*[1 3 3 1]/8;
A(4,1:5)=2*[7 32 12 32 7]/45; A(5,1:6)=5*[19,75,50,50,75,19]/288;
A(6,1:7)=[41,216,27,272,27,216,41]/140;
for i=1:m, k=6*(i-1)+1;
    if i==m, y1=y(k:end); else, y1=y(k:k+6); end
    n=length(y1)-1; if n~=0, S=S+A(n,1:n+1)*y1*h; end
end
```

The syntax of the function is S=num_integral($\boldsymbol{y}, h$), where $\boldsymbol{y}$ stores the equally spaced samples, h is the step-size, the returned S is the approximate value of the definite integral. Since closed formulas are used, this function is not suitable for improper integral problems. Improper integrals can be evaluated instead with Newton–Cotes open or semi-open formulas,[13] or other algorithms.

Example 8.4. Consider again the integral problem in Example 8.2. If there are about 500 equally spaced samples, what will be the results? If the error level is expected to be 10^{-16}, how many points should be known?

Solutions. Assume that there are 499 points (multiple of 7 plus 1), the samples can be generated with the following statements, and the approximate integral can be computed. The result obtained is $S_1 = 0.066666666678740$, the step-size is about $h = 0.0095$, similar to $h = 0.01$ in Table 8.1, while the accuracy is much higher, about 10^8 times, than that of the trapezoidal algorithm.

```
>> x=linspace(0,3*pi/2,499); y=cos(15*x);
   h=x(2)-x(1), S1=num_integral(y,h)
```

If the number of sample points is selected as 500, the result is about $S_{1a} = 0.066665549702459$.

Why is the difference here so large? According to the grouping rules in (8.1.5), there are only two samples left in the final group, where trapezoidal algorithm is used, therefore the error is relatively large. Although the trapezoidal method is used in the final group, the total error is still smaller than that of a typical trapezoidal algorithm, since most of the groups are computed accurately. If one more sample is used, the solution becomes $S_{1b} = 0.066666672566053$.

```
>> x=linspace(0,3*pi/2,500); y=cos(15*x);
   h=x(2)-x(1), S1a=num_integral(y,h)
   x=linspace(0,3*pi/2,501); y=cos(15*x);
   h=x(2)-x(1), S1b=num_integral(y,h)
```

Now let us answer the second problem. It can be seen from trials that if 1 603 samples are used, the definite integral obtained is $S_2 = 0.066666666666667$, and the step-size is about $h \approx 0.0029$, and time requirement is of only 0.0010 seconds. It can be seen that the efficiency of the new algorithm is much higher than that of the trapezoidal method. In real applications, if equally spaced samples are known, num_integral() function is recommended.

```
>> x=linspace(0,3*pi/2,1603); y=cos(15*x);
   tic, h=x(2)-x(1), S2=num_integral(y,h), toc
```

8.2 Numerical integrals of univariate functions

If the integrand $f(x)$ is theoretically not integrable, even if powerful computer mathematics languages are used, the analytical solutions of the integrals cannot be evaluated. The function vpa() can be used then to find more accurate solutions. There are indeed special integration problems, such as that in Example 5.23, where the integrand is not integrable, so not even when finding high precision solutions, numerical method might be the only choice.

There are many algorithms to evaluate numerical integrals, for instance, trapezoidal, Newton–Cotes, and Romberg methods are the commonly introduced algorithms in numerical analysis courses.

The ideas of numerical integrals are introduced first in this section, then some typical algorithms are presented with MATLAB implementations. Improper integrals, integrals with parameters, and integral functions are also presented.

8.2.1 Simple numerical integral problems

Integrals of univariate functions can be evaluated numerically with commonly used algorithms. Here, the ideas of Simpson formula are presented.

Theorem 8.1. *Simpson formula can be used to find the approximate integral s_i in the interval $[x_i, x_{i+1}]$ which can be evaluated with*

$$s_i \approx \frac{h_i}{12}\left[f(x_i) + 4f\left(x_i + \frac{h_i}{4}\right) + 2f\left(x_i + \frac{h_i}{2}\right) + 4f\left(x_i + \frac{3h_i}{4}\right) + f(x_i + h_i)\right], \qquad (8.2.1)$$

where $h_i = x_{i+1} - x_i$.

In fact, the algorithm can be evaluated with a loop structure in MATLAB

```
function S=simpson_int(f,x)
S=0; n=length(x)-1; h=diff(x);
for i=1:n,
    h0=h(i); x0=x(i)+[0:h0/4:h0]; S=S+h0*f(x0)*[1; 4; 2; 4; 1]/12;
end
```

The syntax of the function is S=simpson_int(f,x), where f is the function handle of the integrand. It can be an anonymous function or M function. The vector x defines the subintervals of the integration interval, with $a = x_1$, $b = x_{n+1}$, and n is the number of subintervals. The returned S is a numerical approximation of the integral. The function can be used directly to compute the numerical integral.

Example 8.5. Compute the integral again in Example 8.2 with Simpson formula. Larger intervals can be tried.

Solutions. The integrand can be described with an anonymous function. The number of samples should be selected, and the definite integral can be evaluated directly. The results are listed in Table 8.3. It can be seen that if the number of points is sufficient,

Table 8.3: Simpson formula results.

number of points	100	1 000	10 000	100 000
result	0.066667044	0.0666666667	0.066666666666670	0.066666666666667
time needed (s)	0.0015	0.002797	0.0259	0.2335

the accuracy is relatively high, and the speed is high. At least, it is much better than that of a typical trapezoidal algorithm.

```
>> f=@(x)cos(15*x); n0=[100,1000,10000,100000];
   for i=1:length(n0)
      tic, x=linspace(0,3*pi/2,n0(i)); S=simpson_int(f,x), toc
   end
```

If the interval is enlarged to $t \in (0, 1\,000)$, the analytical solution becomes

$$I = \sin(15\,000)/15 \approx 0.05956191052641859089450785303996,$$

with 0.093 seconds required.

```
>> syms x; tic, I=int(cos(15*x),x,0,1000), vpa(I), toc % analytical
```

With Simpson formula, 1 000 000 samples can be selected, and within 2.21 seconds, the integral obtained is $S = 0.059561910526479$. Compared with theoretical result, the accuracy is 10^{-13}.

```
>> tic, x=linspace(0,1000,1000000); S=simpson_int(f,x), toc
```

If the above mentioned num_integral() function is used, and the number of samples selected is 240 001, the numerical solution of $S_1 = 0.059561910526419$ can be obtained within 0.083 seconds, which is significantly more efficient than the Simpson formula.

```
>> x=linspace(0,1000,240001); y=f(x);
   tic, h=x(2)-x(1), S1=num_integral(y,h), toc
```

In fact, even more efficient high precision algorithm should be considered in actually finding definite integrals. The functions provided in MATLAB are also tried to evaluate definite integrals.

8.2.2 MATLAB solutions of numerical integral problems

An adaptive step-size algorithm is introduced in MATLAB in the function integral(), whose syntax is I=integral(f,a,b,option pairs), where f is used to describe the integrand, which can be an anonymous function or Fun.m file (with @Fun or 'Fun' to describe it). A typical call of the function is y=Fun(x). Arguments a and b are also used to describe the upper and lower boundaries in the definite integrals. The "option pairs" can be used to control the integral precision, as shown in Table 8.4. The following examples are given to demonstrate the numerical integration processes.

Table 8.4: Options for numerical integrals.

options	explanation of the options
'RelTol'	relative error tolerance, which can be used to specify the precision
'AbsTol'	absolute error, to work with 'RelTol' option
'ArrayValued'	vector flag. If the integrand contains parameters other than the independent variable, the parameter can be set to 1, to use vector or mesh grid data
'waypoint'	key points setting, and it can be set to discontinuous or singular points. This option is useful in solving improper integrals

There are yet several other solvers in the earlier versions of MATLAB, such as quad(), quadl(), quadgk(), and quadv(), for finding numerical integrals, the syntaxes of them are almost the same as that in integral(). The efficiency of integral() is much higher and recommended in real applications. Besides, functions quadgk() and quadv() are no longer supported.

Example 8.6. Consider the analytically non-integrable $\mathrm{erf}(x) = \dfrac{2}{\sqrt{\pi}} \int\limits_0^x e^{-t^2} dt$. Evaluate this integral with numerical methods.

Solutions. Before finding the numerical solutions, the integrand should be described first, by either of the three ways:

(1) M function. Establish a MATLAB function to save it in a file, with

```
function y=c8ffun(x)
y=2/sqrt(pi)*exp(-x.^2); % integrand M file description
```

Therefore, file c8ffun.m can be created. Since the independent variable x may contain values of many points, dot operations should be used in evaluating the function values in vector y.

(2) Anonymous function. An anonymous function can also be created with the syntax of f=@(x)2/sqrt(pi)*exp(-x.^2). The command describes the function dynamically, and there is no need to establish a stand-alone file. This method is more suitable to solve simple problems directly. In the function, the argument enclosed by parentheses after @ includes the independent variable. The function expression follows.

(3) `inline()` function. Similar to the anonymous function, the function `inline()` can also be used, with the command

```
>> f=inline('2/sqrt(pi)*exp(-x.^2)','x'); % inline() function description
```

Similarly, there is no need to create a MATLAB stand-alone file. The function `inline()` format is an old-style description, and is not recommended.

When the integrand is defined, the function `integral()` can be used to evaluate numerical integrals under double precision framework. The integral obtained is 0.966105146475311, which is the most accurate one under double precision systems.

```
>> f=@(x)2/sqrt(pi)*exp(-x.^2);
   tic, y=integral(f,0,1.5), toc % double precision solution
```

If Simpson integral formula discussed earlier is used for the same problem, 100 000 points can be generated, and the result is 0.966105146475322 within 0.26 seconds. It can be seen that `integral()` function is much more efficient than the traditional Simpson formula.

```
>> x=linspace(0,1.5,100000); simpson_int(f,x)
```

In fact, for the integrals of univariate functions, more accurate result under Symbolic Math Toolbox can be obtained as

$$I_0 = 0.96610514647531071393693372994991.$$

It can be seen that the numerical solutions are the most accurate possible under double precision framework.

```
>> syms x, I0=vpa(int(2/sqrt(pi)*exp(-x^2),0,1.5)) % high precision
```

Although the three methods can be used to describe the integrand, they all have different characteristics. The M function method can deal with problems with intermediate variables, while the other two cannot. Also in the problems presented later, if more returned variables are involved, anonymous and `inline()` functions can be used. From the viewpoint of computation speed, anonymous functions are faster than M functions. In this book, anonymous functions are used, while if anonymous functions fail, M functions can be used instead.

Example 8.7. Solve again the problem in Example 8.2 with `integral()` function for

$$I = \int_0^{3\pi/2} \cos 15x \, dx.$$

Solutions. With the fixed-step algorithm in Example 8.2, even if the step-size is selected extremely small, 11 correct digits in the results can be obtained, while it is quite time consuming. In fact, with a variable-step algorithm, an accurate result of $S = 0.06666666666667$ can easily be found within 0.0035 seconds. It can be seen that the time consumption is significantly reduced.

```
>> f=@(x)cos(15*x);
   tic, S=integral(f,0,3*pi/2,'RelTol',1e-20), toc
```

If larger integration interval is considered, the result $S = 0.059561910526150$ can be obtained, with the error of 10^{-14} and time requirement of 0.032 seconds. The efficiency is much higher than using Simpson formula.

```
>> tic, S=integral(f,0,1000,'RelTol',1e-20), toc
```

With functions quad() and quadl() in the old versions of MATLAB, one may receive erroneous results such as

$$S_1 = 175.6383838547378 \quad \text{and} \quad S_2 = -205.8148036931392.$$

Singularity phenomenon is witnessed.

```
>> S1=quad(f,0,1000,1e-20), S2=quadl(f,0,1000,1e-20)
```

It can be concluded that, in real applications, fixed-step algorithms commonly addressed in numerical analysis courses are not practical, and the accuracy cannot be maintained. If the step-size is small, the computation load is rather large, and the accuracy cannot be ensured. If variable step-size algorithms are used, this kind of problem can be solved easily. With function integral(), such troubles can be avoided easily, and efficient numerical solutions can be reached.

Example 8.8. Solve the integral problem for the piecewise function[9]

$$I = \int_0^4 f(x)dx, \quad \text{where } f(x) = \begin{cases} e^{x^2}, & 0 \leqslant x \leqslant 2, \\ 80/[4 - \sin(16\pi x)], & 2 < x \leqslant 4. \end{cases}$$

Solutions. The filled curve of the piecewise function can easily be drawn as shown in Figure 8.3. Special treatment is made by introducing a small quantity ϵ. It can be seen that there is a jump at $x = 2$ in the integrand.

```
>> x=[0:0.01:2, 2+eps:0.01:4,4];
   y=exp(x.^2).*(x<=2)+80./(4-sin(16*pi*x)).*(x>2);
   y(end)=0; x=[eps, x]; y=[0,y]; fill(x,y,'g') % draw the filled region
```

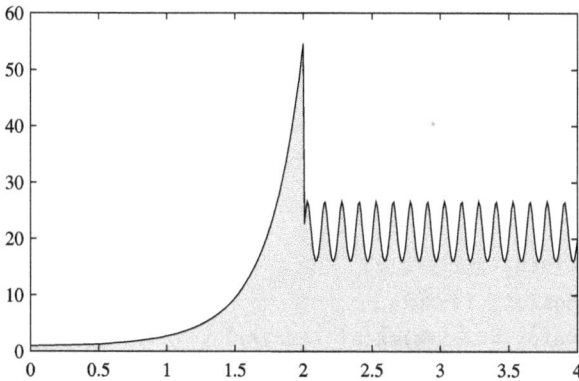

Figure 8.3: Filled region of the integrand.

The integrand can be expressed by an anonymous function, and numerical integral can be obtained with `integral()` function, $I_1 = 57.764450125048505$.

```
>> f=@(x)exp(x.^2).*(x<=2)+80./(4-sin(16*pi*x)).*(x>2);
   I=integral(f,0,4)
```

In fact, the integral can also be decomposed into $\int_0^2 + \int_2^4$, and with `int()` function, high-precision is obtained using

```
>> syms x; I0=vpa(int(exp(x^2),0,2)+int(80/(4-sin(16*pi*x)),2,4))
```

with the result being $I = 57.764450125053010333315235385 18$.

The analytical solution to this example exists, and the accuracy of the numerical results can be compared. In real applications where the analytical solutions are not known, how to validate the numerical results? A more strict error tolerance can be used to see whether consistent results can be obtained. One can set a different `Rel-Tol` value. For instance, when the error tolerance is set to 10^{-20}, the result obtained is $I_2 = 57.764450125053010$. It can be seen that the result is the most accurate one under double precision data type.

```
>> I2=integral(f,0,4,'RelTol',1e-20) % double precision solution
```

With a piecewise function specification for symbolic variables, the following statements can be used, giving exactly the same result as I_0.

```
>> f=piecewise(x<=2,exp(x^2), x>2,80/(4-sin(16*pi*x)));
   syms x; I=vpa(int(f,x,0,4)) % high precision solution
```

Example 8.9. Compute the complex integral $\displaystyle\int_{2}^{6-j5} e^{-x^2-jx} \sin(7+j2)x \, dx$.

Solutions. Complex integral problems can be solved directly with the following state-ments, yielding $I = -0.9245 + j25.792$. The results can be validated with theoretical solutions.

```
>> f=@(x)exp(-x.^2-1i*x).*sin((7+2i)*x);
   I=integral(f,2,6-5i,'RelTol',1e-20)
   syms x; i=sqrt(-1); F=exp(-x^2-i*x)*sin((7+2i)*x);
   I0=vpa(int(F,2,6-5i))
```

8.2.3 Numerical computation of improper integrals

The function `integral()` can be used to evaluate improper integrals directly, and the syntax of the function is exactly the same as for those introduced earlier. If an infinite integral is evaluated, the quantities `-inf` or `inf` can be used directly. If there exist sin-gularities in the integrand, within the integration interval, the same function can still be used, regardless of the singularity issues. Or a tiny offset can be introduced around the singularities and the integral can be evaluated again. To ensure sufficient accu-racy, the option `RelTol` can be set to a very small quantity. If a small value of 10^{-20} is used, the requirement cannot be reached under double precision framework, and the most accurate solution can be expected. Examples are introduced later to demonstrate the solutions of improper problems.

Example 8.10. Evaluate numerically the infinite integral $\displaystyle\int_{0}^{\infty} e^{-x^2} dx$.

Solutions. For such an infinite integral problem, regular statements can be used, in-volving `integral()` function, with the upper bound set to `inf`. The result obtained is $I = 0.886226925452758$, which is very close to the theoretical result $I_1 = \sqrt{\pi}/2 \approx 0.88622692545275801365$. The error is around 10^{-16}.

```
>> f=@(x)exp(-x.^2); I=integral(f,0,inf,'RelTol',1e-20) % numerical
   syms x; I1=int(exp(-x^2),0,inf), vpa(I1) % analytical solution
```

Example 8.11. Solve the volume problem studied in Example 5.23.

Solutions. It has been shown in Example 5.23 that, with `int()` and `vpa()` functions, the analytical solution cannot be obtained. Numerical solution should be tried. Ex-pressing first the radius of the rotation with an anonymous function f, the integrand can be declared with another anonymous function F. Finally, the numerical integral can be obtained with `integral()`, with the result of 57.5928.

```
>> f=@(x)1+x.*sin(4./x); F=@(x)pi*f(x).^2;
   integral(F,0,pi,'RelTol',1e-10)
```

A more strict error tolerance can be tried, however, it is found that singularity issues may appear. Therefore, for this particular example, the maximum relative error can only be assigned to 10^{-10}.

Example 8.12. Compute the numerical improper integral studied in Example 5.12, namely

$$\int_1^{2e} \frac{1}{x\sqrt{1-\ln^2 x}}\,dx.$$

Solutions. An anonymous function can be used to describe the integrand, and there is no need to worry about the singularity at $x = e$. The improper integral found is $I = 1.5708687 - 1.11822j$, with the accuracy of around 10^{-5}.

```
>> f=@(x)1./(x.*sqrt(1-log(x).^2));
   I=integral(f,1,2*exp(1),'RelTol',1e-15)
```

High-precision solution to the original problem can be computed with

```
>> syms x; f(x)=1/x/sqrt(1-log(x)^2);
   I=int(f,x,1,2*exp(sym(1))), vpa(I) % compute improper integral
```

yielding the result of

$$I = 1.5707963267948966192313216916398 - 1.1182308528192447293713675895252j.$$

Example 8.13. Compute the improper integral in Example 5.12.

Solutions. The numerical approach is tried to solve the integral problem, where there is a singularity in the integrand $f(x)$ at $x = e$. The interval can be divided into two subintervals $(1, e-\epsilon)$ and $(e+\epsilon, 2e)$, where ϵ can be selected very small, e. g., $\epsilon = 10^{-15}$. The integrals in the two subintervals can be evaluated independently, and the final result is $I = 1.570796326794387 - 1.118230852817493j$. It can be seen that the error is as low as 10^{-12}, which is smaller than the direct one obtained earlier.

```
>> f=@(x)1./(x.*sqrt(1-log(x).^2)); ex=exp(1);
   I1=integral(f,1,ex-1e-15,'RelTol',1e-15);
   I2=integral(f,ex+1e-15,2*ex,'RelTol',1e-15); I=I1+I2
```

8.2.4 Numerical integrals for integrands with parameters

Consider the definite integral for the independent variable x. If there exist other variables in the integrand besides x, for instance, another parameter α, or other parameters, the integral problems are referred to as integrals with parameters. If the integrand is not integrable analytically with respect to x, numerical solutions can only be expected. Some samples of the parameters can be selected, and, for each sample, the numerical integral can be evaluated. Therefore, graphical representation of the results as curves or surfaces can be obtained.

In normal situations, the fundamental treatment is to use loops to evaluate the integrals. For integrals of univariate functions, the option `ArrayValue` in the function `integral()` can alternatively be employed to bypass loops. This will be demonstrated through examples.

Example 8.14. For the integral $I(\alpha) = \displaystyle\int_0^\infty e^{-\alpha x^2} \sin(\alpha^2 x)\mathrm{d}x$ with parameter α, compute the relationship of $I(\alpha)$ and α, where the interval of the parameter is $\alpha \in (0,4)$.

Solutions. The definite integrals discussed earlier were those for univariate integrands. Here a series of samples for α can be selected, and the following MATLAB statements can be used to compute the integrals with vectorized parameters. The curve of the integrals can be obtained as shown in Figure 8.4. In the earlier versions, a loop structure should be used for the same problem.

```
>> a=0:0.1:4; f=@(x)exp(-a*x.^2).*sin(a.^2*x); %integral of vectors
   I=integral(f,0,inf,'RelTol',1e-20,'ArrayValued',true);
   plot(a,I)
```

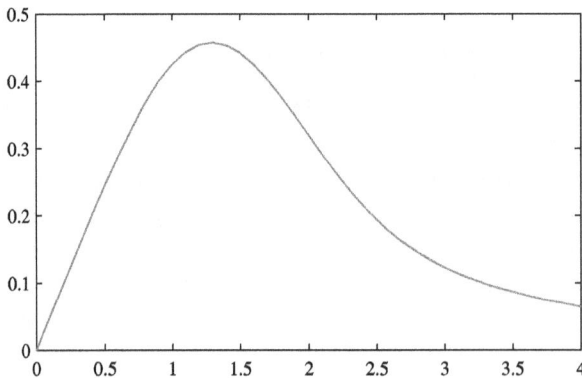

Figure 8.4: Integral $I(\alpha)$ as a function of α.

In practical situations, where there exists more than one parameter, similar procedures should be adopted. For instance, if there are two parameters, mesh grid data for the samples of the two parameters can be generated first, then for each pair of parameters, the integrals can be evaluated numerically. The surface representation of the integrals can be obtained. Normally, double loop structures should be used, however, with function integral(), one command is sufficient to solve the problem with vectorized syntax.

Example 8.15. For the integral $I(\alpha, \beta) = \int_0^\infty e^{-\alpha x^2} \sin(\beta x)dx$, the relationship of the integral $I(\alpha, \beta)$ and parameters α and β can be obtained, where the intervals of the parameters are $\alpha \in (1, 4)$ and $\beta \in (-2, 2)$.

Solutions. Similar to the case in Example 8.14, mesh grid data for the parameters α and β should be generated first. Then a vectorized integral can be evaluated for each pair of values in the mesh grid of α and β, to compute infinite integrals. The size of matrix **I** is exactly the same as those of the parameters **α** and **β**. The surface of the integral can be obtained with function surf(), as shown in Figure 8.5.

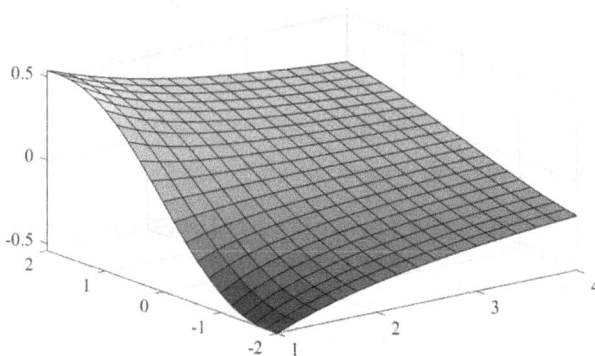

Figure 8.5: Integral $I(\alpha, \beta)$ as a function of parameters α and β.

```
>> [a,b]=meshgrid(1:0.2:4,-2:0.2:2);   % mesh grid data
   f=@(x)exp(-a*x.^2).*sin(b*x);       % integral with parameters
   I=integral(f,0,inf,'RelTol',1e-20,'ArrayValued',true);
   surf(a,b,I)
```

8.2.5 Numerical solutions of integral functions

The topics considered so far involved the evaluation of definite integrals in the given interval (a, b). There is one question remaining. How can we evaluate an integral function

$$F(x) = \int_a^x f(\tau)\, d\tau? \tag{8.2.2}$$

Similar to the case of numerical integral evaluation, the interval (a, b) can be divided into n equal subintervals. Let $x_1 = a$, $x_2 = a + h$, ..., $x_{n+1} = b$, where $h = (b-a)/n$, the integral function at point a is 0, i.e., the definite integral in the interval (a, a) is zero. Denoting $F_1 = 0$, the integral function can be evaluated recursively from

$$F_{k+1} = F_k + \int_{x_k}^{x_{k+1}} f(\tau) d\tau, \quad k = 1, 2, \ldots, n - 1. \tag{8.2.3}$$

The following MATLAB function can be written:

```
function [x,f1]=intfunc(f,a,b,n)
if nargin<=3, n=100; end;
x=linspace(a,b,n); f1=0; F=0; % default parameters
for i=1:n-1,
    F=F+integral(f,x(i),x(i+1),'RelTol',1e-20); f1=[f1, F];
end
```

with the syntax $[x, f_1] = \texttt{intfunc}(f, a, b, n)$, and the default value of n being 100.

Example 8.16. Draw the integral curve for the piecewise function in Example 8.8.

Solutions. Since function e^{x^2} in the piecewise function is not analytically integrable, analytical solutions cannot be obtained to draw the surface of the integral. Numerical method should be used instead. An anonymous function can be used to describe the integrand, then function $\texttt{intfunc}()$ can be used to evaluate the integral function, and the curve obtained is shown in Figure 8.6. It can be seen that the definite integral in Example 8.8 is only the right terminal of the curve.

```
>> f=@(x)exp(x.^2).*(x<=2)+80./(4-sin(16*pi*x)).*(x>2); % integrand
   [x1,f1]=intfunc(f,0,4,100); plot(x1,f1,x1(end),f1(end),'o'),
   f1(end)
```

8.3 Numerical computation of double integrals

Consider the double integral problems studied so far. If the inner integral cannot be obtained, there might be difficulties in finding analytical solutions of the outer integral. Therefore, numerical integration might be a good choice in evaluating double integrals. Numerical methods will be tried in this section to evaluate double integrals.

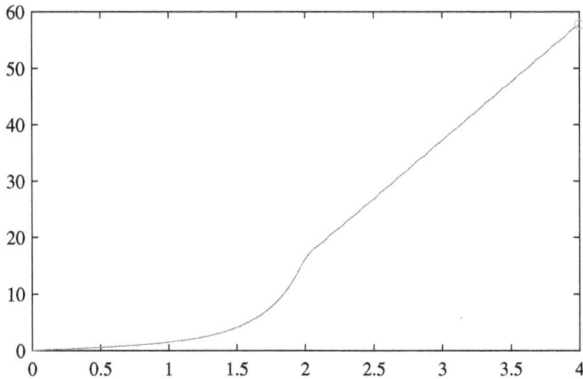

Figure 8.6: Integral curve of the function.

8.3.1 Computing double integrals

Definition 8.5. The standard form of double integrals is

$$I = \int_{x_m}^{x_M} \int_{y_m(x)}^{y_M(x)} f(x,y)\,dy\,dx. \tag{8.3.1}$$

A MATLAB function `integral2()` can be used to solve directly the evaluate an integral given in the standard form. The syntax of the function is

I=integral2$(f, x_m, x_M, y_m, y_M,$ option pairs$)$

where "option pairs" specifications are similar to those presented in function `integral()`, however, the `'ArrayValued'` is not supported. Besides, y_m and y_M can be function handles for the inner boundaries.

It should be noted that function `integral()` can only be used to evaluate double integrals in the order of y and then x.

Example 8.17. Evaluate numerically the double integral

$$J = \int_{-1}^{1} \int_{-2}^{2} e^{-x^2/2} \sin(x^2 + y)\,dx\,dy.$$

Solutions. For integrals in rectangular regions, the order of integration can be interchanged, without affecting the final results. If one can pair the independent variables and boundaries correctly, the results can be obtained directly. An anonymous function can be used to describe the integrand. Also it is noted that the boundaries for x and y are respectively $[-2, 2]$ and $[-1, 1]$, so the following MATLAB statements can be used to evaluate the double integral, and the result obtained is 1.574498159218786.

```
>> f=@(x,y)exp(-x.^2/2).*sin(x.^2+y);
   J=integral2(f,-2,2,-1,1,'RelTol',1e-20)
```

If the analytical method is used, the following statements can be used:

```
>> syms x y; clear f; f(x,y)=exp(-x^2/2)*sin(x^2+y);
   I=int(int(f,y,-1,1),x,-2,2), vpa(I)
```

and the result is $I = 1.57449815921736052274208452227$. It can be seen that the error in the numerical solution is around 10^{-12}, which is acceptable in most of the applications.

It should be noted that the command `clear f` was used here, before the new integrand can be specified. This is because the variable f was defined earlier as an anonymous function with one independent variable. If the command "$f(x, y) =$" is used to update f, errors may occur. Therefore the easiest way is to clear f first, before it can be redefined.

Example 8.18. Compute the double integral

$$J = \int_{-1/2}^{1} \int_{-\sqrt{1-x^2/2}}^{\sqrt{1-x^2/2}} e^{-x^2/2} \sin(x^2 + y)\mathrm{d}y\mathrm{d}x.$$

Solutions. The order of integration here is y first and then x, which is exactly the same as that in the standard form. Therefore, the following statements can be used, and the result of $I = 0.411929546176295$ is found.

```
>> fh=@(x)sqrt(1-x.^2/2); fl=@(x)-sqrt(1-x.^2/2); % inner bounds
   f=@(x,y)exp(-x.^2/2).*sin(x.^2+y);                    % integrand
   I=integral2(f,-1/2,1,fl,fh)                           % direct computation
```

The analytical solution cannot be obtained, however, high-precision solution can be obtained instead, $I_1 = 0.41192954617629511965175994$.

```
>> syms x y   % analytical solutions
   i1=int(exp(-x^2/2)*sin(x^2+y),y,-sqrt(1-x^2/2),sqrt(1-x^2/2));
   int(i1,x,-1/2,1), I1=vpa(ans) % warning is given
```

8.3.2 Computation of double integral functions

According to the ideas illustrated in Figure 8.6, the numerical solution function can be written in MATLAB, for equally spaced integral regions. An integral surface can then be shown. The syntax of the function is

$[x,y,F]$=intfunc2(f,x_m,x_M,y_m,y_M,n,m)

where f is an anonymous function or an M function, (x_m, x_M) and (y_m, y_M) are the bounds of the rectangular region. Arguments n and m are respectively the number of subintervals in axes x and y, with the default values of 50. The returned F(end,end) is the value of the definite integral on the mesh grid points.

```
function [yv,xv,F]=intfunc2(f,xm,xM,varargin)
[ym,yM,n,m]=default_vals({xm,xM,50,50},varargin{:}); % default pars
xv=linspace(xm,xM,n); yv=linspace(ym,yM,m); d=yv(2)-yv(1);
[x y]=meshgrid(xv,yv); F=zeros(n,m); % mesh grid setting
for i=2:n, for j=2:m,               % loop for each mesh grid point
    F(i,j)=integral2(f,xv(1),xv(i),yv(1),yv(j),'RelTol',1e-20);
end, end
```

Example 8.19. Draw the integral function surface for the problem in Example 8.17.

Solutions. An anonymous function can be used again to describe the integrand. The integral surface can then be obtained with the following statements, as shown in Figure 8.7. The upper-right corner value should be the same as that in Example 8.17, $I = 1.574498159218787$. The function is quite time-consuming, requiring 5.44 seconds, and more efficient algorithms are expected for such problems.

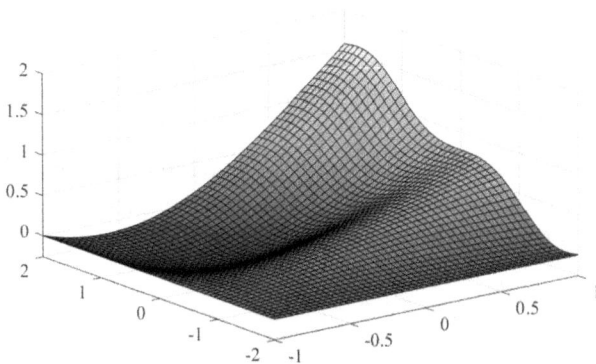

Figure 8.7: Integral surface for 2D functions.

```
>> f=@(x,y)exp(-x.^2/2).*sin(x.^2+y); % integrand
   tic, [x,y,z]=intfunc2(f,-2,2,-1,1); toc,
   surf(x,y,z), I=z(end,end)
```

8.3.3 Evaluations of double integrals in different order

It is a pity that there is no MATLAB function provided directly for solving double integral problems, in the order of x first and then y as in

$$I = \int_{y_m}^{y_M} \int_{x_m(y)}^{x_M(y)} f(x,y)\,dxdy. \tag{8.3.2}$$

Variable substitution method can be introduced so as to convert it directly into the standard form in (8.3.1). Then function `integral2()` can be used to solve the original problem.

Letting $\hat{x} = y$ and $\hat{y} = x$, (8.3.2) can be converted into an equivalent form

$$I = \int_{\hat{x}_m}^{\hat{x}_M} \int_{\hat{y}_m(\hat{x})}^{\hat{y}_M(\hat{x})} f(\hat{y},\hat{x})\,d\hat{y}d\hat{x}. \tag{8.3.3}$$

It can be seen that the best way is to swap the interface in the integrand $f(x,y)$ to $f(y,x)$, without touching other parts in the whole problem. The integrand should be written as $f=@(y,x)$. An example will be given below to illustrate the solutions of such problems.

Example 8.20. Compute numerically the double integral

$$J = \int_{-1}^{1} \int_{-\sqrt{1-y^2}}^{\sqrt{1-y^2}} e^{-x^2/2}\sinh(x^2 + y)\,dxdy.$$

Solutions. Theoretically speaking, there is no analytical solution in this example; with variable precision solution schemes, the high-precision numerical solution is $I = 0.704121334903356899478003120225l7$, and the time requirement is as high as 123.57 seconds.

```
>> syms x y, tic,
   i1=int(exp(-x^2/2)*sinh(x^2+y),x,-sqrt(1-y^2),sqrt(1-y^2));
   I=int(i1,y,-1,1), vpa(I), toc % warding appears
```

For the problem in this example, the order of the integration is x first and then y, therefore, the integrand can be expressed by just swapping the interface, without changing other parts in the syntaxes. The numerical solution can be found to be 0.704121334903362 with the following statements, and it took only 0.0195 seconds, which is extremely efficient.

```
>> tic, f=@(y,x)exp(-x.^2/2).*sinh(x.^2+y); % swap the orders
   fh=@(y)sqrt(1-y.^2); fl=@(y)-sqrt(1-y.^2);
   I=integral2(f,-1,1,fl,fh,'RelTol',1e-20), toc
```

8.4 Numerical computation of multiple integrals

Triple integrals in arbitrary integration regions are explored first in this section, followed by the computation triple integrals with parameters. Then multiple integrals in hyper-rectangular regions are explored.

8.4.1 Numerical triple integrals

Definition 8.6. The standard form of the triple integral is defined as

$$I = \int_{x_m}^{x_M} \int_{y_m(x)}^{y_M(x)} \int_{z_m(x,y)}^{z_M(x,y)} f(x,y,z) dz dy dx. \tag{8.4.1}$$

Please pay attention to the order of the integrals in the standard form. MATLAB function integral3() can be used to evaluate numerically triple integrals given in standard form, with the syntax

I=integral3$(f, x_m, x_M, y_m, y_M, z_m, z_M,$ optoin pairs$)$

where f describes the integrand with either an M function or anonymous function format; "option pairs" is quite similar to that in integral2(). The arguments y_m, y_M, z_m, and z_M can be function handles or constants. If the orders of the integrals are different from those in the standard form, similar manipulation should be made as shown for integral2() before final results can be obtained.

Example 8.21. Compute numerically the triple integral in Example 5.18, that is,

$$\int_0^2 \int_0^\pi \int_0^\pi 4xze^{-x^2y-z^2} dz dy dx.$$

Solutions. The integrand can be described by an anonymous function. Then the following statements can be used to evaluate the triple integral, with the result of 3.108079402085465, and 0.42 seconds needed. Compared with the analytical solutions in Example 5.18, it can be seen that the error is about 10^{-14}.

```
>> f=@(x,y,z)4*x.*z.*exp(-x.*x.*y-z.*z); % integrand
   tic, I=integral3(f,0,2,0,pi,0,pi,'RelTol',1e-20), toc % integral
```

Example 8.22. Compute numerically the triple integral with functional boundaries

$$I = \int_0^1 \int_0^{\sqrt{1-x^2}} \int_{\sqrt{x^2+y^2}}^{\sqrt{2-x^2-y^2}} z^2 e^{-(x+y^2)} \, dz \, dy \, dx.$$

Solutions. The result of $I = 0.237902335517189$ can be obtained with the following statements, needing 0.16 seconds:

```
>> tic, f=@(x,y,z)z.^2.*exp(-(x+y.^2));              % integrand
   yM=@(x)sqrt(1-x.^2); zm=@(x,y)sqrt(x.^2+y.^2);    % boundaries
   zM=@(x,y)sqrt(2-x.^2-y.^2);
   I=integral3(f,0,1,0,yM,zm,zM,'RelTol',1e-20); toc % triple integral
```

If an analytical solution is expected, the following statements can be tried. However, after 43.9 seconds of waiting, it may be prompted that there is no analytical or high-precision solution. Numerical solution becomes the only choice.

```
>> syms x y z, zm=sqrt(x^2+y^2); zM=sqrt(2-x^2-y^2); tic % analytical
   I=int(int(int(z^2*exp(-(x+y^2)),z,zm,zM),y,0,sqrt(1-x^2)),x,0,1)
   vpa(I), toc
```

Example 8.23. Compute numerically the triple integral problem with functional bounds

$$I = \int_0^1 \int_0^{\sqrt{1-z^2}} \int_{\sqrt{y^2+z^2}}^{\sqrt{2-y^2-z^2}} z^2 e^{-(x+y^2)} \, dx \, dy \, dz.$$

Solutions. It can be seen that the integrand here is the same as that in Example 8.22, however, the order of integration is different. The following statements can be used directly to find the analytical solution of the integral:

```
>> syms x y z, xm=sqrt(y^2+z^2); xM=sqrt(2-y^2-z^2); tic % exact one
   I=int(int(int(z^2*exp(-(x+y^2)),x,xm,xM),y,0,sqrt(1-z^2)),z,0,1)
   vpa(I), toc
```

It is a pity that after 59.1 seconds of waiting, an error prompt is given, indicating there is no analytical solution, while vpa() may also yield numerical solutions. With a less demanding command vpa(I, 16), the approximate solution 0.02420459178538355 can be obtained within 34.59 seconds. Also the function int() can be converted into the following results, where the triple integral problem can be simplified to a double integral problem:

$$I = \int\limits_0^1 \int\limits_0^{\sqrt{1-z^2}} z^2 e^{-y^2} \left(e^{-\sqrt{y^2+z^2}} - e^{-\sqrt{-y^2-z^2+2}}\right) dy dz.$$

Considering the double integral problem, variable z can be regarded as x, the order of the integration is exactly the same as that defined in the standard form in Definition 8.5. Therefore, the following statements can be used to evaluate the simplified numerical solution $I = 0.024204591785398$, which took 0.038 seconds. The first few digits are the same as the those obtained using vpa() function.

```
>> tic, yM=@(z)sqrt(1-z.^2);
   f=@(z,y)z.^2.*exp(-y.^2).*(exp(-sqrt(y.^2+z.^2))-...
            exp(-sqrt(-y.^2-z.^2+2)));
   I=integral2(f,0,1,0,yM,'RelTol',1e-20), toc
```

The numerical solution can also be evaluated with integral3() function. Compared with the standard form in (8.4.1), it can be seen that the arguments of the integrand can be converted to (z, y, x). With the command in Example 8.22, the problem can be solved. However, it is better to unify the variable names in the inner integral bounds. The result obtained is $I = 0.024204591786321$, and it took 0.1 seconds.

```
>> tic, f=@(z,y,x)z.^2.*exp(-(x+y.^2));            % integral
   yM=@(z)sqrt(1-z.^2); xm=@(y,z)sqrt(y.^2+z.^2); % boundaries
   xM=@(y,z)sqrt(2-y.^2-z.^2);
   I=integral3(f,0,1,0,yM,xm,xM,'RelTol',1e-20); toc % triple integral
```

8.4.2 Triple integrals of integrands with parameters

If a certain integrand comes with other parameters besides the necessary independent variables, the parameters must be sampled, and for the samples, the values of the integrals can be evaluated. Finally, the curve and surface of the integrals with respect to the changes in the parameters can be obtained. For the functions integral2() and integral3(), the vectorized calling option ArrayValued cannot supported, loops must be used to evaluate integrals for the parameter samples. Finally, the relationship between the integral values and the parameters can be established.

Example 8.24. Compute the triple integrals with integrand parameters

$$I(\alpha, \beta) = \int\limits_0^1 \int\limits_0^{\sqrt{1-x^2}} \int\limits_{\sqrt{x^2+y^2}}^{\sqrt{2-x^2-y^2}} \sinh z^2 e^{-(\alpha \sin 40x + \beta \cos^2 y^2)} dz dy dx.$$

Solutions. For the convenience of graphical representation, the parameters a and b should be selected as mesh grids. Then, double loop structure can be used to find the integral matrix from the samples. Function surf() can be used to draw the surface. The surface can be obtained as shown in Figure 8.8. It should be noted that this structure is rather time consuming, and the time used by the code is 335.7 seconds.

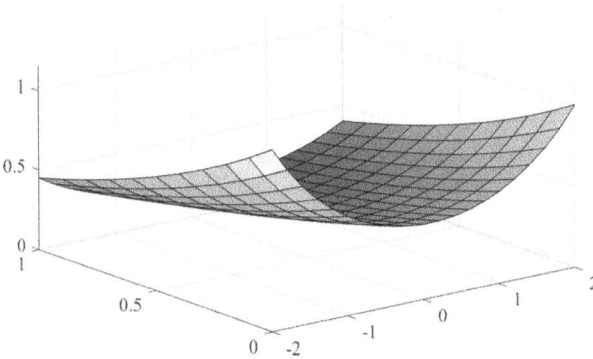

Figure 8.8: The surface of the triple integral with respect to parameters.

```
>> [a0,b0]=meshgrid(-2:0.2:2,0:0.1:1); tic
   zm=@(x,y)sqrt(x.^2+y.^2); zM=@(x,y)sqrt(2-x.^2-y.^2);% the bounds
   yM=@(x)sqrt(1-x.^2); [n,m]=size(a0); I=zeros(n,m);
   for i=1:n, for j=1:m, a=a0(i,j); b=b0(i,j);
       f=@(x,y,z)sinh(z.^2).*exp(-(a*sin(40*x)+b*cos(y.^2).^2));
       I(i,j)=integral3(f,0,1,0,yM,zm,zM,'RelTol',1e-20); i,j
   end, end
   toc, surf(a0,b0,I) % surface of the triple integral
```

8.4.3 Multiple integrals

Multiple integral with multiplicity higher than three cannot be evaluated numerically with the current MATLAB functions. The Numerical Integral Toolbox (NIT) by Wilson and Gardner[26] can be used to evaluate multiple integrals with hyper-rectangular bounds, for instance, with quadndg() function. However, there is no existing function for evaluating multiple integrals with functional bounds.

Definition 8.7. The multiple integrals with hyper-rectangular bounds can be expressed as

$$I = \int_{x_{1m}}^{x_{1M}} \int_{x_{2m}}^{x_{2M}} \cdots \int_{x_{pm}}^{x_{pM}} f(x_1, x_2, \ldots, x_p) dx_p \cdots dx_2 dx_1. \tag{8.4.2}$$

The problem can be solved with the following command:

$$I=\text{quadndg}(f, [x_{1m}, x_{2m}, \ldots, x_{pm}], [x_{1M}, x_{2M}, \ldots, x_{pM}], \epsilon)$$

where f can be used to describe the integrand with M or anonymous functions. The argument ϵ is the error tolerance, and can be omitted. The independent variables are specified in the vector x, such that more independent variables can be stored.

Example 8.25. Triple integral in Example 8.21 can be evaluated with multiple integral solvers, namely we compute

$$\int_0^2 \int_0^\pi \int_0^\pi 4xze^{-x^2y-z^2}\, dz\, dy\, dx.$$

Solutions. Letting $x_1 = x$, $x_2 = y$, $x_3 = z$, the original problem can be rewritten for

$$f(x) = 4x_1x_3e^{-x_1^2x_2-x_3^2}.$$

An anonymous function can be used to describe the integrand, and then the original integral can be evaluated from the following statements, where $I = 3.108079402085409$, which is the same as that in Example 8.21. The efficiency of quadndg() is higher than that of integral3(). For instance, the time requirement is only 1/10 of that in Example 8.21.

```
>> f=@(x)4*x(1)*x(3)*exp(-x(1)^2*x(2)-x(3)^2);  % integrand description
   tic, I=quadndg(f,[0 0 0],[2,pi,pi]), toc     % triple integral
```

Example 8.26. Find the numerical and analytical solutions of the following quintuple integral problem:

$$I = \int_0^5 \int_0^4 \int_0^1 \int_0^2 \int_0^3 \sqrt[3]{v}\sqrt{w}x^2y^3z\, dz\, dy\, dx\, dw\, dv.$$

Solutions. For this specific problem, the analytical solution can be found as $120\sqrt[3]{5}$, needing 0.133 seconds.

```
>> syms x y z w v; F=v^(1/3)*sqrt(w)*x^2*y^3*z; tic % analytical solution
   I=int(int(int(int(int(F,z,0,3),y,0,2),x,0,1),w,0,4),v,0,5), toc
```

In fact, there is no analytical solution for most of the quintuple integral problems. Numerical solutions should be evaluated instead. Letting $x_1 = v$, $x_2 = w$, $x_3 = x$, $x_4 = y$, and $x_5 = z$, the integrand can be rewritten as

$$f(x) = \sqrt[3]{x_1}\sqrt{x_2}x_3^2x_4^3x_5.$$

The integrand can be described by an anonymous function, and the following statements can be used to find the solution, $I = 205.2205 \approx 120\sqrt[3]{5}$. Since the algorithm uses non-vectorized format, the speed of the function is rather slow. The time used is 4.92 seconds.

```
>> f=@(x)(x(1))^(1/3)*sqrt(x(2))*x(3)^2*x(4)^3*x(5); %integrand
   tic, I=quadndg(f,[0 0 0 0 0],[5,4,1,2,3]), toc %numerical integral
```

Example 8.27. Compute numerically the following quintuple integral problem:

$$I = \int_0^5 \int_0^4 \int_0^1 \int_0^2 \int_0^3 (e^{-\sqrt[3]{v}} \sin\sqrt{w} + e^{-x^2 y^3 \cos z^2}) \, dz\,dy\,dx\,dw\,dv.$$

Solutions. The analytical solution for the problem does not exist, therefore, numerical algorithms must be employed.

Again letting $x_1 = v$, $x_2 = w$, $x_3 = x$, $x_4 = y$, and $x_5 = z$, the integrand can be rewritten as

$$f(x) = e^{-\sqrt[3]{x_1}} \sin\sqrt{x_2} + e^{-x_3^2 x_4^3 \cos x_5^2}.$$

The integrand can be expressed as an anonymous function. With the following statements, the result obtained is $I = 281.1757867591371$. Although the integrand is much more complicated than that in the previous example, the time needed is almost the same. It can be seen that the time consumption does not depend upon the complexity of the integrand.

```
>> f=@(x)exp(-(x(1))^(1/3))*sin(sqrt(x(2)))+...
         exp(-x(3)^2*x(4)^3*cos(x(5)^2));
   tic, I=quadndg(f,[0 0 0 0 0],[5,4,1,2,3]), toc %direct computation
```

8.4.4 Numerical solutions of some multiple integrals with variable bounds

The function quadndg() can only be used in finding multiple integrals with cuboid bounds, and those with functional bounds cannot be handled. Therefore, the universality of the methods is restricted. An alternative method can be explored for some specific multiple integrals.

Example 8.28. Evaluate the following quadruple integral with functional bounds:

$$J = \int_0^2 \int_{-2}^2 \int_{-\sqrt{1-z^2}}^{\sqrt{1-z^2}} \int_{-\sqrt{4-y^2-z^2}}^{\sqrt{4-y^2-z^2}} u^2 \sin u \, dx\,dy\,dz\,du.$$

Solutions. Since functional bounds are involved, the function `quadndg()` cannot be used for finding the results. Therefore, the following method can be explored. The function `int()` is used to simplify the original problem:

```
>> syms x y z u; clear f; f(x,y,z,u)=u^2*sin(u);
   I=int(int(int(int(f,x,-sqrt(4-y^2-z^2),sqrt(4-y^2-z^2)),...
       y,-sqrt(1-z^2),sqrt(1-z^2)),z,-2,2),u,0,2)
   vpa(I)   % however, the solution cannot be obtained directly
```

Although an analytical solution of the original problem cannot be obtained, the problem is simplified as the following double integral problem:

$$J = \int_0^2 \int_{-2}^2 -2u^2 \sin u \left(2\mathrm{asin}\left(\frac{\sqrt{1-z^2}}{\sqrt{4-z^2}} \right)\right)\left(\frac{z^2}{2} - 2 \right) - \sqrt{3}\sqrt{1-z^2}\,\bigg)dzdu.$$

Further, the original problem can be expressed as the product of two independent integrals of univariate integrands:

$$J = \int_0^2 -2u^2 \sin u \, du \int_{-2}^2 \left(2\mathrm{asin}\left(\frac{\sqrt{1-z^2}}{\sqrt{4-z^2}} \right)\right)\left(\frac{z^2}{2} - 2 \right) - \sqrt{3}\sqrt{1-z^2}\,\bigg)dz.$$

The numerical solution J = 23.280912550946475 + 31.326203493421794j can be found with the following statements:

```
>> g=@(u)-2*u.^2.*sin(u);
   f=@(z)2*asin(sqrt(1-z.^2)/sqrt(4-z.^2)).*(z.^2/2-2)-sqrt(3-3*z.^2);
   J=integral(g,0,2,'RelTol',1e-15)*integral(f,-2,2,'RelTol',1e-15)
```

It should be noted that there are limitations in the method. If the integral can be simplified to triple or lower-dimensional integrals with `int()` function, the numerical solutions can be found. Otherwise, numerical solutions cannot be found.

8.5 Other numerical methods for multiple integrals

The numerical algorithms are regular numerical integral methods. In some specific situations, some problems which cannot be evaluated with regular algorithms also appear. Specific methods must be employed. Two kinds of specific algorithm will be introduced here. The first is the Monte Carlo method – a statistical test approach, which can be used in the approximate evaluation of integrals with irregular regions. The other, interpolation-based algorithm is introduced afterwards, which can be used in evaluating integrals with sparsely distributed or irregular samples.

8.5.1 Numerical integral approximations with Monte Carlo method

Monte Carlo is a commonly used approach to obtain approximations of certain quantities through many statistical experimental data. For problems which are hard to model mathematically in modern scientific research, examples are given to show the use of Monte Carlo method in integral evaluations.

Definition 8.8. Consider the integral problem $\int_a^b f(x)\,dx$, where $f(x) \geqslant 0$, as shown in Figure 8.9. Selecting an upper bound of y as $y = M$, the shaded area is the definite integral expected. Throw N samples randomly in the rectangle $(a, 0) \sim (b, M)$, at points (x_i, y_i), which are random numbers in the intervals $[a, b]$ and $(0, M)$, respectively. Counting the number of samples satisfying the inequality $y_i \leqslant f(x_i)$ as N_1, the approximate integral can be obtained from[12]

$$\frac{N_1}{N} \approx \frac{1}{M(b-a)} \int_a^b f(x)\,dx. \tag{8.5.1}$$

Therefore, the approximate integral can be found with

$$\int_a^b f(x)\,dx \approx \frac{M(b-a)N_1}{N}. \tag{8.5.2}$$

Example 8.29. Compute the following integral with Monte Carlo method:

$$I = \int_1^3 \left[1 + e^{-0.2x} \sin(x + 0.5)\right] dx.$$

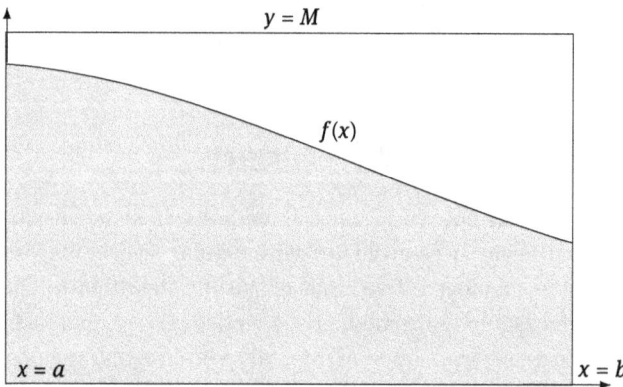

Figure 8.9: Illustration of Monte Carlo method.

Solutions. The illustration in Figure 8.9 is, in fact, the integrand in this example. The upper bound can be selected as $M = 2$. The approximate result can be found as $p = 2.74476$, with the following statements. The exact result can also be found as $I = 2.74393442001011808$.

```
>> f=@(x)1+exp(-0.2*x).*sin(x+0.5); a=1; b=3; M=2; % parameters
   N=100000; x=a+(b-a)*rand(N,1); y=M*rand(N,1);     % random numbers
   i=y<=f(x); N1=sum(i); p=M*N1*(b-a)/N              % approximation
   syms x; I=vpa(int(1+exp(-0.2*x)*sin(x+0.5),x,a,b)) % exact one
```

If the value of N is further increased, better solutions might be obtained, however, it is also possible to have worse approximations. It should be noted that an accurate solution with more digits can never be expected using Monte Carlo methods.

It should also be noted that the method can only be used to handle problems like $f(x) \geqslant 0$ or $f(x) \leqslant 0$ in the entire interval. Otherwise, improved version of Monte Carlo method should be introduced. For instance, let $f_1(x) = f(x) + C$ such that $f_1(x)$ does not change its sign in the interval of interest.

Another important issue when using Monte Carlo method is that the amount of data must be large, otherwise the statistical result may be useless. It should also be noted that the increase of the amount of data may increase the accuracy of computation, which for a specific experiment may not be true.

Example 8.30. Compute the volume in Example 5.24 again with Monte Carlo method, if $r = 2$.

Solutions. From the computation of the volume of the hemisphere in Example 5.24, the difficulty is that the integration region must be found explicitly. The original problem can be converted into a triple integral problem, so that the problem can be solved in that way. If Monte Carlo method is to be used, another kind of thinking must be carried out. Let us try to demonstrate the problem with this example.

A cuboid can be defined as $-2 \leqslant x \leqslant 2$, $-2 \leqslant y \leqslant 2$, and $0 \leqslant z \leqslant 2$. A set of N random points can be thrown into the cuboid, some of them will fall into the hemisphere, some will be outside. If the number of the random points falling into the hemisphere is N_1, then

$$\frac{N_1}{N} \approx \frac{V}{\text{volume of the cuboid}} = \frac{V}{4 \cdot 4 \cdot 2} = \frac{V}{32}.$$

Therefore the volume of the hemisphere can be evaluated as $V \approx 32N_1/N$.

The question and key point now become: How to check if a random point is inside or outside a hemisphere?

It is known that the mathematical expression of a sphere of radius 2 is $f(x, y, z) = x^2 + y^2 + z^2 - 4$. Substituting a random point (x_0, y_0, z_0) into the function, if the value

is less than zero, the point is inside the hemisphere, otherwise it is outside. Therefore, a set of 100 000 random points can be generated, from which the volume of the
hemisphere can be computed as $V = 16.72192$, which is close to the theoretical result
16.755160819145562. If N is increased, solutions with higher accuracy may be obtained,
but the result is random.

```
>> N=100000; V0=2*pi*2^3/3 % statistical method
   x=-2+4*rand(N,1); y=-2+4*rand(N,1); z=2*rand(N,1);
   F=x.^2+y.^2+z.^2-4; N1=sum(F<=0); V=32*N1/N
```

In fact, the method can be extended easily to higher-dimensional integral problems, even functional bounds are also allowed. The main idea may be different from
those when computing regular integrals. The problem must be described by other
meaningful quantities such as area and volume, or even multidimensional "volume".
The integral model should be reconstructed again, rather than using the original forms
of the mathematical models.

8.5.2 Spline-based integral evaluations

The spline interpolation technique was introduced earlier. However, before considering its applications in numerical integration, the following example is to be considered. Comparisons will be made to show the benefit of using interpolation techniques
in finding the integrals.

Example 8.31. Consider the set of sparsely distributed samples in Example 7.15. Use
the trapezoidal method to approximate the integral.

Solutions. The samples can be input to MATLAB, and then the integrand can be
drawn, where the shaded area is the approximation of the integral by the trapezoidal
method, shown in Figure 8.10. The trapezoidal method can be used to approximate
the definite integral, yielding $I_0 = 1.84155830$. The error is rather large.

```
>> x=[0,0.4,1 2,pi]; y=sin(x); I0=trapz(x,y)
   fplot(@(t)sin(t),[0,pi]), hold on, plot(x,y,'o'), fill(x,y,'g')
```

Let us review the regular integral algorithms. Since the samples known are too
few, and the step-sizes are not regular, the trapezoidal method seems to be the only
choice in solving this kind of problems.

Is it possible to solve this problem with higher accuracy, from only five sparsely
distributed samples? Cubic and B-splines were introduced in Chapter 7, with numerical evaluation of differentials. Similar tools can be used to numerically explore the
integrals.

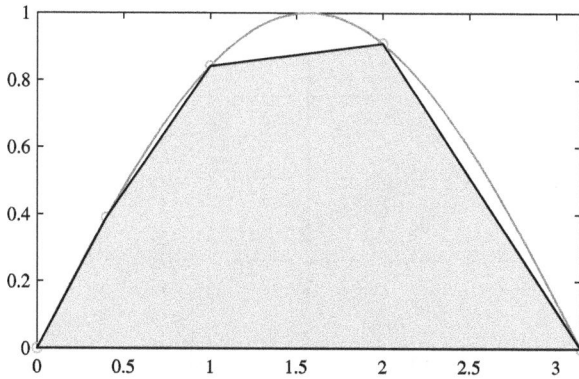

Figure 8.10: Illustration of the trapezoidal algorithm.

The function `fnint()` provided in MATLAB can be used to evaluate an integral function. Also a definite integral can be evaluated. The spline object of the integral can be evaluated with S_1=`fnint(S,F0)`, where F_0 is the initial value, with a default of zero. The integral function can be evaluated from y=`fnval(S1,x)`, where y is the vector, and the returned arguments are the values of the integral function at x. Since an indefinite integral can have an added constant, the actual integral function is, in fact, a simple translation of the given result. The initial value of the integral function is expressed by F_0, while the initial values of multiple integral functions may also be nonzero.

With the integral function, the definite integral over the interval $[a, b]$ can also be evaluated using `fnval()` function, by computing the function values at terminals $[a, b]$, F=`fnval(fnint(S),[a,b])`, then Newton–Leibniz formula can be used to evaluate the definite integral I=$F(2)$–$F(1)$.

Example 8.32. Compute the definite integral and integral function with splines for the problem in Example 8.31.

Solutions. The following statements can be used to set up two splines, then with `fnint()` function, the integral function can be obtained, from which the definite integral can also be obtained. The results obtained by using the two splines are I_1 = 2.01905 and I_2 = 1.999942. It can be seen that the results are far more accurate for finding the definite integrals.

```
>> x=[0,0.4,1 2,pi]; y=sin(x);
   a=fnint(csapi(x,y)); F=fnval(a,[0,pi]); I1=F(2)-F(1)
   b=fnint(spapi(5,x,y)); F=fnval(b,[0,pi]); I2=F(2)-F(1)
```

It is known that the indefinite integral of the sine function is $F(x) = -\cos x + C$. Let $F(0) = 0$, then $C = 1$, i. e., $F(x) = -\cos x + 1$. It can be seen that the integral function obtained is shown in Figure 8.11, where the error of the cubic spline is large, and

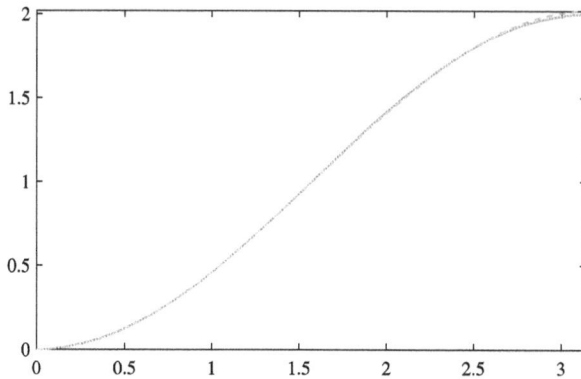

Figure 8.11: Integral function reconstruction from splines.

B-spline result is quite close to the analytical result, the difference cannot be distinguished from the curves.

```
>> fplot(@(t)[-cos(t)+1],[0,pi]); hold on;
   fnplt(a,'--'); fnplt(b,':') % integral function
```

As it was described earlier, if a cubic spline is used, the result from fnint() function is also a cubic spline, therefore the fitting quality may not be satisfactory, and not suitable in evaluating definite integrals. B-spline is recommended for this type of problems.

Example 8.33. Compute the multiple integral problem in Example 8.31 with B-splines.

Solutions. It is quite natural that the following method can be used in finding directly the multiple integral function, as shown in Figure 8.12. Only when $n = 1$, the result is correct, the others are wrong.

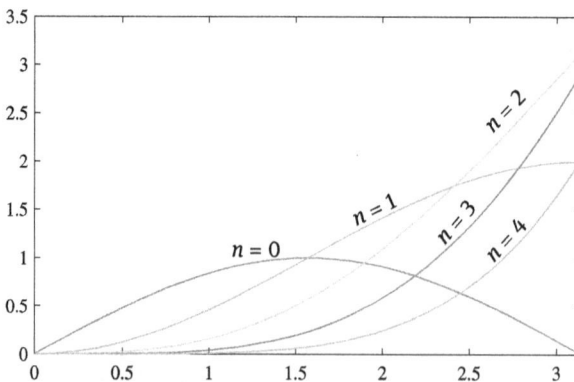

Figure 8.12: Multiple integral evaluation with B-splines.

```
>> x=[0,0.4,1 2,pi]; y=sin(x);
   b=spapi(5,x,y); fnplt(b), hold on
   for i=1:4, b=fnint(b); fnplt(b), F=fnval(b,[0,pi]); end
```

Many orders in B-spline were explored, and the number of samples were also tried. It seems that the evaluation of multiple integrals of univariate functions from samples using B-splines may not be possible.

8.5.3 Numerical evaluations of multiple integrals

If the samples are sparsely or unevenly distributed, the numerical integral approaches discussed so far cannot be used. Interpolation based algorithms should be introduced to solve related problems.

It should be noted that function fnint() provided in MATLAB can only be used to evaluate integrals of univariate functions, and cannot be used for solving multiple integral problems. To solve multiple integral problems, function fnder() can be considered, by setting the orders of differentiation to negative integers.

Example 8.34. Compute the double integral in Example 8.17 with B-splines, where

$$J = \int_{-1}^{1} \int_{-2}^{2} e^{-x^2/2} \sin(x^2 + y) \mathrm{d}x \mathrm{d}y.$$

Solutions. B-spline model can be established with function fnder(), and then the orders of differentiation with respect to x and y to -1, such that the integral function can be obtained. The surface of the integral function obtained is shown in Figure 8.13, which is exactly the same as that in Example 8.19. The definite integral obtained is $I = 1.5744966$, being very close to the theoretical result $I_0 = 1.57449815921736052274208$ of Example 8.17.

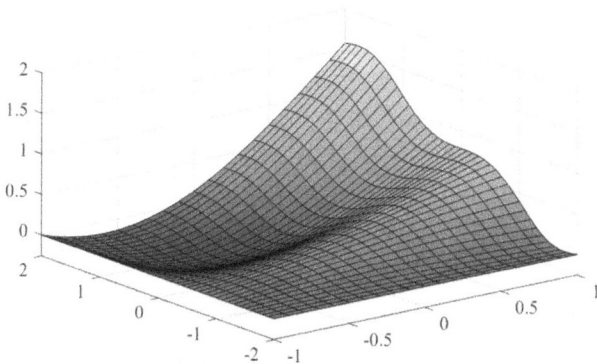

Figure 8.13: Surface of integral function from splines.

```
>> x0=-2:0.1:2; y0=-1:0.1:1; [x y]=ndgrid(x0,y0);          % samples
   z=exp(-x.^2/2).*sin(x.^2+y); S=spapi({5,5},{x0,y0},z); % B-spline
   S1=fnder(S,[-1 -1]); S2=fnval(S1,{x0,y0});
   surf(y0,x0,S2), I=S2(end,end) % compute integrals
```

Example 8.34 was solved based on assumptions that the mesh grid samples are known. In real applications, may be only scattered data are known. How can we compute the integral or integral function from these samples? A new method is conceived, i. e., `griddata()` function can be used to reconstruct mesh grid data. Then we establish a B-spline object from the mesh grid data, and compute the integral function surface. An example is given below to demonstrate the idea and implementation of the solution process.

Example 8.35. If only a set of scattered samples are known, evaluate the integral in Example 8.34 from the samples, and draw the surface of the integral function.

Solutions. A set of 80 pseudorandom numbers can be generated in vectors x_0, y_0 and z_0, then mesh grid matrices x and y can be generated, finally, interpolation can be obtained with `griddata()` function to reconstruct mesh grid values to build B-spline object S. Negative orders of differentiation from B-spline can be taken to find integral functions and create the surface, which is the same as that in Figure 8.13. The definite integral obtained is $I = 1.575368459080101$, and it is close to the theoretical value.

```
>> x0=-2+4*rand(80,1); y0=-1+2*rand(80,1);
   z0=exp(-x0.^2/2).*sin(x0.^2+y0);
   x1=-2:0.1:2; y1=-1:0.1:1; [x y]=ndgrid(x1,y1); % mesh grid data
   z=griddata(x0,y0,z0,x,y,'v4');                 % build samples
   S=spapi({5,5},{x1,y1},z); S1=fnder(S,[-1 -1]); % create B-spline
   S2=fnval(S1,{x1,y1}); surf(y1,x1,S2) % evaluate integrals
   I=S2(end,end)                        % compute the definite integral
```

For the scattered data interpolation function `griddata()`, the interpolation results at the boundaries may not be satisfactory. Therefore this may lead to bias in the integral computation. Well, in practical situations, reconstruction of integrals from scattered samples may be the only choice.

8.6 Exercises

8.1   Evaluate the definite integral for the data in Table 8.5. If high precision algorithm is used, what will be the result?

Table 8.5: Data in Problem 8.1.

x_i	0	0.1	0.2	0.3	0.4	0.5	0.6	0.7	0.8	0.9	1	1.1	1.2
y_i	0	2.2077	3.2058	3.4435	3.241	2.8164	2.311	1.8101	1.3602	0.9817	0.6791	0.4473	0.2768

8.2 Modify the function num_integral(), such that the final two sets of samples can be fairly manipulated, to ensure trapezoidal method may not be selected. Also, the number of points can be automatically selected.

8.3 Newton–Cotes open and closed formulas of different orders are given in [13], together with Maple language implementation. Rewrite the code with MATLAB.

8.4 Evaluate numerically $\displaystyle\int_0^\pi (\pi-t)^{1/4} f(t) dt$, where, $f(t) = e^{-t}\sin(3t+1)$. If the samples are selected at $t = 0.1, 0.2, \ldots, \pi$, find the function values of

$$F(t) = \int_0^t (t - \tau)^{1/4} f(\tau) \, d\tau,$$

and draw the $F(t)$ curve.

8.5 Compute numerical integral,[9] assess the accuracy, and draw integral function for

$$I = \int_0^1 \left[\frac{1}{(x - 0.3)^2 + 0.01} + \frac{1}{(x - 0.9)^2 + 0.04} + 6 \right] dx.$$

8.6 Draw the curve $\displaystyle I(s) = \int_0^s \frac{e^x \sqrt{e^x - 1}}{e^x + 3} dx, \ s \in (0, 10).$

8.7 For a set of samples of a, draw the integral curve $\displaystyle I(a) = \int_0^\infty \frac{\cos ax}{1 + x^2} dx.$

8.8 Compute numerically the integrals, and compare with theoretical results:

$$(1) \iint_{|x|+|y|\leqslant 1} (|x| + |y|) dx dy, \quad (2) \iint_{\pi^2 \leqslant x^2 + y^2 \leqslant 4\pi^2} \sin\sqrt{x^2 + y^2} dx dy.$$

8.9 Compute the following numerical and improper integrals:

$$(1) I = \int_0^\infty \frac{\cos x}{\sqrt{x}} dx, \quad (2) I = \int_0^1 \frac{1 + x^2}{1 + x^4} dx, \quad (3) \int_{e^{-2\pi n}}^1 \left|\cos\left(\ln\frac{1}{x}\right)\right| dx.$$

8.10 Evaluate numerically the triple integral $\displaystyle\iiint_V x^3 y^2 z \, dx dy dz$, where V is the region given by $0 \leqslant x \leqslant 1, 0 \leqslant y \leqslant x$, and $0 \leqslant z \leqslant xy$.

8.11 Evaluate numerically the following multiple integrals:

(1) $\displaystyle\int_0^2 \int_0^{\sqrt{4-x^2}} \sqrt{4 - x^2 - y^2}\, dydx,$ (2) $\displaystyle\int_0^3 \int_0^{3-x} \int_0^{3-x-y} xyz\, dzdydx,$

(3) $\displaystyle\int_0^2 \int_0^{\sqrt{4-x^2}} \int_0^{\sqrt{4-x^2-y^2}} z(x^2 + y^2)\, dzdydx.$

8.12 Evaluate the following multiple integrals with numerical methods. It should be noted that the analytical solutions do not exist. Validate the numerical results.

(1) $\displaystyle\int_0^2 \int_0^{e^{-x^2/2}} \sqrt{4 - x^2 - y^2}\, e^{-x^2-y^2}\, dydx,$ (2) $\displaystyle\int_0^2 \int_0^2 \int_0^2 z(x^2 + y^2)e^{-x^2-y^2-z^2-xz}\, dzdydx,$

(3) $\displaystyle\int_0^{7/10} \int_0^{4/5} \int_0^{9/10} \int_0^1 \int_0^{11/10} \sqrt{6 - x^2 - y^2 - z^2 - w^2 - u^2}\, dwdudzdydx.$

8.13 Assume that a set of samples in the function $z = f(x,y)$ are obtained, as shown in Table 8.6. The mathematical expression of the function is not known. Compute the integral $\displaystyle\int_{-0.2}^{0.2} \int_{-0.1}^{0.3} f(x,y)dydx,$ and draw the integral surface.

Table 8.6: Scattered samples for Exercise 8.13.

x_i	y_i	z_i	x_i	y_i	z_i	x_i	y_i	z_i
0.06	−0.03	−0.0018	0.09	−0.01	−0.0009	−0.16	0.28	−0.044785
0.16	−0.02	−0.0032	−0.14	0.09	−0.0126	−0.08	−0.08	0.0064
−0.14	0.05	−0.0070	0.05	−0.09	−0.0045	−0.17	0.15	−0.025497
−0.2	0.03	−0.006	−0.18	0.15	−0.026997	−0.2	0.1	−0.020
−0.03	0.22	−0.0066	0.08	0.16	0.0128	−0.12	0.08	−0.0096
−0.02	0.23	−0.0046	0.06	0.3	0.018	−0.02	0.14	−0.0028
0.08	−0.09	−0.0072	0.03	0.27	0.0081	0.15	0.16	0.023998
0.11	−0.03	−0.0033	0.1	0.13	0.013	−0.05	0.22	−0.011
−0.07	−0.04	0.0028	0.09	0.3	0.026997	0.07	0.2	0.014
−0.08	0.26	−0.0208	−0.1	0.06	−0.006	−0.1	0.14	−0.014

9 Integral transforms

Integral transform techniques play a very important role in science and engineering. One of the applications of integral transform technique is that they can be used to map expressions from one domain into another such that simple manipulations are made possible. For instance, Laplace transform can be used to map time domain functions into complex domain functions, and it can be used to map ordinary differential equations into algebraic equations, which are much easier to handle. Also, complex domain-based techniques, such as stability analysis, are made possible. This kind of transform established the foundation of classical control theory.

In real applications, Fourier, Mellin, and Hankel transforms are also very useful in different fields. The major topic of the chapter is to show how to solve integral transform problems by the use of computers. Even though the reader may know nothing about integral transforms, he/she can still use MATLAB to directly solve integral transform problems.

In Section 9.1, definitions and properties are presented first, and then we will concentrate on how to use MATLAB to solve Laplace transform problems. For functions where analytical expressions of Laplace transforms and their inverses do not exist, numerical Laplace transforms are presented in Section 9.2. A universal MATLAB function is written to perform numerical Laplace transforms for a wide variety of problems. In Section 9.3, various Fourier transforms and their inverses are presented. In Section 9.4, Mellin and Hankel transforms are introduced, and universal solvers with numerical applications are also written for these transforms. The z transform is another category of transforms, which is the foundation of digital signal processing and digital control systems. In Section 9.5, z transforms are also presented.

9.1 Laplace transforms and their inverses

French mathematician Pierre-Simon Laplace (1749–1827) introduced integral transforms to map linear ordinary differential equations into algebraic ones. Laplace transform technique established foundations of many fields, such as electric circuit analysis and automatic control systems. In this section, the definitions and properties in Laplace transform and its inverse are introduced, then we will focus on the Laplace transform problem solutions with MATLAB Symbolic Math Toolbox. In the next section, numerical approaches will be introduced for Laplace transform problems.

https://doi.org/10.1515/9783110666977-009

9.1.1 Definition and properties of Laplace transform

Definition 9.1. Laplace transform of a time domain function $f(t)$ is defined as

$$\mathcal{L}[f(t)] = \int_0^\infty f(t)e^{-st}dt = F(s), \qquad (9.1.1)$$

where $\mathcal{L}[f(t)]$ is a simplified notation for Laplace transform.

With the Laplace transform, a t-domain (if t is time, it can be regarded as time domain) signal $f(t)$ can be transformed to an s-domain signal $F(s)$, and s can be regarded as a complex variable.

Theorem 9.1. *The following properties are listed below without further proofs:*
(1) *Linearity property. For scalars a and b, $\mathcal{L}[af(t) \pm bg(t)] = aF(s) \pm bG(s)$.*
(2) *Time domain translation. $\mathcal{L}[f(t-a)] = e^{-as}F(s)$.*
(3) *s-domain translation. $\mathcal{L}[e^{-at}f(t)] = F(s+a)$.*
(4) *Differentiation. $\mathcal{L}[df(t)/dt] = sF(s)-f(0^+)$. More generally, the nth order derivative can be evaluated from*

$$\mathcal{L}[d^n f(t)/dt^n] = s^n F(s) - s^{n-1}f(0^+) - s^{n-2}f'(0^+) - \cdots - f^{(n-1)}(0^+). \qquad (9.1.2)$$

If the initial values of $f(t)$ and its derivatives are all zero, then (9.1.2) can be simplified as

$$\mathcal{L}\left[\frac{d^n f(t)}{dt^n}\right] = s^n F(s). \qquad (9.1.3)$$

With such a property, differential equations can be mapped into algebraic equations.

(5) *Integral. If the initial conditions are zero, $\mathcal{L}\left[\int_0^t f(\tau)\,d\tau\right] = \dfrac{F(s)}{s}$, more generally, the nth multiple integral of $f(t)$ can be derived from*

$$\mathcal{L}\left[\int_0^t \cdots \int_0^t f(\tau)d\tau^n\right] = \frac{F(s)}{s^n}. \qquad (9.1.4)$$

(6) *Initial value. $\lim_{t\to 0} f(t) = \lim_{s\to\infty} sF(s)$.*
(7) *Final value. If $F(s)$ has no poles at $s \geq 0$, then $\lim_{t\to\infty} f(t) = \lim_{s\to 0} sF(s)$.*
(8) *Convolution. $\mathcal{L}[f(t) * g(t)] = \mathcal{L}[f(t)]\mathcal{L}[g(t)]$, where the convolution operator $*$ is defined as*

$$f(t) * g(t) = \int_0^t f(\tau)g(t-\tau)d\tau = \int_0^t f(t-\tau)g(\tau)d\tau. \qquad (9.1.5)$$

(9) *Other properties:*

$$\mathcal{L}[t^n f(t)] = (-1)^n \frac{d^n F(s)}{ds^n}, \quad \mathcal{L}\left[\frac{f(t)}{t^n}\right] = \int_s^\infty \cdots \int_s^\infty F(s) ds^n. \qquad (9.1.6)$$

Definition 9.2. If the Laplace transform expression is $F(s)$, the inverse Laplace transform is defined as

$$f(t) = \mathcal{L}^{-1}[F(s)] = \frac{1}{2\pi j} \int_{\sigma-j\infty}^{\sigma+j\infty} F(s) e^{st} ds, \qquad (9.1.7)$$

where σ is larger than any of the poles in $F(s)$.

9.1.2 Computer solutions of Laplace transforms

It might be complicated to derive manually Laplace transforms for given functions. Computers should be employed to find the transforms for complicated functions. For instance, MATLAB can be used directly in solving these problems.

In Example 5.27, the direct integral method was used to evaluate Laplace transform of a given signal, however, it was not quite successful. A dedicated solver in MATLAB is introduced to solve such problems.

Laplace transform problems can easily be solved with MATLAB Symbolic Math Toolbox, with the following procedures:
(1) Command syms can be used to declare t as a symbolic variable, then the time domain expression f can be entered into MATLAB.
(2) Call laplace() function to find the Laplace transform

   F=laplace(f), %with default variables
   F=laplace(f,v,u), %the related variables v and u can be specified

(3) Function simplify() can be used to simplify the results.

Besides, functions like pretty() and latex() can further be used to manipulate the final results.

If the Laplace transform expression is given in F, function ilaplace() can be used to find the inverse Laplace transform, with the syntaxes

f=ilaplace(F), %with default variables
f=ilaplace(F,u,v), %the related variables v and u can be specified

the expression f can further be simplified and displayed.

Example 9.1. For the given function $f(t) = t^2 e^{-2t} \sin(t + \pi)$, find its Laplace transform.

Solutions. Based on the original problem, t should be declared as a symbolic variable, then $f(t)$ can be expressed with MATLAB commands, and then the following commands to find the Laplace transform can be used:

```
>> syms t; f=t^2*exp(-2*t)*sin(t+pi);
   F=simplify(laplace(f)) % direct transform
```

where Laplace transform is obtained as

$$F(s) = \frac{2}{((s+2)^2 + 1)^2} - \frac{2(2s+4)^2}{((s+2)^2 + 1)^3}.$$

Example 9.2. Assume that the function $f(x) = x^2 e^{-2x} \sin(x + \pi)$ is given. Find its Laplace transform, and then take the inverse Laplace transform of the result and see whether the original function can be restored.

Solutions. Again the function `laplace()` can be used to solve the problem

```
>> syms x w; f=x^2*exp(-2*x)*sin(x+pi);
   F=laplace(f,x,w), g=simplify(ilaplace(F))
```

It can be seen that the results are the same, if variable substitution is made. If the inverse Laplace transform is computed with `ilaplace(F)`, the returned function is $-t^2 e^{-2t} \sin t$, since $\sin(t + \pi) = -\sin t$.

Example 9.3. Compute the Laplace transform of the following rational function:

$$G(x) = \frac{-17x^5 - 7x^4 + 2x^3 + x^2 - x + 1}{x^6 + 11x^5 + 48x^4 + 106x^3 + 125x^2 + 75x + 17}.$$

Solutions. For the original problem, the following statements can be used to solve it:

```
>> syms x t;              % declare symbolic variable and input the function
   G=(-17*x^5-7*x^4+2*x^3+x^2-x+1)...
     /(x^6+11*x^5+48*x^4+106*x^3+125*x^2+75*x+17);
   f=ilaplace(G,x,t) % find Laplace transform
```

It can be seen that the readability is rather poor. In fact, since there is no analytical solution for the equation $x^6 + 11x^5 + 48x^4 + 106x^3 + 125x^2 + 75x + 17 = 0$, the Laplace transform problem has no analytical solution. With command `vpa(f)`, high precision solutions can be obtained as

$$y(t) = -556.2565e^{-3.2617t} + 1.7589e^{-1.0778t} \cos 0.6021t + 10.9942e^{-1.0778t} \sin 0.6021t$$
$$+ 0.2126e^{-0.5209t} + 537.2850e^{-2.5309t} \cos 0.3998t - 698.2462e^{-2.5309t} \sin 0.3998t.$$

Example 9.4. For the function $f(t)$ given in Example 9.1, find the relationship between $\mathscr{L}\left[\dfrac{\mathrm{d}^5 f(t)}{\mathrm{d}t^5}\right]$ and $s^5 \mathscr{L}[f(t)]$.

Solutions. To solve this type of problem, function `diff()` can be used to compute the fifth order derivative of $f(t)$, and then apply Laplace transform

```
>> syms t s; f=t^2*exp(-2*t)*sin(t+pi);
   F=simplify(laplace(diff(f,t,5)))
```

The difference between the two quantities can be measured with the following MAT-LAB statements, and it can be seen that the difference is $6s - 48$.

```
>> F0=laplace(f); simplify(F-s^5*F0)
```

Since the difference between them is not zero, it seems that (9.1.3) is not satisfied. This is correct, since the initial conditions of $f(t)$ are not zero. Since $f(0) = f'(0) = 0$, while the initial values of high-order derivatives are not zero, (9.1.3) is not satisfied, while (9.1.2) is satisfied instead. It can be seen from (9.1.2) that when initial conditions are satisfied, the difference obtained above can fully be explained.

```
>> ss=0; f1=f;
   for i=4:-1:0, ss=ss-s^i*subs(f1,t,0); f1=diff(f1,t); end, ss
```

Example 9.5. Derive the differentiation formula for $\mathscr{L}\left[\dfrac{\mathrm{d}^2 f(t)}{\mathrm{d}t^2}\right]$.

Solutions. Some of the properties of Laplace transform can be derived directly with MATLAB Symbolic Math Toolbox. Assume that Laplace transform of the second-order derivative of $f(t)$ is expected, the function $f(t)$ should be defined first, then the following statements can be employed to solve the problem:

```
>> syms t f(t); laplace(diff(f,t,2))
```

and the result is `s^2*laplace(f(t),t,s)-s*f(0)-D(f)(0)`. The mathematical expression is $s^2 F(s) - sf(0) - f'(0)$. It can be seen that the result is exactly the same as that in (9.1.2). The facility can further be extended, for instance, the Laplace transform of the eighth-order derivative can be formulated as

```
>> Y=collect(laplace(diff(f,t,8))) % Laplace transform property
```

Example 9.6. For the given function $f(t) = e^{-5t} \cos(2t + 1) + 5$, compute $\mathscr{L}\left[\dfrac{\mathrm{d}^5 f(t)}{\mathrm{d}t^5}\right]$.

Solutions. This example is an extension to the previous one. For a specific function $f(t)$, functions `diff()` and `laplace()` can be used together to solve the problem, with

```
>> syms t; f=exp(-5*t)*cos(2*t+1)+5;
   F=laplace(diff(f,t,5)); F=simplify(F) % solutions
```

the result obtained is

$$F = \frac{1\,475\cos 1s - 1\,189\cos 1 - 24\,360\sin 1 - 4\,282\sin 1s}{s^2 + 10s + 29}.$$

Fine tuning can be applied to the simplified results. For instance, if one wants to have collected terms for the numerator, the following commands can be used:

```
>> syms s; F1=collect(F) % collect the terms
```

the results obtained is

$$F_1(s) = \frac{(1\,475\cos 1 - 4\,282\sin 1)s - 1\,189\cos 1 - 24\,360\sin 1}{s^2 + 10s + 29}.$$

Example 9.7. For the given function $f(t) = \dfrac{1}{\sqrt{t}\,(at + b)}$, $a, b > 0$, find its Laplace transform.

Solutions. The problem can be input into MATLAB, then solved directly with MATLAB statements:

```
>> syms t; syms a b positive
   f(t)=1/sqrt(t)/(a*t+b); simplify(laplace(f))
```

The result obtained is

$$F(s) = \frac{\pi e^{bs/a}}{\sqrt{ab}}[1 - \operatorname{erf}(\sqrt{bs/a})].$$

9.1.3 Solving differential equations with Laplace transform

Solving ordinary differential equations will be fully elaborated in Volume V of the series. Here, only a simple approach is introduced to solve linear differential equations with Laplace transform facilities.

Observing (9.1.2), since the nth order derivative of a signal gets mapped to the Laplace transform of the signal, $Y(s)$, just multiplied by s^n, the differential equation can be converted into an algebraic equation. Solving it, the Laplace transform expression of certain signals can be obtained. The inverse Laplace transform can be applied to find the analytical solutions to the original differential equations. This idea will be demonstrated through the following examples.

Example 9.8. Assume that the input signal is $u(t) = e^{-5t}\cos(2t + 1) + 5$, and $y(0) = 3$, $y'(0) = 2$, $y''(0) = y'''(0) = 0$, find the analytical solutions of the following differential equations:

$$y^{(4)}(t) + 10y'''(t) + 35y''(t) + 50y'(t) + 24y(t) = 5u''(t) + 4u'(t) + 2u(t).$$

Solutions. Taking Laplace transform of both sides of the equation, the right-hand side expression can easily be evaluated with `laplace()` function.

```
>> syms t s; u=exp(-5*t)*cos(2*t+1)+5;   % input signal description
   R=simplify(laplace(diff(u,2)+4*diff(u)+2*u))  % right hand side
```

Taking Laplace transform of the left-hand side of the equation, the first term can be written manually as

$$s^4 Y(s) - y(0)s^3 - y'(0)s^2 - y''(0)s - y'''(0) = s^4 Y(s) - 3s^3 - 2s^2,$$

and the subsequent terms are respectively

$$10s^3 Y(s) - 30s^2 - 20s, \quad 35s^2 Y(s) - 105s - 70, \quad 50sY(s) - 150, \quad 24Y(s),$$

the sum of the terms can be obtained, and the following equation is derived:

$$(s^4 + 10s^3 + 35s^2 + 50s + 24)Y(s) - 3s^3 - 32s^2 - 125s - 220 = R(s).$$

It can immediately be found that the analytical solution of Laplace transform $Y(s)$ of the output can be written as

$$Y(s) = \frac{R(s) + 3s^3 + 32s^2 + 125s + 220}{s^4 + 10s^3 + 35s^2 + 50s + 24}.$$

The analytical solution $y(t)$ can be obtained by taking the inverse Laplace transform of $Y(s)$, with the following statements, and the graphical representation of the solution can be drawn as shown in Figure 9.1.

```
>> Y=(R+3*s^3+32*s^2+125*s+220)/(s^4+10*s^3+35*s^2+50*s+24);
   y=ilaplace(Y), fplot(y,[0,10])  % Laplace transform
```

The analytical solution of the equation is

$$y(t) = e^{-t}\left(\frac{3\cos 1}{10} + \frac{7\sin 1}{20} + 19\right) - e^{-4t}\left(\frac{9\cos 1}{10} + \frac{\sin 1}{5} + \frac{25}{4}\right)$$
$$+ e^{-3t}\left(\frac{15\cos 1}{8} + \frac{9\sin 1}{8} + \frac{73}{3}\right) - e^{-2t}\left(\frac{33\cos 1}{26} + \frac{15\sin 1}{13} + \frac{69}{2}\right)$$
$$- e^{-5t}\left(\cos 2t + \frac{\sin 2t(21\cos 1 - \sin 1)}{\cos 1 + 21\sin 1}\right)\left(\frac{3\cos 1}{520} + \frac{63\sin 1}{520}\right) + \frac{5}{12}.$$

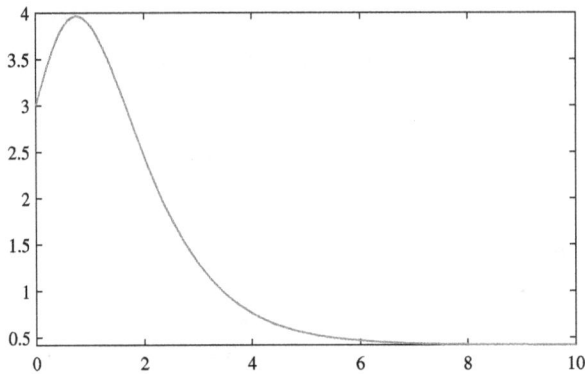

Figure 9.1: Solution of an ordinary differential equation.

Example 9.9. Solve the following ordinary differential equations:

$$\begin{cases} x'' - x + y + z = 0, \\ x + y'' - y + z = 0, \\ x + y + z'' - z = 0, \end{cases} \quad x(0) = 1, \quad y(0) = z(0) = x'(0) = y'(0) = z'(0) = 0.$$

Solutions. The equations can be solved by taking Laplace transform. Considering the nonzero term $x(0) = 1$, the following three equations can be set up:

$$s^2X - s - X + Y + Z = 0, \quad X + s^2Y - Y + Z = 0, \quad X + Y + s^2Z - Z = 0.$$

Solving the above simultaneous equations, and taking the inverse Laplace transform of the results, the analytical solution of the equation can be found

```
>> syms s X Y Z
   S=solve(s^2*X-s-X+Y+Z==0,X+s^2*Y-Y+Z==0,X+Y+s^2*Z-Z==0,[X,Y,Z]);
   x=ilaplace(S.X), y=ilaplace(S.Y), z=ilaplace(S.Z)
```

and it is given by

$$x(t) = \frac{2\cosh\sqrt{2}t}{3} + \frac{\cos t}{3}, \quad y(t) = \frac{\cos t}{3} - \frac{\cosh\sqrt{2}t}{3}, \quad z(t) = \frac{\cos t}{3} - \frac{\cosh\sqrt{2}t}{3}.$$

9.2 Numerical solutions of Laplace transform problems

It was demonstrated earlier that with function `laplace()`, Laplace transform of some time domain functions can be obtained directly. However, there are quite a few time domain functions whose Laplace transforms do not exist analytically at all. In this case, numerical methods should be considered to find Laplace transforms.

9.2.1 Numerical inverse Laplace transform

One of the numerical inverse Laplace transform algorithms was proposed by Juraj Valsa, with MATLAB function named INVLAP(),[23, 22] with the following syntax:

$[t,y]$=INVLAP$(f,t_0,t_n,N,$other parameters$)$

where character s can be used in describing the string of Laplace transform expression, (t_0, t_n) is the time interval of interest, $t_0 \neq 0$, and N is the number of points to compute. Different values of N can be tried to test computation results. "Other parameters" option can be a reference from the online help of the original function, and it is suggested here that, if it is not absolutely necessary, do not change these parameters.

An extension to the INVLAP() function is written in this section, as a new function INVLAP_new(), where the original function can be fully restored. Besides, more facilities are introduced in the new function. Examples are introduced to demonstrate them.

Example 9.10. Use the numerical inverse Laplace transform method to solve again the problem in Example 9.3.

Solutions. It can be seen that in Example 9.3 that, although the analytical solution to the inverse Laplace transform is not known, numerical solutions to the problem can easily be obtained. The original function can be expressed as a string of s, and with the following statements, the numerical solution can be obtained. The relative error in the problem is 1.83×10^{-5} %, which satisfies most of the requirements in application problems.

```
>> syms x t; % declare necessary symbolic variables
   G=(-17*x^5-7*x^4+2*x^3+x^2-x+1) ...
      /(x^6+11*x^5+48*x^4+106*x^3+125*x^2+75*x+17);
   f=ilaplace(G,x,t);        % theoretical Laplace transform
   fun=char(subs(G,x,'s'));  % convert x into s in the string
   [t1,y1]=INVLAP_new(fun,0,5,100); y0=subs(f,t,t1);
   norm(vpa((y1-y0)./y0))
```

It can be seen from the computation load that, when the number of points is increased from 100 to 5 000, the time required with INVLAP_new() function is 0.61 seconds, while with ilaplace() and subs() functions, one needs 261 seconds. It can be seen that the numerical algorithm is more efficient.

9.2.2 Ideas of feedback control systems

Consider the typical feedback control system structure shown in Figure 9.2. The forward path is constructed in series of two transfer function blocks, $P(s)$ and $G_c(s)$, with

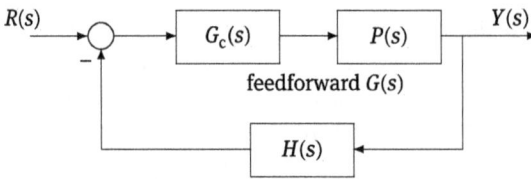

Figure 9.2: Block diagram of a typical feedback control system.

the overall transfer function of the forward path, $P(s)G_c(s)$, denoted as $G(s)$. If the Laplace transform of the input signal is $R(s)$, under its excitation, how to evaluate the Laplace transform of the output $Y(s)$? How to compute the time domain response $y(t)$?

Theorem 9.2. *It is known from control theory that the Laplace transform of the output signal $Y(s)$ can be evaluated from*

$$Y(s) = \frac{G(s)}{1 + G(s)H(s)} U(s), \tag{9.2.1}$$

where the time domain response of the output signal is $y(t) = \mathscr{L}^{-1}[Y(s)]$.

In traditional control systems, all transfer functions are expressed as rational functions, therefore, it is easy to find the output signal $Y(s)$. With the inverse Laplace transform, the analytical solution $y(t)$ can be obtained. If the transfer functions have fractional or irrational orders, the typical method cannot be used to evaluate time domain responses. Numerical inverse Laplace transform methods should be used instead.

9.2.3 Numerical Laplace transforms

Based on the above mentioned ideas, the function INVLAP() is extended and converted as follows. The closed-loop system response under any input signal can be evaluated from the new function.

```
function [t,y]=INVLAP_new(G,t0,tn,N,H,tx,ux)
G=add_dots(G);
if nargin<=5, tx='1'; end, if nargin<=4, H=0; end
if ischar(H), H=add_dots(H); end
if ischar(tx), tx=add_dots(tx); end
a=6; ns=20; nd=19; t=linspace(t0,tn,N);
if t0==0, t=t(2:N); N=N-1; end,
n=1:ns+1+nd; alfa=a+(n-1)*pi*j;
bet=-exp(a)*(-1).^n; n=1:nd; bet(1)=bet(1)/2;
bdif=fliplr(cumsum(gamma(nd+1)./gamma(nd+2-n)./gamma(n)))./2^nd;
bet(ns+2:ns+1+nd)=bet(ns+2:ns+1+nd).*bdif;
```

```
if isnumeric(H), H=num2str(H); end
for i=1:N   % inverse Laplace transform for any point
    tt=t(i); s=alfa/tt; bt=bet/tt; sG=eval(G); sH=eval(H);
    if ischar(tx), sU=eval(tx); % with known input Laplace transform
    else                        % compute Laplace transform
        if isnumeric(tx),
            f=@(x)interp1(tx,ux,x,'spline').*exp(-s.*x); % interpolation
        else, f=@(x)tx(x).*exp(-s.*x); end       % input signal
        sU=integral(f,t0,tn,'ArrayValued',true); % numerical Laplace
    end
    btF=bt.*sG./(1+sG.*sH).*sU; y(i)=sum(real(btF)); % closed-loop
end
function F=add_dots(F) % sub function: unified dot operation
F=strrep(strrep(strrep(F,'.*','*'),'./','/'),'.^','^'); % delete dots
F=strrep(strrep(strrep(F,'*','.*'),'/','./'),'^','.^'); % add back
```

The syntaxes of the function are listed below

$[t,y]$=INVLAP_new(G,t_0,t_n,N), %inverse Laplace transform
$[t,y]$=INVLAP_new(G,t_0,t_n,N,H), %impulse response of G and H
$[t,y]$=INVLAP_new(G,t_0,t_n,N,H,u), %u input signal
$[t,y]$=INVLAP_new(G,t_0,t_n,N,H,t_x,u_x), %samples of the inputs t_x, u_x

There are many syntaxes in the function, where G is the string expression of the Laplace transform of the forward path. If H is also provided, G is the forward path transfer function string, H is the feedback string. If u is a string, then it should be the Laplace transform of the input signal; u can also be a function handle of the input; the input signal can also be expressed as samples (t_x, u_x). If only G is provided, H is automatically set to 0.

Apart from the above mentioned extensions, two bugs are fixed: the first is that the initial instance t_0 can be set to zero, however, the zero instance is excluded automatically in the function; the second is that in describing the string, dot operation should be made. Modification is made to the new function to add back dot operations automatically even if the user forgot to add them.

In practical applications, if the system model $G(s)$ is given, and Laplace transform of the input signal is $R(s)$, the Laplace transform of the output signal can be obtained from $Y(s) = G(s)U(s)$. Numerical solutions of the output signal can be evaluated through the inverse Laplace transform of $Y(s)$.

Example 9.11. Considering the fractional-order transfer function

$$G(s) = \frac{(s^{0.4} + 0.4s^{0.2} + 0.5)}{\sqrt{s}\,(s^{0.2} + 0.02s^{0.1} + 0.6)^{0.4}(s^{0.3} + 0.5)^{0.6}},$$

compute the inverse Laplace transform for $t \in (0,1)$, and draw the output signal.

Solutions. Function `ilaplace()` cannot be used to evaluate the inverse Laplace transform of $G(s)$, since there is no analytical solution. Numerical solution is the only choice. Selecting the number of points as $N = 1\,000$, the inverse Laplace transform of $G(s)$ can be obtained easily, as shown in Figure 9.3. If the number of points is increased to $N = 5\,000$, then the same results can be obtained, indicating the solution obtained is correct.

```
>> G=['(s^0.4+0.4*s^0.2+0.5)/sqrt(s)/',...
       '(s^0.2+0.02*s^0.1+0.6)^0.4/(s^0.3+0.5)^0.6'];
   [t,y]=INVLAP_new(G,0,1,1000); plot(t,y) %numerical inverse Laplacian
```

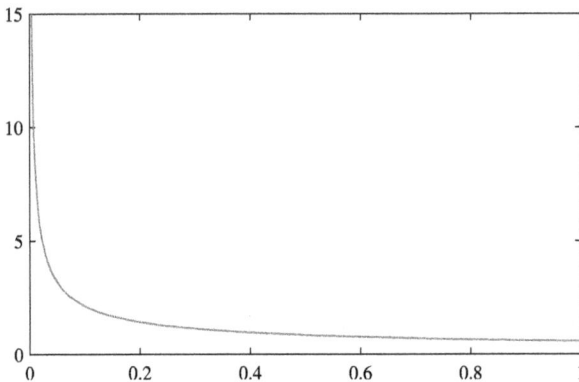

Figure 9.3: Numerical inverse Laplace transform.

It should be noted that, since the fractional-order term $p^y(x)$ is, in fact, an infinite series, it is not possible to find its analytical solution. Numerical approach should be used instead. Since the function is fast, the time needed is around 0.3 seconds.

If the Laplace transform of the input signal $u(t)$ is not known, the Laplace transform of the input signal can only be evaluated with numerical approaches. Analyzing the source code in the function `INVLAP()`, a loop structure is used, and in each step, a vector s is generated with the mathematical formula, where the numerical Laplace transform is computed as

$$\mathscr{L}[u(t)] = \int_0^\infty u(t)e^{-st}dt = U(s),\tag{9.2.2}$$

with s being a vector. Since the e^{-st} factor exists, when the interval $(0, t_n)$ is large enough, a finite interval can be used to replace the infinite interval and compute the numerical Laplace transform.

If the input signal is described by a set of samples, in vectors x_0 and u_0, the input $u(t)$ can be evaluated with an interpolation method, resulting in vector t.

Example 9.12. Assuming $G(s)$ is a fractional-order transfer function, and the input is $u(t) = e^{-0.3t} \sin t^2$, compute the output signal and draw the response.

Solutions. The input of the transfer function is the same as the input earlier. The input signal can be described as an anonymous function, and the output signal can be evaluated numerically. The curve of the output signal is shown in Figure 9.4. Inside the solver, numerical Laplace transform computation is embedded, and the total time required is 9 seconds.

```
>> f=@(t)exp(-0.3*t).*sin(t.^2); % given input signal
   G=['(s^0.4+0.4*s^0.2+0.5)/sqrt(s)',...
      '/(s^0.2+0.02*s^0.1+0.6)^0.4/(s^0.3+0.5)^0.6'];
   tic, [t,y]=INVLAP_new(G,0,15,400,0,f); toc
   plot(t,y) % numerical inverse Laplace transform
```

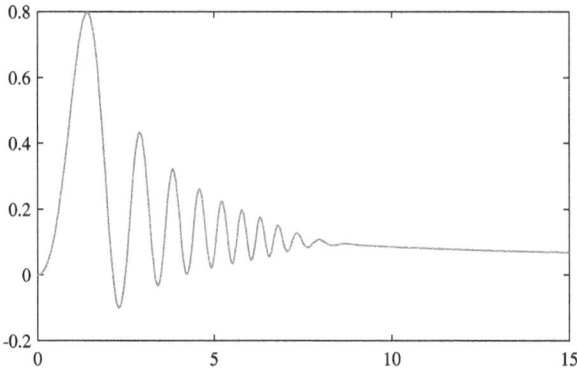

Figure 9.4: Output of a complicated fractional-order system.

Consider the input signal and assume that some samples are computed in the interval $t \in (0, 15)$. Then, based on the samples, the output of the system can be evaluated, and the curve is the same as those given earlier, meaning the signal obtained is correct.

Since interpolation is used internally in the function, it is quite time demanding, which takes about 552 seconds. It can be seen that the efficiency is rather low.

```
>> x0=0:0.1:15; u0=exp(-0.3*x0).*sin(x0.^2); % input samples
   tic, [t,y]=INVLAP_new(G,0,15,400,0,x0,u0); toc, plot(t,y)
```

9.2.4 Computation of irrational systems

If the transfer function model contains transcendental components, the fractional-order model is no longer rational, and it can be regarded as an irrational system model.

With the use of numerical Laplace transforms, an irrational system response can easily be obtained. This can be demonstrated through examples.

Example 9.13. For the complicated irrational model given below,[2] compute the output of the system in unity negative feedback system.

$$G(s) = \left[\frac{\sinh(0.1\sqrt{s})}{0.1\sqrt{s}} \right]^2 \frac{1}{\sqrt{s}\sinh(\sqrt{s})}.$$

Solutions. The open-loop irrational transfer function is described as a string, and the Laplace form of the input is $1/s$, where the output signal of the closed-loop system can be obtained as shown in Figure 9.5.

```
>> G='(sinh(0.1*sqrt(s))/0.1/sqrt(s))^2/sqrt(s)/sinh(sqrt(s))';
   [t,y]=INVLAP_new(G,0,10,1000,1,'1/s'); plot(t,y) % step response
```

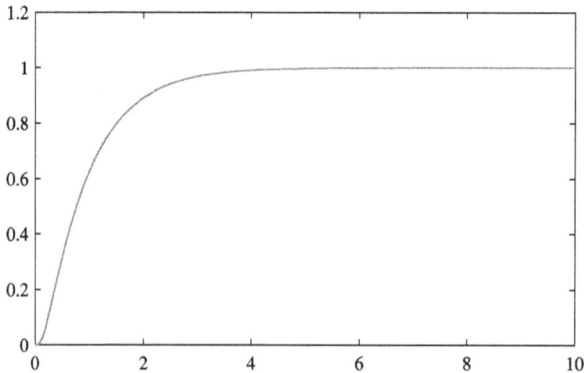

Figure 9.5: Closed-loop response of an irrational system.

9.3 Fourier transforms and their inverses

Laplace transforms can be used to map time domain signals into complex domain signals, while Fourier transforms can be used to convert time domain signals into frequency domain signals. Fourier transforms are useful in the areas such as circuit analysis and digital signal processing.

Definitions and properties of Fourier transforms are presented in this section first, followed by the MATLAB solution methodology. Finally, various Fourier transforms, including fast Fourier transforms, are presented.

9.3.1 Definition and properties of Fourier transforms

In Laplace transforms, the operator e^{-st} is introduced. If s is substituted by $j\omega$, where ω is also known as the frequency, Fourier transform can be introduced.

Definition 9.3. The general form of the Fourier transform is

$$\mathscr{F}[f(t)] = \int_{-\infty}^{\infty} f(t)e^{-j\omega t} dt = F(\omega). \tag{9.3.1}$$

Definition 9.4. If the Fourier transform is $F(\omega)$, the inverse Fourier transform is defined as

$$f(t) = \mathscr{F}^{-1}[F(\omega)] = \frac{1}{2\pi} \int_{-\infty}^{\infty} F(\omega)e^{j\omega t} d\omega. \tag{9.3.2}$$

As with Laplace transforms, several similar properties of Fourier transforms are listed without proofs.

Theorem 9.3. *Fourier transform has the following properties:*
(1) *Linearity. For scalars a and b, $\mathscr{F}[af(t) \pm bg(t)] = a\mathscr{F}[f(t)] \pm b\mathscr{F}[g(t)]$.*
(2) *Translation. $\mathscr{F}[f(t \pm a)] = e^{\pm j a\omega} F(\omega)$.*
(3) *Complex domain translation. $\mathscr{F}[e^{\pm jat}f(t)] = F(\omega \mp a)$.*
(4) *Differentiation. $\mathscr{F}\left[\dfrac{df(t)}{dt}\right] = j\omega F(\omega)$. More generally, the nth order derivative can be transformed as*

$$\mathscr{F}\left[\frac{d^n}{dt^n}f(t)\right] = (j\omega)^n \mathscr{F}[f(t)]. \tag{9.3.3}$$

(5) *Integral. $\mathscr{F}\left[\displaystyle\int_{-\infty}^{t} f(\tau)d\tau\right] = \dfrac{F(\omega)}{j\omega}$. Normally, the nth multiple integral of a function $f(t)$ can be evaluated from*

$$\mathscr{F}\left[\int_{-\infty}^{t} \cdots \int_{-\infty}^{t} f(\tau)d\tau^n\right] = \frac{\mathscr{F}[f(t)]}{(j\omega)^n}. \tag{9.3.4}$$

(6) *Scaling. $\mathscr{F}[f(at)] = F(\omega/a)/a$.*
(7) *Convolution. $\mathscr{F}[f(t)*g(t)] = \mathscr{F}[f(t)]\mathscr{F}[g(t)]$, where convolution is defined in (9.1.5).*

9.3.2 Computer evaluation of Fourier transform

As in the case in Laplace transform, symbolic variables should be declared first, the function can be expressed in variable f, and the function `fourier()` can be used to evaluate Fourier transform, with the following syntaxes

F=fourier(*f*), %compute Fourier transform
F=fourier(*f*,*v*,*u*), %the variables *v* and *u* are specified

The inverse Fourier transform can be evaluated from ifourier() function, with the syntaxes

f=ifourier(*F*), %default variables
f=ifourier(*F*,*u*,*v*), %the variables *u* and *v* are specified

It can be seen from the above syntaxes that, if the function is known, Fourier transform can be evaluated directly with a single command. If Fourier transform expression is known, ifourier() function can be used to compute the inverse Fourier transform.

Example 9.14. Considering $f(t) = 1/(t^2 + a^2)$, $a > 0$, compute its Fourier transform.

Solutions. The original function can be input into MATLAB environment, then Fourier transform can be obtained with the following statements, and the result obtained is $F = \pi e^{-a|\omega|}/a$. Taking the inverse Fourier transform of the result, the original function can be restored in f_1.

```
>> syms t w; syms a positive
   f(t)=1/(t^2+a^2); F=fourier(f,t,w) % direct transform
   f1=ifourier(F,w,t)                 % inverse transform
```

Example 9.15. For the function $f(t) = \sin^2(at)/t$, $a > 0$, find its Fourier transform.

Solutions. For the given expression, the following statements can be used to evaluate its Fourier transform

```
>> syms t w; syms a positive
   f(t)=sin(a*t)^2/t; F=fourier(f,t,w) % direct transform
```

with the results

$$F = -\pi j\, \text{heaviside}(-2a - \omega)/2 - \pi j\, \text{heaviside}(2a - \omega)/2 + \pi j\, \text{heaviside}(-\omega),$$

where heaviside(x) is a step function of x, also known as Heaviside function. If $x > 0$, the function is 1; $x = 0$ for 0.5; otherwise it is 0. If $\omega > 2a$, the three heaviside() functions are all 1, which makes $F(\omega) = 0$; if $\omega \leqslant -2a$, the three functions are all zero, thus $F(\omega) = 0$; If $0 < \omega < 2a$, the second and third heaviside() functions are both 1, with $F(\omega) = -j\pi/2$; if $0 > \omega > -2a$, then $F(\omega) = j\pi/2$. It can be concluded that the original problem can be simplified manually as

$$\mathscr{F}[f(t)] = \begin{cases} 0, & |\omega| > 2a, \\ -j\pi\, \text{sign}(\omega)/2, & |\omega| < 2a. \end{cases}$$

Example 9.16. For the given function $f(t) = e^{-a|t|}/\sqrt{|t|}$, compute its Fourier transform.

Solutions. With the existing function `fourier()`, the Fourier transform function can be obtained from

```
>> syms w t; syms a positive;
   f(t)=exp(-a*abs(t))/sqrt(abs(t)); F=fourier(f,t,w)
```

and the result obtained is $\sqrt{2\pi}/(2\sqrt{|w - ja|}) + \sqrt{2\pi}/(2\sqrt{|w + ja|})$.

9.3.3 Fourier sine and cosine transforms

Since in the traditional Fourier transform, a complex function $e^{-j\omega t}$ is introduced, while it is known from Euler formula that $e^{-j\omega t} = \cos\omega t - j\sin\omega t$, Fourier sine and cosine transforms can also be defined.

Definition 9.5. Fourier sine transform is defined as

$$\mathscr{F}_{\mathscr{S}}[f(t)] = \int_0^\infty f(t)\sin(\omega t)dt = F_s(\omega). \tag{9.3.5}$$

Definition 9.6. Fourier cosine transform is defined as

$$\mathscr{F}_{\mathscr{C}}[f(t)] = \int_0^\infty f(t)\cos(\omega t)dt = F_c(\omega). \tag{9.3.6}$$

Since the latter two transforms are not directly supported in MATLAB Symbolic Math Toolbox, integrals can be used to compute them directly. Examples will be shown later how to compute sine and cosine Fourier transforms.

Example 9.17. Compute cosine Fourier transforms for $f(t) = t^n e^{-at}$, $a > 0$, $n = 1, 2, \ldots, 8$.

Solutions. Loop structure can be used to compute cosine Fourier transform, for different values of i, and Fourier cosine transforms can be obtained with direct integration. Simplified results can be obtained as shown in Table 9.1.

```
>> syms t; syms w real; syms a positive % declare symbolic variables
   for i=1:8,
       f=t^i*exp(-a*t); F=simplify(int(f*cos(w*t),t,0,inf)),
   end
```

In fact, according to the formulas given in the Mathematics Handbook,[14] for integer n, one has

$$\mathscr{F}_{\mathscr{C}}[t^n e^{-at}] = n!\left(\frac{a}{a^2+\omega^2}\right)^{n+1}\sum_{m=0}^{[n/2]}(-1)^m C_{n+1}^{2m+1}\left(\frac{\omega}{a}\right)^{2m+1}. \tag{9.3.7}$$

Table 9.1: Fourier cosine transforms for different n.

n	$\mathscr{F}_{\mathscr{C}}[f(t)]$
1~4	$\dfrac{a^2 - \omega^2}{(a^2 + \omega^2)^2}, \quad -2\dfrac{(-a^2 + 3\omega^2)a}{(a^2 + \omega^2)^3}, \quad 6\dfrac{(-a^2 + 2a\omega + \omega^2)(-a^2 - 2a\omega + \omega^2)}{(a^2 + \omega^2)^4},$
4,5	$24\dfrac{a(a^4 - 10a^2\omega^2 + 5\omega^4)}{(a^2 + \omega^2)^5}, \quad -120\dfrac{(-a + \omega)(a + \omega)(a^2 - 4\omega a + \omega^2)(a^2 + 4\omega a + \omega^2)}{(a^2 + \omega^2)^6},$
6	$-720\dfrac{(-a^6 + 21a^4\omega^2 - 35\omega^4a^2 + 7\omega^6)a}{(a^2 + \omega^2)^7}$
7	$5\,040\dfrac{(a^4 + 4a^3\omega - 6a^2\omega^2 - 4a\omega^3 + \omega^4)(a^4 - 4a^3\omega - 6a^2\omega^2 + 4a\omega^3 + \omega^4)}{(a^2 + \omega^2)^8}$
8	$40\,320\dfrac{a(-a^2 + 3\omega^2)(-a^6 + 33a^4\omega^2 - 27a^2\omega^4 + 3\omega^6)}{(a^2 + \omega^2)^9}$

Example 9.18. Compute Fourier cosine transform for the piecewise function

$$f(t) = \begin{cases} \cos t, & 0 < x < a, \\ 0, & \text{otherwise.} \end{cases}$$

Solutions. It can be seen that Fourier cosine transform can be computed from the definition that, the integrand in (9.3.6) is zero in the interval $t \in (a, \infty)$. Therefore the integral is also zero. The whole integral can be evaluated in the interval $t \in (0, a)$. The following statements can be issued to evaluate Fourier cosine transform

```
>> syms t w; syms a positive; f=cos(t);
   F=simplify(int(f*cos(w*t),t,0,a))
```

the result is also a piecewise function

$$F(\omega) = \begin{cases} a/2 + \sin 2a/4, & \omega \in \{-1, 1\}, \\ \sin[a(\omega - 1)]/[2(\omega - 1)] + \sin[a(\omega + 1)]/[2(\omega + 1)], & \omega \notin \{-1, 1\}, \\ 0, & a = \pi/2 \,\&\, \omega = 3. \end{cases}$$

If f is described by a piecewise function, the same results can be obtained.

```
>> f=piecewise(t<a & t>=0,cos(t), t>=a,0);
   F=int(f*cos(w*t),t,0,inf)
```

9.3.4 Discrete Fourier sine and cosine transforms

Discrete Fourier sine and cosine transforms are also known as finite Fourier sine and cosine transforms. Compared with Fourier sine and cosine transforms discussed earlier, the integral interval is changed from $t \in (0, \infty)$ into $t \in (0, a)$.

Definition 9.7. Finite Fourier sine and cosine transforms are defined as

$$F_s(k) = \int_0^a f(t) \sin \frac{k\pi t}{a} dt, \quad F_c(k) = \int_0^a f(t) \cos \frac{k\pi t}{a} dt. \qquad (9.3.8)$$

Definition 9.8. The inverse finite Fourier sine and cosine transforms are defined as

$$f(t) = \frac{2}{a} \sum_{k=1}^{\infty} F_s(k) \sin \frac{k\pi t}{a}, \quad f(t) = \frac{1}{a} F_c(0) + \frac{2}{a} \sum_{k=1}^{\infty} F_c(k) \cos \frac{k\pi t}{a}. \qquad (9.3.9)$$

The inverse transform is no longer an integral. Instead, it is an infinite series expression. Examples will be introduced to show how to solve discrete Fourier transform problems.

Example 9.19. For the following piecewise function, with $a > 0$, compute its discrete Fourier sine transform

$$f(t) = \begin{cases} t, & t \leqslant a/2, \\ a - t, & t > a/2. \end{cases}$$

Solutions. The finite discrete Fourier sine transform can be evaluated directly with the following MATLAB statements:

```
>> syms t k; assumeAlso(k,'integer');
   syms a positive, f1=t; f2=a-t;
   Fs=int(f1*sin(k*pi*t/a),t,0,a/2)+int(f2*sin(k*pi*t/a),t,a/2,a);
   simplify(Fs)
```

and the result is $((-1)^{k/2} a^2 ((-1)^{k+1/2} - j))/(k^2 \pi^2)$.

If a piecewise function is used to describe f, the same results can be obtained.

```
>> f=piecewise(t<=a/2,t, t>a/2,a-t); % piecewise function input
   Fs=simplify(int(f*sin(k*pi*t/a),t,0,a)) % Fourier sine transform
```

9.3.5 Fast Fourier transform

It can be seen that only some of the simple functions have analytical Fourier transforms. For most of the functions, numerical solutions to Fourier transform problems should be introduced. Discrete Fourier transforms are very simple examples of numerical Fourier transforms.

Definition 9.9. For the discrete samples x_i, $i = 1, 2, \ldots, N$, discrete Fourier transform is defined as

$$X(k) = \sum_{i=1}^{N} x_i e^{-2\pi j(k-1)(i-1)/N}, \quad \text{where } 1 \leqslant k \leqslant N. \qquad (9.3.10)$$

Definition 9.10. Discrete inverse Fourier transform is defined as

$$x(k) = \frac{1}{N} \sum_{i=1}^{N} X(i)e^{2\pi j(k-1)(i-1)/N}, \quad \text{where } 1 \leqslant k \leqslant N. \tag{9.3.11}$$

The fast Fourier transform (FFT) technique can be used to evaluate discrete Fourier transform in an efficient way. It is so far the most practical and universal way to evaluate the transforms. Function fft() is provided in MATLAB as a built-in function, with the syntax f=fft(x) for direct FFT, and $\hat{x}$=ifft(f) for inverse FFT. Another important property of fft() function is that the vector x can be of any length, it is not necessary to satisfy the $2^n - 1$ constraint.

Example 9.20. Assume that a function is expressed as

$$x(t) = 12\sin(2\pi \cdot t + \pi/4) + 5\cos(2\pi \cdot 4t).$$

For the step-size h selected, FFT can be taken, and the results can be shown as a magnitude curve. For the obtained FFT, take inverse FFT and see whether the original signal can be restored.

Solutions. For the step-size h, also known as the sampling period, the time interval is $t \in (0, t_n)$, and a total of L samples t_i are generated. Compute the function values x_i. Also the corresponding frequency can be computed as $f_0 = 1/(ht_n), 2f_0, 3f_0, \ldots$ The following statements can be used to compute FFT.

```
>> h=0.01; t=0:h:10; x=12*sin(2*pi*t+pi/4)+5*cos(2*pi*4*t);
   X=fft(x); f=t/h/10; N=floor(length(f)/2);
   stem(f(1:N),abs(X(1:N))), xlim([0,10])
```

The relationship between FFT magnitude and frequency can be obtained as shown in Figure 9.6.

Here, half of the data can be used to draw the FFT display, in order to avoid aliasing phenomenon. It can be seen that there are two peaks in the magnitude plot, at 1 and 4 Hz, respectively, which are in fact the two frequencies in the original function.

Inverse FFT can be evaluated with ifft(), and it can be seen that the original signal can be restored, with the norm of the error vector being $e = 1.029 \times 10^{-13}$.

```
>> ix=real(ifft(X)); plot(t,x,t,ix,':'); xlim([0,1]); e=norm(x-ix)
```

The inverse FFT thus obtained is exactly the same as that in Figure 9.7, meaning the original function can be fully restored. Since the samples are sparsely distributed, the curves obtained are not quite smooth.

Two-dimensional or even multidimensional FFT and inverse FFT can be evaluated with the functions fft2(), ifft2(), fftn(), and ifftn().

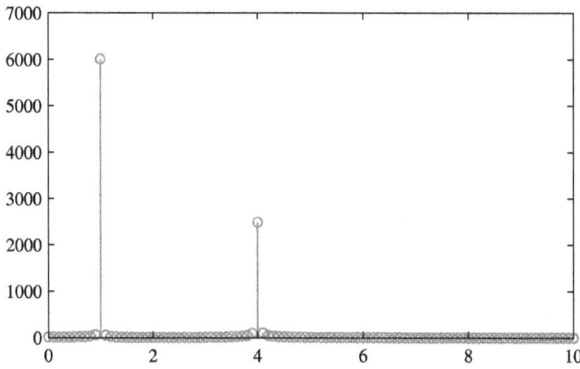

Figure 9.6: FFT analysis of the data.

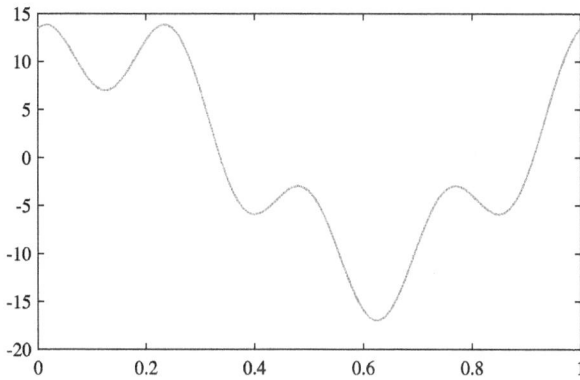

Figure 9.7: Inverse FFT results.

9.4 Computation of other integral transforms

Apart from Laplace and various Fourier transforms, other integral transforms, such as Mellin and Hankel transforms, are also commonly used. Unfortunately, these transforms cannot be well evaluated with MATLAB Symbolic Math Toolbox. Direct integrals can be tried, while numerical transforms may also be used in evaluating the transforms.

9.4.1 Mellin transform

Definition 9.11. Mellin transform is defined as

$$\mathscr{M}[f(x)] = \int_0^\infty f(x)x^{z-1}\mathrm{d}x = M(z). \tag{9.4.1}$$

Definition 9.12. Inverse Mellin transform is defined as

$$f(x) = \mathcal{M}^{-1}[M(z)] = \frac{1}{2\pi j} \int_{c-j\infty}^{c+j\infty} M(z)x^{-z}dz. \qquad (9.4.2)$$

There is no MATLAB function for Mellin transforms. Direct integrals can be tried from definition to find the expected Mellin transforms or their inverses. Examples will be given to show how to evaluate Mellin transforms.

Example 9.21. Consider the time domain function $f(t) = \ln t/(t + a)$, where $a > 0$. Compute its Mellin transform.

Solutions. One can try to apply the definition with the following statements, and Mellin transform can be found.

```
>> syms t z; syms a positive;
   f=log(t)/(t+a); M=simplify(int(f*t^(z-1),t,0,inf))
```

Simplification can be carried out to find the result

$$\mathcal{M}[f(t)] = a^{z-1}\pi(\ln a \sin \pi z - \pi \cos \pi z) \csc^2 \pi z.$$

Example 9.22. For the given function $f(t) = 1/(t + a)^n$ ($a > 0$), compute the Mellin transforms for different n, and check whether some conclusion can be made for different n.

Solutions. Letting $n = 1, 2, \ldots, 8$, Mellin transform can be evaluated with

```
>> syms t z; syms a positive % declare symbolic variables
   for i=1:8, f=1/(t+a)^i; disp(int(f*t^(z-1),t,0,inf)), end
```

It can be found from the loop structure that the transforms are:

$$a^{z-1}\pi \csc \pi z,$$

$$- a^{-2+z}\pi(z - 1) \csc \pi z,$$

$$1/2a^{-3+z}\pi(-2 + z)(z - 1) \csc \pi z,$$

$$- 1/6a^{z-4}\pi(z - 1)(-2 + z)(-3 + z) \csc \pi z,$$

$$1/24a^{-5+z}\pi(z - 4)(-3 + z)(-2 + z)(z - 1) \csc \pi z,$$

$$- 1/120a^{-6+z}\pi(-5 + z)(z - 4)(-3 + z)(-2 + z)(z - 1) \csc \pi z,$$

$$1/720a^{-7+z}\pi(-6 + z)(-5 + z)(z - 4)(-3 + z)(-2 + z)(z - 1) \csc \pi z,$$

$$- 1/5\,040a^{-8+z}\pi(-7 + z)(-6 + z)(-5 + z)(z - 4)(-3 + z)(-2 + z)(z - 1) \csc \pi z.$$

Therefore it can be concluded that the Mellin transform for arbitrary n is

$$\mathcal{M}\left[\frac{1}{(t+a)^n}\right] = \frac{(-1)^{k-1}\pi}{(n-1)!}a^{z-n}\prod_{i=1}^{n-1}(z-i)\csc\pi z.$$

Like with other integral transforms, most functions do not have their analytical Mellin transforms. Numerical integrals can be considered to compute them. A universal function is written below to compute the Mellin integral for any given function with the syntax

F=mellin_trans $(f,z,$ parameter pairs$)$

where "parameter pairs" are exactly the same as those for function integral().

```
function F=mellin_trans(f,z,varargin)
f1=@(x)f(x).*x.^(z-1); % anonymous description for the integrand
F=integral(f1,0,Inf,'ArrayValued',true,varargin{:}); % numerical integral
```

Example 9.23. Compute the Mellin integral of the function $f(x) = \sin(3x^{0.8})/(x+2)^{1.5}$ with numerical methods.

Solutions. It is not possible to find the analytical solution to the Mellin transform of the original function. Numerical solution becomes the only choice. The original function can be described by an anonymous function, and the Mellin transform can be evaluated, with the results shown in Figure 9.8.

```
>> f=@(x)sin(3*x.^0.8)./(x+2).^1.5;              % original function
   z=0:0.05:1; F=mellin_trans(f,z); plot(z,F) % draw the transform
```

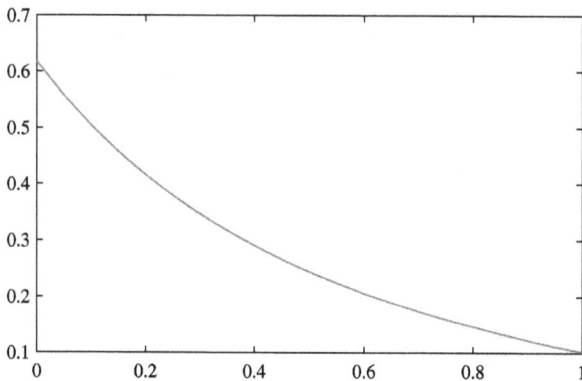

Figure 9.8: Numerical Mellin transform.

9.4.2 Computation of Hankel transforms

Definition 9.13. The vth order Hankel transform is defined as

$$\mathscr{H}[f(t)] = \int_0^\infty tf(t)J_v(\omega t)dt = H_v(\omega), \tag{9.4.3}$$

where $J_v(z)$ is the vth order Bessel function, which can be evaluated in MATLAB using J=besselj(v,z).

Definition 9.14. The vth order inverse Hankel transform is defined as

$$\mathscr{H}^{-1}[H(\omega)] = \int_0^\infty \omega H_v(\omega)J_v(\omega t)d\omega. \tag{9.4.4}$$

It can be seen from definition that Henkel transform is, in fact, an infinite integral, which can be evaluated analytically by int() function.

Example 9.24. Compute the zeroth order Hankel transform for the function $f(t) = e^{-a^2t^2/2}$.

Solutions. With the following MATLAB statements, Hankel transform of the original function can be evaluated directly

```
>> syms t w a positive; f=exp(-a^2*t^2/2); % original function
   F=int(f*t*besselj(0,w*t),t,0,inf); F=simplify(F) % compute transform
   f1=int(w*F*besselj(0,w*t),w,0,Inf)  % inverse Hankel transform
```

and the result obtained is $e^{-\omega^2/(2a^2)}/a^2$. With the above command, the original function can be restored through the inverse Hankel transform.

The direct integral approach can only be used to solve Hankel transform problems of very few functions, involving low-order Bessel functions such as $v = 0$. For most functions, numerical Hankel transforms should be employed. A universal numerical Hankel transform function is written as follows:

```
function H=hankel_trans(f,w,nu,varargin)
F=@(t)t.*f(t).*besselj(nu,w*t); % describing integrand for Hankel transform
H=integral(F,0,Inf,'ArrayValued',true,varargin{:}); % numerical integral
```

with the following syntax:

H=hankel_trans(f,w,v,parameter pairs)

where "parameter pairs" are exactly the same as in function integral().

Example 9.25. Consider again the previous example. If $a = 2$, draw Hankel transform curves of different orders.

Solutions. The numerical solution to the problem can easily be obtained with the following statements, as shown in Figure 9.9. The analytical result for $v = 0$ is also superimposed on the plot, which validates the numerical solution.

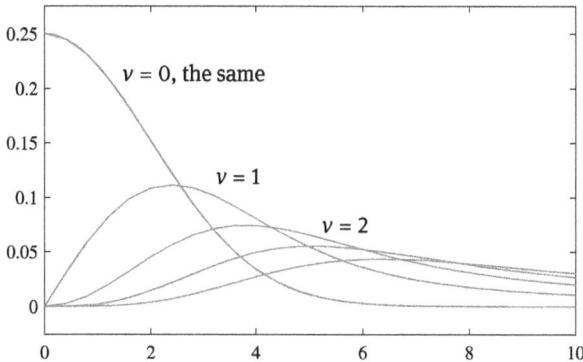

Figure 9.9: Numerical Hankel transforms of different orders.

Besides, it can be seen that in high-order Hankel transforms, the decay speed is rather low, therefore, larger interval should be used in the integral to get better results in the inverse Hankel transform evaluations. If function $F(\omega)$ is known, the following commands can be used to evaluate the inverse Hankel transforms.

```
>> syms t w a positive; f=exp(-a^2*t^2/2);   % original function
   F=int(f*t*besselj(0,w*t),t,0,inf); F=simplify(F) % Hankel transform
   F1=subs(F,a,2); ezplot(F1,[0,10]);          % theoretical
   a=2; f=@(t)exp(-a^2*t.^2/2); w=0:0.4:10; % input original function
   for i=0:4, H=hankel_trans(f,w,i); line(w,H); end % different orders
```

9.5 The *z* transform and its inverse

Strictly speaking, the *z* transform is not a genuine integral transform. Since its definition and properties are all very close to those of other integral transforms, the *z* transform is also presented in this chapter. In this section, definition and properties of the *z* transform will be given, followed by MATLAB solution strategies.

9.5.1 Definition and properties of z transforms

Definition 9.15. The z transform of a discrete signal $f(k)$, $k = 1, 2, \ldots$ is defined as

$$\mathscr{Z}[f(k)] = \sum_{k=0}^{\infty} f(k)z^{-k} = F(z). \tag{9.5.1}$$

Like for Laplace and Fourier transforms discussed earlier, there are also many properties of the z transform. They are listed below without further proofs.

Theorem 9.4. *The z transform has the following properties:*
(1) *Linearity. For scalars a and b, $\mathscr{Z}[af(k) \pm bg(k)] = a\mathscr{Z}[f(k)] \pm b\mathscr{Z}[g(k)]$.*
(2) *Backward translation. $\mathscr{Z}[f(k-n)] = z^{-n}F(z)$.*
(3) *Forward translation. For nonzero forward shifts, the z transform can be evaluated from*

$$\mathscr{Z}[f(k+n)] = z^n F(z) - \sum_{i=0}^{n-1} z^{n-i}f(i). \tag{9.5.2}$$

Specifically, for zero initial condition problems, $\mathscr{Z}[f(k+n)] = z^n F(z)$.
(4) *s-Domain scaling. $\mathscr{Z}[r^{-k}f(k)] = F(rz)$.*
(5) *Frequency domain differentiation. $\mathscr{Z}[kf(k)] = -zdF(z)/dz$.*
(6) *Frequency domain integrals. $\mathscr{Z}[f(k)/k] = \int_z^{\infty} \dfrac{F(\omega)}{\omega} d\omega$.*
(7) *Initial values. $\lim_{k\to 0} f(k) = \lim_{z\to\infty} F(z)$.*
(8) *Final values. If $F(z)$ has no poles outside of the unit circle, $\lim_{k\to\infty} f(k) = \lim_{z\to 1}(z-1)F(z)$.*
(9) *Convolution. $\mathscr{Z}[(f*g)(k)] = \mathscr{Z}[f(k)]\mathscr{Z}[g(k)]$, where the convolution operator $*$ is defined as*

$$(f*g)(k) = \sum_{l=0}^{\infty} f(k)g(k-l). \tag{9.5.3}$$

Definition 9.16. For the z transform $F(z)$, the inverse z transform is defined as

$$f(k) = \mathscr{Z}^{-1}[f(k)] = \frac{1}{2\pi j} \oint F(z)z^{k-1}dz. \tag{9.5.4}$$

9.5.2 Computer evaluation of z transforms

With the Symbolic Math Toolbox in MATLAB, the z transform can easily be evaluated with ztrans() function, while iztrans() function can be used to evaluate the inverse z transforms. The syntaxes of the two functions are:

F=ztrans(f,k,z), %z transform
F=iztrans(f,z,k), %inverse z transform with specified variables

If there is only one independent variable in the function, there is no need to further specify k or z in the function call.

Example 9.26. Evaluate the z transform of $f(kT) = akT - 2 + (akT + 2)e^{-akT}$.

Solutions. The z transform for the given signal can be obtained directly with the MATLAB statements

```
>> syms a T k; f=a*k*T-2+(a*k*T+2)*exp(-a*k*T);
   F=ztrans(f) %z transform
```

and the result obtained is

$$\mathscr{L}[f(kT)] = \frac{aTz}{(z-1)^2} - \frac{2z}{z-1} + \frac{aTze^{-aT}}{(z-e^{-aT})^2} + 2ze^{aT}\left(\frac{z}{e^{-aT}} - 1\right)^{-1}.$$

Example 9.27. Considering the z expression $F(z) = q/(z^{-1} - p)^m$, find its inverse z transforms for different m, and formulate a universal formula.

Solutions. One can select $m = 1, 2, \ldots, 8$, then with a loop structure, the inverse z transforms can be evaluated directly with

```
>> syms p q z;
   for i=1:8, disp(simplify(iztrans(q/(1/z-p)^i))), end
```

For different values of m, the following results can be obtained:

$$-q/p(1/p)^n,$$
$$q/p^2(1+n)(1/p)^n,$$
$$-1/2q(1/p)^n(1+n)(2+n)/p^3,$$
$$1/6q(1/p)^n(3+n)(2+n)(1+n)/p^4,$$
$$-1/24q(1/p)^n(4+n)(3+n)(2+n)(1+n)/p^5,$$
$$1/120q(1/p)^n(5+n)(4+n)(3+n)(2+n)(1+n)/p^6,$$
$$-1/720q(1/p)^n(6+n)(5+n)(4+n)(3+n)(2+n)(1+n)/p^7,$$
$$1/5\,040q(1/p)^n(7+n)(6+n)(5+n)(4+n)(3+n)(2+n)(1+n)/p^8.$$

From the above results, it can be concluded that the inverse z transform is

$$\mathscr{L}^{-1}\left[\frac{q}{(z^{-1}-p)^m}\right] = \frac{(-1)^m q}{(m-1)!p^{n+m}}\prod_{i=1}^{m-1}(n+i).$$

Example 9.28. Compute the z transform for the Laplace expression

$$F(s) = \frac{bs + c}{s^2(s + a)}.$$

Solutions. Since $F(s)$ is the Laplace transform, function `ilaplace()` can be called first to get the time domain expression, then `ztrans()` function can be used to apply the z transform

```
>> syms a b c s; F=(b*s+c)/s^2/(s+a);
   f=ilaplace(F); F1=simplify(ztrans(f))
```

It can be seen that the z transform of the original function becomes

$$F_1 = \frac{cz}{a(z-1)^2} + \frac{(c-ab)z}{a^2(z-e^{-a})} - \frac{(c-ab)z}{a^2(z-1)}.$$

9.5.3 Bilateral z transform

The typical z transform deals with a sequence signal for $n \geqslant 0$. Therefore it is referred to as a unilateral z transform. If n is extended to the whole set of integers, bilateral z transform can be defined.

Definition 9.17. Bilateral z transform is defined as

$$\mathscr{Z}[f(k)] = \sum_{k=-\infty}^{\infty} f(k)z^{-k} = F(z). \tag{9.5.5}$$

There is no existing MATLAB function for bilateral z transforms. It can be found from definition that bilateral z transform can easily be evaluated with MATLAB using

$F=\text{symsum}(f*z^{\wedge}(-k),k,0,\text{inf})+\text{symsum}(f*z^{\wedge}(-k),k,-\text{inf},-1)$

since `symsum()` function does not support directly the sum in the interval $(-\infty, \infty)$.

Example 9.29. Compute the z transform for the following piecewise function:[20]

$$f(n) = \begin{cases} 2^n, & n \geqslant 0, \\ -3^n, & n < 0. \end{cases}$$

Solutions. With the idea discussed earlier, the sum of the bilateral z transform can be divided into two parts and evaluated with the following statements. The result obtained is $F = z/(z-2) + z/(z-3)$.

```
>> syms z n;
   F=symsum(2^n*z^(-n),n,0,inf)+symsum(-3^n*z^(-n),n,-inf,-1)
```

9.5.4 Numerical inverse z transforms for rational functions

For many functions, the analytical expression of the inverse z transforms cannot be evaluated with `iztrans()` function. Therefore, numerical approaches should be explored. Here, the numerical inverse z transform can be evaluated for rational functions.

Definition 9.18. A typical z transform expression is given by

$$F(z^{-1}) = z^{-d} \frac{b_0 + b_1 z^{-1} + b_2 z^{-2} + \cdots + b_{m-1} z^{-(m-1)} + b_m z^{-m}}{a_0 + a_1 z^{-1} + a_2 z^{-2} + \cdots + a_{n-1} z^{-(n-1)} + a_n z^{-n}}, \tag{9.5.6}$$

and it can also be converted into a power series of z^{-k} as

$$F(z^{-1}) = f_0 + f_1 z^{-1} + f_2 z^{-2} + \cdots = \sum_{k=0}^{\infty} f_k z^{-k}. \tag{9.5.7}$$

The above equation is, in fact, the definition of the z transform. It can be seen that by applying the long division algorithm to $F(z^{-1})$, the numerical solution of its inverse can be obtained.

The long division algorithm can be implemented in MATLAB with the following syntax

```
num=[b₀,b₁,b₂,...,bₘ];  den=[a₀,a₁,a₂,...,aₙ];
y=inv_z(num,den,d,N)
```

where d is the number of delays, and N is the number of points with a default of $N = 10$. The inverse z transform can be returned in the sequence **y**.

```
function y=inv_z(num,den,d,N)
if nargin==2, d=0; end, if nargin<=3, N=10; end, num(N)=0;
for i=1:N-d, y(d+i)=num(1)/den(1); % loop structure
    if length(num)>1, ii=2:length(den);
        if length(den)>length(num); num(length(den))=0; end
        num(ii)=num(ii)-y(end)*den(ii); num(1)=[]; % remove the first term
end, end
```

Example 9.30. Find the inverse z transform for the rational function

$$G(z) = \frac{z^2 + 0.4}{z^5 - 4.1z^4 + 6.71z^3 - 5.481z^2 + 2.2356z - 0.3645}.$$

Solutions. Multiplying both the numerator and denominator by z^{-5}, the expected standard form can be obtained as

$$F(z^{-1}) = z^{-3} \frac{1 + 0.4z^{-2}}{1 - 4.1z^{-1} + 6.71z^{-2} - 5.481z^{-3} + 2.2356z^{-4} - 0.3645z^{-5}}.$$

Therefore, the following statements can be used to evaluate the inverse z transform, and the result obtained is shown in Figure 9.10.

```
>> num=[1 0 0.4]; den=[1 -4.1 6.71 -5.481 2.2356 -0.3645];
   N=50; y=inv_z(num,den,3,N); t=0:(N-1); stem(t,y) % long division
```

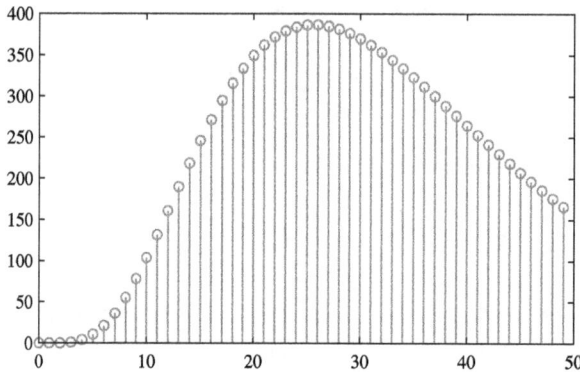

Figure 9.10: Numerical inverse z transform.

9.6 Exercises

9.1 Find Laplace transforms for the following functions:

(1) $f_a(t) = \sin at/t$, (2) $f_b(t) = t^5 \sin \alpha t$, (3) $f_c(t) = t^8 \cos \alpha t$,
(4) $f_d(t) = t^6 e^{\alpha t}$, (5) $f_e(t) = 5e^{-at} + t^4 e^{-at} + 8e^{-2t}$,
(6) $f_f(t) = e^{\beta t} \sin(\alpha t + \theta)$, (7) $f_g(t) = e^{-12t} + 6e^{9t}$.

9.2 Find the inverse Laplace transform for the above results and see whether the original function can be retained.

9.3 The formulas below are the properties of Laplace transform. Verify the two properties for the selected samples of n.

(1) $\mathscr{L}[t^n f(t)] = (-1)^n \dfrac{d^n \mathscr{L}[f(t)]}{ds^n}$, (2) $\mathscr{L}[t^{n-1/2}] = \dfrac{\sqrt{\pi}(2n-1)!}{2^n} s^{-n-1/2}$.

9.4 Compute inverse Laplace transforms for the following $F(s)$:

(1) $F_a(s) = \dfrac{1}{\sqrt{s}(s^2 - a^2)(s + b)}$, (2) $F_b(s) = \sqrt{s-a} - \sqrt{s-b}$,

(3) $F_c(s) = \ln\dfrac{s-a}{s-b}$, (4) $F_d(s) = \dfrac{1}{\sqrt{s}(s+a)}$, (5) $F_e(s) = \dfrac{3a^2}{s^3 + a^3}$,

(6) $F_f(s) = \dfrac{(s-1)^8}{s^7}$, (7) $F_g(s) = \ln\dfrac{s^2 + a^2}{s^2 + b^2}$,

(8) $F_h(s) = \dfrac{s^2 + 3s + 8}{\prod\limits_{i=1}^{8}(s + i)}$,   (9) $F_i(s) = \dfrac{1}{2}\dfrac{s + \alpha}{s - \alpha}$.

9.5   Show that the two Laplace transform properties hold:

(1) for any non-integer y, $\mathscr{L}[t^y] = \dfrac{\Gamma(y + 1)}{s^{y+1}}$,

(2) for any $a > 0$, $\mathscr{L}\left[\dfrac{1}{\sqrt{t}\,(1 + at)}\right] = \dfrac{\pi}{a}\,e^{s/a}\mathrm{erfc}(\sqrt{s/a})$.

9.6   One of the most important applications of Laplace transform is that it can be used in solving linear ordinary differential equations. If the initial values of the output and all its derivatives are zeros, $\mathscr{L}[d^n f(t)/dt^n] = s^n \mathscr{L}[f(t)]$ property can be used to map differential equations into algebraic equations. If the initial values are not zero, Laplace transform can also be used in solving differential equations. Solve the following differential equations, with Laplace transform properties:

(1) $y''(t) + 3y'(t) + 2y(t) = e^{-t}$, $y(0) = y'(0) = 0$,

(2) $y'' - y = 4\sin t + 5\cos 2t$, $y(0) = -1$, $y'(0) = -2$,

(3) $\begin{cases} x'' - x + y + z = 0, \\ x + y'' - y + z = 0, \\ x + y + z'' - z = 0, \end{cases}$

$x(0) = 1$, $x'(0) = y(0) = y'(0) = z(0) = z'(0) = 0$.

9.7   Assume that a fractional-order system is composed of two blocks, $G_1(s)$ and $G_2(s)$, in parallel connection. The overall model can be expressed as $G(s) = G_1(s) + G_2(s)$. Draw the step response of the overall system if

$$G_1(s) = \frac{(s^{0.4} + 2)^{0.8}}{\sqrt{s}(s^2 + 3s^{0.9} + 4)^{0.3}}, \quad G_2(s) = \frac{s^{0.4} + 0.6s + 3}{(s^{0.5} + 3s^{0.4} + 5)^{0.7}}.$$

9.8   If blocks $G_1(s)$ and $G_2(s)$ are connected in series, the overall system can be expressed as $G(s) = G_2(s)G_1(s)$. Draw the step response of the overall system.

9.9   For the functions given below, compute their Fourier transforms. Then for the results, compute their inverse Fourier transforms and see whether the original functions can be restored.

(1) $f(x) = x^2(3\pi - 2|x|)$, $0 \leqslant x \leqslant 2\pi$,   (2) $f(t) = t^2(t - 2\pi)^2$, $0 \leqslant t \leqslant 2\pi$,

(3) $f(t) = e^{-t^2}$, $-l \leqslant t \leqslant l$,   (4) $f(t) = te^{-|t|}$, $-\pi \leqslant t \leqslant \pi$.

9.10 Compute Fourier sine and cosine transforms for the following functions, and then take corresponding inverse transforms, and see whether the original functions can be restored.

(1) $f(t) = e^{-t}\ln t$,   (2) $f(x) = \dfrac{\cos x^2}{x}$,   (3) $f(x) = \ln\dfrac{1}{\sqrt{1 + x^2}}$,

(4) for any $a > 0$, $f(x) = x(a^2 - x^2)$,   (5) $f(x) = \cos kx$.

9.11 Find the discrete Fourier sine and cosine transforms for the following functions:
(1) $f(x) = e^{kx}$, (2) $f(x) = x^3$.

9.12 Compute Mellin transform for the following piecewise function:

$$f(x) = \begin{cases} \sin(a \ln x), & x \leqslant 1, \\ 0, & \text{otherwise.} \end{cases}$$

9.13 Compute the z transform for the following sequences $f(kT)$, then take the inverse z transform and see whether the original sequences can be restored:

(1) $f_a(kT) = \cos(kaT)$, (2) $f_b(kT) = (kT)^2 e^{-akT}$,

(3) $f_c(kT) = \dfrac{1}{a}(akT - 1 + e^{-akT})$, (4) $f_d(kT) = e^{-akT} - e^{-bkT}$,

(5) $f_e(kT) = \sin(akT)$, (6) $f_f(kT) = 1 - e^{-akT}(1 + akT)$.

9.14 For the following z transform expressions $F(z)$, take their inverse z transforms:

(1) $F_a(z) = \dfrac{10z}{(z-1)(z-2)}$, (2) $F_b(z) = \dfrac{z^2}{(z-0.8)(z-0.1)}$,

(3) $F_c(z) = \dfrac{z}{(z-a)(z-1)^2}$, (4) $F_d(z) = \dfrac{z^{-1}(1-e^{-aT})}{(1-z^{-1})(1-z^{-1}e^{-aT})}$,

(5) $F_e(z) = \dfrac{Az[z \cos \beta - \cos(\alpha T - \beta)]}{z^2 - 2z \cos(\alpha T) + 1}$.

9.15 Compute the z transforms for the given Laplace expressions and validate the results:

(1) $G(s) = \dfrac{b}{s^2(s+a)}$, (2) $G(s) = \dfrac{b}{s^2(s+a)^2} \cdot \dfrac{1 - e^{-2s}}{s}$.

9.16 For the given function $G(s) = 1/(s+1)^3$, if $s = 2(z-1)/[T(z+1)]$ is used to substitute into $G(s)$, the function $H(z)$ can be obtained. Such a transform is referred to as a bilinear transform. If $T = 1/2$, compute $H(z)$. If a new transform $z = (1+Ts/2)/(1-Ts/2)$ is used, can the original function be restored?

9.17 Show with MATLAB that

$$\mathscr{L}\left\{1 - e^{-akT}\left[\cos(bkT) + \frac{a}{b}\sin(bkT)\right]\right\} = \frac{z(Az + B)}{(z-1)(z^2 - 2e^{-aT}\cos(bT)z + e^{-2aT})},$$

where

$$A = 1 - e^{-aT}\cos(bT) - \frac{a}{b}e^{-aT}\sin(bT) \quad \text{and}$$

$$B = e^{-2aT} + \frac{a}{b}e^{-aT}\sin(bT) - e^{-aT}\cos(bT).$$

10 Introduction to fractional calculus

Calculus problems were presented extensively in the previous chapters, and MATLAB-based solution strategies were given for finding their analytical and numerical solutions. In the traditional calculus courses, the orders of derivatives and integrals are all assumed to be integers. Hence the traditional calculus we are familiar with is referred to as the integer-order calculus.

It is known that the notation $\mathrm{d}^n y / \mathrm{d}x^n$ represents the nth order derivative of y with respect to x. What if $n = 1/2$? This was the question asked 300 years ago by French mathematician Guillaume François Antoine L'Hôpital to Leibniz,[25] one of the inventors of calculus. Leibniz replied, "Thus it follows that $\mathrm{d}^{1/2}x$ will be equal to $x\sqrt[2]{\mathrm{d}x : x}$, an apparent paradox, from which one day useful consequences will be drawn". This was recognized as the starting point of a branch called fractional calculus.

Strictly speaking, "fractional-order" is a misused word, accurate ones should be "non-integer-order", or "arbitrary-order", since $\sqrt{2}$ can be used as the order, while it is not a fractional number. While the key word "fractional" is widely used in fractional calculus communities, this key word is used throughout the book to really mean "arbitrary real order".

Fractional calculus is a mathematics branch with more than 300 years of history. Earlier research in this area was all pure theoretical research. It is not until the recent few decades that the application research on fractional calculus was introduced to a variety of fields. For instance, in control systems, the fractional-order control is an active research field. In this chapter, the definitions of fractional calculus are presented first. Then we concentrate on how to find high precision numerical solutions to fractional calculus problems.

A dedicated MATLAB toolbox – FOTF Toolbox[29] – was written by the author, which can be used in dealing with fractional calculus and fractional-order control systems. The toolbox can be downloaded directly from[30]

```
http://cn.mathworks.com/matlabcentral/fileexchange/60874-fotf-toolbox
```

A systematic presentation of fractional-order derivatives and integrals is given this chapter. In Section 10.1, various definitions are given, followed by the commonly used properties in Section 10.2. In Section 10.3, numerical algorithms and implementation of Grünwald–Letnikov definition are proposed, with high precision MATLAB functions. In Section 10.4, another important definition – Caputo definition – is fully addressed, with high precision numerical algorithms and MATLAB solvers. For signals whose mathematical expressions are not known before computation, Oustaloup filter approximation to their fractional-order derivatives and integrals is addressed in Section 10.5.

https://doi.org/10.1515/9783110666977-010

10.1 Definitions in fractional calculus

In the development of fractional calculus, there were quite a few definitions. Some of the widely used definitions are Cauchy integral formula, Grünwald–Letnikov, Riemann–Liouville, and Caputo definitions, and so on. The definitions and properties in fractional calculus are presented in this section.

10.1.1 Why fractional calculus?

Integer-order calculus is well established, and already widely used in applications. It is quite natural to ask the question: If the world can be modeled well with integer-order calculus, why we bother to introduce fractional calculus? The following example should be given to show the benefit of fractional calculus.

Example 10.1. Consider a sinusoidal signal $\sin t$. It is known that the first-order derivative of the signal is $\cos t$. Taking high-order derivatives of the signal, the results have only four variations: $\pm \sin t$ and $\pm \cos t$. Other signals cannot be obtained. What will happen if fractional calculus is introduced?

Solutions. It is known from Cauchy's integral formula that

$$\frac{d^n}{dt^n} \sin t = \sin\left(t + n\frac{\pi}{2}\right).$$

In fact, the above formula holds even if n is an arbitrary real number. Therefore, with the following statements, the derivatives of the function for different orders can be obtained, as shown in Figure 10.1.

```
>> n0=0:0.1:1.5; t=0:0.2:2*pi; Z=[];
   for n=n0, Z=[Z; sin(t+n*pi/2)]; end, surf(t,n0,Z)
```

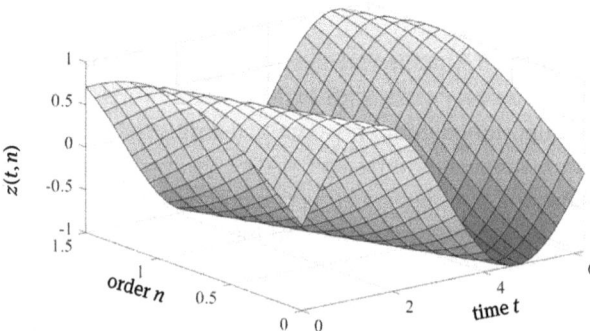

Figure 10.1: Derivative surface of different orders.

It can be seen that, apart from the four functions $\pm \sin t$ and $\pm \cos t$, more information can also be witnessed. Gradual changes are provided in fractional calculus, indicating that fractional calculus may provide more information than ordinary integer-order calculus. If the viewpoint of fractional calculus is employed, more hidden information invisible from the viewpoint of integer-order calculus may be revealed.

10.1.2 Definitions

Various definitions exist in the area of fractional calculus. Some of the commonly used ones are given below.

Definition 10.1 (Fractional Cauchy integral formula). The formula is directly extended from integer-order calculus as

$$\mathscr{D}^{\gamma} f(t) = \frac{\Gamma(\gamma + 1)}{2\pi j} \oint_{C} \frac{f(\tau)}{(\tau - t)^{\gamma+1}} d\tau, \tag{10.1.1}$$

where C is a smooth closed-path such that $f(t)$ is single-valued and analytic, and $\Gamma(\cdot)$ is the Gamma function.

Definition 10.2 (Riemann–Liouville definition). Fractional integral is defined as

$$_{t_0}^{RL}\mathscr{D}_t^{-\alpha} f(t) = \frac{1}{\Gamma(\alpha)} \int_{t_0}^{t} \frac{f(\tau)}{(t - \tau)^{1-\alpha}} d\tau, \tag{10.1.2}$$

where $\alpha > 0$ and t_0 is the initial instance.

If $t_0 = 0$, the notation can be simplified as $^{RL}\mathscr{D}_t^{-\alpha} f(t)$. If there is no conflict, the subscript RL can also be omitted. The subscripts on the two sides of the symbol $\mathscr{D}$ denote the lower and upper bounds in the integral.[11]

Riemann–Liouville derivatives can also be defined.

Definition 10.3. If $m = \lceil \alpha \rceil$, then the fractional Riemann–Liouville derivative is defined as

$$_{t_0}^{RL}\mathscr{D}_t^{\alpha} f(t) = \frac{1}{\Gamma(m - \alpha)} \frac{d^m}{dt^m} \left[\int_{t_0}^{t} \frac{f(\tau)}{(t - \tau)^{1+\alpha-m}} d\tau \right]. \tag{10.1.3}$$

Definition 10.4. Grünwald–Letnikov definition. The fractional Grünwald–Letnikov derivative is

$$_{t_0}^{GL}\mathscr{D}_t^{\alpha} f(t) = \lim_{h \to 0} \frac{1}{h^{\alpha}} \sum_{j=0}^{[(t-t_0)/h]} (-1)^j \binom{\alpha}{j} f(t - jh), \tag{10.1.4}$$

where t_0 is the initial instance, and it is assumed that when $t < t_0, f(t) \equiv 0$, and $(-1)^j \binom{\alpha}{j}$ is the binomial coefficient.

Definition 10.5. Caputo fractional derivative. The Caputo fractional derivative is defined as

$$
{}_{t_0}^{C}\mathscr{D}_t^{\alpha} f(t) = \frac{1}{\Gamma(m-\alpha)} \int_{t_0}^{t} \frac{f^{(m)}(\tau)}{(t-\tau)^{1+\alpha-m}} d\tau, \tag{10.1.5}
$$

where $m = \lceil \alpha \rceil$ is an integer.

The definition of Caputo fractional integral is the same as that of Riemann–Liouville fractional integral, namely

$$
{}_{t_0}^{C}\mathscr{D}_t^{-\alpha} f(t) = \frac{1}{\Gamma(\alpha)} \int_{t_0}^{t} \frac{f(\tau)}{(t-\tau)^{1-\alpha}} d\tau = {}_{t_0}^{RL}\mathscr{D}_t^{-\alpha} f(t), \quad \alpha > 0. \tag{10.1.6}
$$

10.2 Properties and relationships among different fractional calculus definitions

It can be shown that,[18] for a wide class of practical functions, Grünwald–Letnikov definition is equivalent to Riemann–Liouville definition. These two definitions are not distinguished in the book. The difference between Caputo and Riemann–Liouville definitions lies in the way the initial values are treated.

If the initial value of $y(t)$ is not zero, and $\alpha \in (0,1)$, it can be found by comparing the Caputo and Riemann–Liouville definitions that

$$
{}_{t_0}^{C}\mathscr{D}_t^{\alpha} f(t) = {}_{t_0}^{RL}\mathscr{D}_t^{\alpha}(f(t) - f(t_0)), \tag{10.2.1}
$$

where the derivative of constant $f(t_0)$ is

$$
{}_{t_0}^{RL}\mathscr{D}_t^{\alpha} f(t_0) = \frac{f(t_0)}{\Gamma(1-\alpha)}(t-t_0)^{-\alpha}. \tag{10.2.2}
$$

Therefore, the relationship between Caputo and Riemann–Liouville derivatives is

$$
{}_{t_0}^{C}\mathscr{D}_t^{\alpha} f(t) = {}_{t_0}^{RL}\mathscr{D}_t^{\alpha} f(t) - \frac{f(t_0)}{\Gamma(1-\alpha)}(t-t_0)^{-\alpha}. \tag{10.2.3}
$$

Theorem 10.1. *More generally, if $\alpha > 1$, denote $m = \lceil \alpha \rceil$, then*

$$
{}_{t_0}^{C}\mathscr{D}_t^{\alpha} f(t) = {}_{t_0}^{RL}\mathscr{D}_t^{\alpha} f(t) - \sum_{k=0}^{m-1} \frac{f^{(k)}(t_0)}{\Gamma(k-\alpha+1)}(t-t_0)^{k-\alpha}. \tag{10.2.4}
$$

It can be seen that the relationship for $0 \leqslant \alpha \leqslant 1$ is only a special case of (10.2.4).

If $\alpha < 0$, it was pointed out that Riemann–Liouville and Caputo integrals are identical. Therefore there is no need to distinguish them.

Theorem 10.2. *The following properties are summarized without proof:*[17]

(1) *The fractional-order derivative ${}_{t_0}\mathscr{D}_t^\alpha f(t)$ of an analytic function $f(t)$ is also analytic as a function of t and α.*

(2) *When $\alpha = n$ is an integer, the fractional-order derivatives are the same as the corresponding integer-order derivatives, and ${}_{t_0}\mathscr{D}_t^0 f(t) = f(t)$.*

(3) *Fractional-order operators are linear, i. e., for any constants a and b,*

$$ {}_{t_0}\mathscr{D}_t^\alpha [af(t) + bg(t)] = a \, {}_{t_0}\mathscr{D}_t^\alpha f(t) + b \, {}_{t_0}\mathscr{D}_t^\alpha g(t). \tag{10.2.5} $$

(4) *For a function $f(t)$ with zero initial conditions, the fractional-order operators satisfy*

$$ {}_{t_0}\mathscr{D}_t^\alpha [{}_{t_0}\mathscr{D}_t^\beta f(t)] = {}_{t_0}\mathscr{D}_t^\beta [{}_{t_0}\mathscr{D}_t^\alpha f(t)] = {}_{t_0}\mathscr{D}_t^{\alpha+\beta} f(t). \tag{10.2.6} $$

Theorem 10.3. *Laplace transform of fractional-order integrals can be evaluated from*

$$ \mathscr{L}[\mathscr{D}_t^{-\gamma} f(t)] = s^{-\gamma} \mathscr{L}[f(t)]. \tag{10.2.7} $$

Theorem 10.4. *Under Riemann–Liouville definition, the Laplace transform of fractional-order derivatives can be evaluated from*

$$ \mathscr{L}[{}_{t_0}^{\mathrm{RL}}\mathscr{D}_t^\alpha f(t)] = s^\alpha \mathscr{L}[f(t)] - \sum_{k=1}^{n-1} s^k \, {}_{t_0}^{\mathrm{RL}}\mathscr{D}_t^{\alpha-k-1} f(t)|_{t=t_0}. \tag{10.2.8} $$

More specifically, if the initial values of $f(t)$ and its fractional-order derivatives are all zero, then

$$ \mathscr{L}[{}_{t_0}\mathscr{D}_t^\alpha f(t)] = s^\alpha \mathscr{L}[f(t)]. \tag{10.2.9} $$

The Laplace transforms of Caputo and Riemann–Liouville integrals are exactly the same.

Theorem 10.5. *The Laplace transform of the Caputo fractional-order derivative satisfies*

$$ \mathscr{L}[{}_{t_0}^{\mathrm{C}}\mathscr{D}_t^\gamma f(t)] = s^\gamma F(s) - \sum_{k=0}^{n-1} s^{\gamma-k-1} f^{(k)}(t_0). \tag{10.2.10} $$

It can be seen from Laplace transform properties that Laplace transforms of Caputo derivatives only involve the initial values of integer-order derivatives, while Riemann–Liouville derivatives involve initial values of fractional-order derivatives, which is difficult to provide in real applications. For fractional-order systems with nonzero initial conditions, Caputo definition is more suitable for describing dynamical systems.

10.3 Numerical implementation of Grünwald–Letnikov derivatives

10.3.1 Simple Grünwald–Letnikov definition evaluation

Since Grünwald–Letnikov and Riemann–Liouville definitions are identical for a wide class of functions, due to its apparent simplicity for numerical computation, the Grünwald–Letnikov definition can be used in numerical computation of fractional calculus. Considering the definition in (10.1.4), if the step-size h is small enough, the limit operation can be omitted. The following numerical computation formula is described below.

Definition 10.6. Numerical implementation of Grünwald–Letnikov definition is

$$_{t_0}\mathscr{D}_t^\alpha f(t) \approx \frac{1}{h^\alpha} \sum_{j=0}^{[(t-t_0)/h]} w_j^{(\alpha)} f(t - jh), \tag{10.3.1}$$

where $w_j^{(\alpha)}$ is the jth polynomial expansion coefficient of $(1 - z)^\alpha$. These coefficients can easily be obtained recursively from

$$w_0^{(\alpha)} = 1, \ w_j^{(\alpha)} = \left(1 - \frac{\alpha + 1}{j}\right) w_{j-1}^{(\alpha)}, \quad j = 1, 2, \dots \tag{10.3.2}$$

Assuming that the step-size h is small enough, (10.3.1) can be used to approximate directly the fractional-order derivatives of given functions. It can be shown that[18] the error bound is $o(h)$. With Grünwald–Letnikov definition, the following MATLAB function can be written to compute the fractional-order derivatives and integrals:

```
function dy=glfdiff(y,t,gam)
if strcmp(class(y),'function_handle'), y=y(t); end % function handle
h=t(2)-t(1); w=1; y=y(:); t=t(:);   % convert to column vectors
for j=2:length(t), w(j)=w(j-1)*(1-(gam+1)/(j-1)); end  % binomial
for i=1:length(t), dy(i)=w(1:i)*[y(i:-1:1)]/h^gam; end % derivatives
```

The syntax of the function is y_1=glfdiff(y,t,y), where t is equally spaced time vector, while y can either be samples of the signal, or the function handle of the signal. The argument y is the order of the derivative. The returned vector y_1 contains the samples of fractional-order derivatives. Here, y can also be negative for integrals.

10.3.2 High precision algorithms and implementation

Compared with the numerical derivative formula discussed earlier, high-precision generating function can be evaluated, and based on it, a high-precision general algorithm for evaluating Grünwald–Letnikov derivatives is proposed and implemented in MATLAB.

Theorem 10.6. *The generating function $g_p(z)$ with $o(h^p)$ precision can be expressed by the following polynomial:*[29]

$$g_p(z) = \sum_{k=0}^{p} g_k z^k, \tag{10.3.3}$$

where the coefficients g_k can be evaluated directly from the following equation, where the coefficient matrix is a Vandermonde matrix:

$$
\begin{bmatrix}
1 & 1 & 1 & \cdots & 1 \\
1 & 2 & 3 & \cdots & p+1 \\
1 & 2^2 & 3^2 & \cdots & (p+1)^2 \\
\vdots & \vdots & \vdots & \ddots & \vdots \\
1 & 2^p & 3^p & \cdots & (p+1)^p
\end{bmatrix}
\begin{bmatrix}
g_0 \\ g_1 \\ g_2 \\ \vdots \\ g_p
\end{bmatrix}
= -
\begin{bmatrix}
0 \\ 1 \\ 2 \\ \vdots \\ p
\end{bmatrix}. \tag{10.3.4}
$$

The generating function algorithm above can be implemented with

```
function g=genfunc(p)
a=1:p+1; A=rot90(vander(a)); g=(1-a)*inv(sym(A'));
```

The syntax of the function is g=genfunc(p), where p specifies the accuracy, and the returned symbolic vector g contains the generating function coefficients. To ensure the accuracy of the coefficients, symbolic computation is used.

Theorem 10.7. *If the fractional-order generating function $g_p^\alpha(z)$ can be written as*

$$g_p^\alpha(z) = (g_0 + g_1 z + \cdots + g_p z^p)^\alpha, \tag{10.3.5}$$

the Taylor series expansion can be written as

$$g_p^\alpha(z) = \sum_{k=0}^{\infty} w_k z^k. \tag{10.3.6}$$

Therefore, when $k = 0$, $w_0 = g_0^\alpha$, and $k > 0$, the subsequent coefficients w_k can be evaluated recursively from

$$w_k = -\frac{1}{g_0} \sum_{i=1}^{p} g_i\left(1 - i\frac{1+\alpha}{k}\right) w_{k-i}, \quad w_k = 0, \text{ if } k < 0. \tag{10.3.7}$$

If the generating function vector g is obtained, the following MATLAB function can be written, so as to compute the vector w. The syntax of the function is w=get_vecw(α,n,g), where α is the order and n is the number of terms.

```
function w=get_vecw(gam,n,g)
p=length(g)-1; b=1+gam; g0=g(1);
```

```
w=zeros(1,n); w(1)=g(1)^gam;
for m=2:p, M=m-1; dA=b/M;
    w(m)=-[g(2:m).*((1-dA):-dA:(1-b))]*w(M:-1:1).'/g0;
end
for k=p+1:n, M=k-1; dA=b/M;
    w(k)=-[g(2:(p+1)).*((1-dA):-dA:(1-p*dA))]*w(M:-1:(k-p)).'/g0;
end
```

Definition 10.7. High-precision Grünwald–Letnikov fractional-order derivatives and integrals can be approximated by

$$\overset{\overline{GL}}{_{t_0}}\mathscr{D}_t^\alpha f(t) = \frac{1}{h^\alpha} \sum_{k=0}^{[(t-t_0)/h]} w_k f(t - kh), \tag{10.3.8}$$

where the coefficient vector $w = [w_0, w_1, w_2, \ldots]$ can be evaluated with Lubich's linear multistep algorithm. The theoretical precision is $o(h^p)$.

Unfortunately, direct use of these formulas may lead to huge errors at the initial stage.[29] Compensations are needed, otherwise it can only be used to deal with functions with zero initial conditions, where the $p + 1$ quantities are all zero.

Theorem 10.8 ([29]). *An ordinary signal $y(t)$ can be decomposed as $y(t) = u(t) + v(t)$, where the first $p + 1$ samples of $v(t)$ are all zero, while $u(t)$ is a polynomial where $u(t) = \sum_{k=0}^{p} c_k(t - t_0)^k$, with*

$$\begin{bmatrix} 1 & 0 & 0 & \cdots & 0 \\ 1 & h & h^2 & \cdots & h^p \\ 1 & 2h & (2h)^2 & \cdots & (2h)^p \\ \vdots & \vdots & \vdots & \ddots & \vdots \\ 1 & ph & (ph)^2 & \cdots & (ph)^p \end{bmatrix} \begin{bmatrix} c_0 \\ c_1 \\ c_2 \\ \vdots \\ c_p \end{bmatrix} = \begin{bmatrix} y(t_0) \\ y(t_0 + h) \\ y(t_0 + 2h) \\ \vdots \\ y(t_0 + ph) \end{bmatrix}. \tag{10.3.9}$$

The unified analytical description of fractional-order derivatives and integrals of signal $u(t)$ can be represented as

$$\overset{RL}{_{t_0}}\mathscr{D}_t^\alpha (t - t_0)^k = \frac{\Gamma(k + 1)}{\Gamma(k + 1 - \alpha)}(t - t_0)^{k-\alpha}, \quad k = 0, 1, 2, \ldots \tag{10.3.10}$$

Based on the above theorem, the following MATLAB function can be written to compute high-precision evaluations of Grünwald–Letnikov derivatives and integrals

```
function dy=glfdiff9(y,t,gam,p)
if strcmp(class(y),'function_handle'), y=y(t); end, y=y(:);
h=t(2)-t(1); t=t(:); n=length(t); u=0; du=0; r=(0:p)*h;
R=sym(fliplr(vander(r))); c=double(inv(R)*y(1:p+1));
```

```
for i=1:p+1, u=u+c(i)*t.^(i-1);
    du=du+c(i)*t.^(i-1-gam)*gamma(i)/gamma(i-gam);
end
v=y-u; g=double(genfunc(p)); w=get_vecw(gam,n,g);
for i=1:n, dv(i)=w(1:i)*v(i:-1:1)/h^gam; end
dy=dv+du'; if abs(y(1))<1e-10, dy(1)=0; end
```

The syntax of the function is y_1=glfdiff9(**y**,**t**,**y**,**p**), the syntax is close to that of glfdiff(), and one more argument, p, is appended, indicating the expected precision $o(h^p)$.[29] It can be seen that Theorem 7.3 is only a special case of the formula, where $n = 1$ – the forward difference algorithm.

Example 10.2. Under the framework of integer-order calculus, the derivative of a constant is zero, and its first-order integral is a straight line. Compute fractional-order derivatives and integrals of a constant.

Solutions. The following commands can be used to construct a constant vector **y**, then function glfdiff() can be used to compute fractional-order derivatives and integrals, as shown in Figure 10.2. It can be seen that there exist differences in the fractional- and integer-order derivatives.

```
>> t=0:0.01:1.5; gam=[-1 -0.5 0.3 0.5 0.7]; y=ones(size(t)); dy=[];
    for a=gam, dy=[dy; glfdiff(y,t,a)]; end, plot(t,dy) % different orders
```

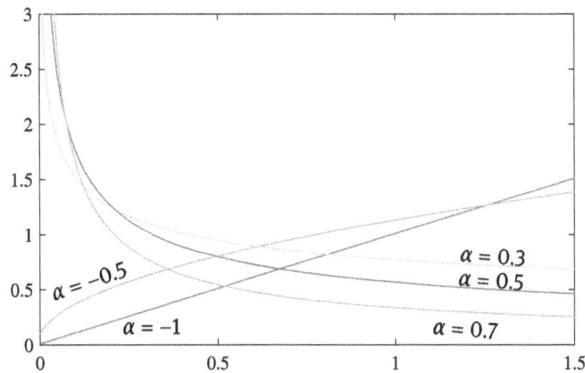

Figure 10.2: Fractional-order derivatives and integrals of a constant.

Example 10.3. Compute fractional-order derivatives of the function $f(t) = \mathrm{e}^{-t}\sin(3t + 1)$, $t \in (0, \pi)$, with nonzero initial conditions.

Solutions. If Grünwald–Letnikov definition is used to compute fractional-order derivative functions, and selecting the step-sizes $T = 0.05$ and $T = 0.001$, the 0.5th order derivative for the two step-sizes can be obtained as shown in Figure 10.3. It can

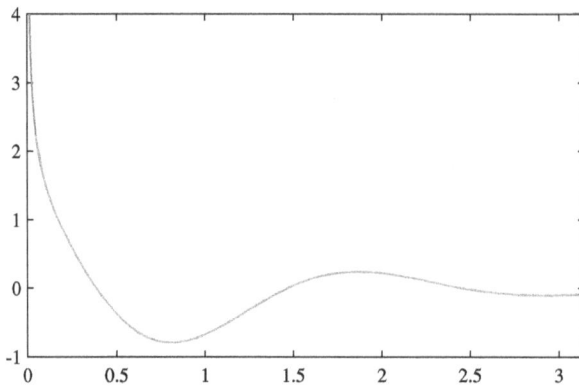

Figure 10.3: Fractional-order derivatives.

be seen that the two curves are very close. Since a high-precision algorithm is selected, very accurate solutions can be found even though $T = 0.05$.

```
>> t1=0:0.001:pi; y=exp(-t1).*sin(3*t1+1); dy=glfdiff9(y,t1,0.5,6);
   t2=0:0.05:pi; y=exp(-t2).*sin(3*t2+1); dy2=glfdiff9(y,t2,0.5,6);
   plot(t1,dy,t2,dy2)
```

For different selections of y, the surface of the fractional-order derivatives can be obtained with the following statements, as shown in Figure 10.4. It can be seen that fractional-order derivatives of different orders can be evaluated directly.

```
>> Z=[]; t=0:0.05:pi; y=exp(-t).*sin(3*t+1); gam0=0:0.1:1;
   for gam=gam0, Z=[Z; glfdiff9(y,t,gam,6)]; end
   surf(t,gam0,Z); axis([0,pi,0,1,-1.2,2.5]) %derivative surface
```

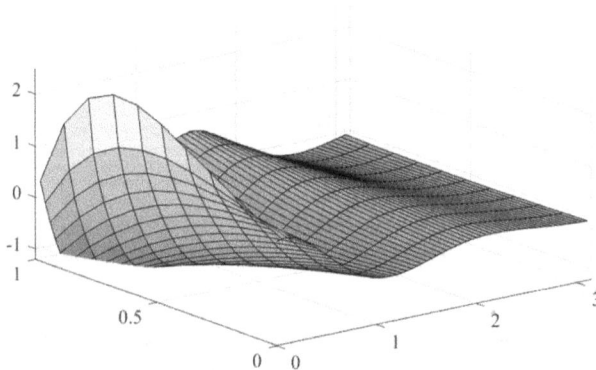

Figure 10.4: Fractional-order derivatives of different orders.

Example 10.4. Compute 0.75th order derivative for $f(t) = \sin(3t + 1)$ under different definitions. Compare the results.

Solutions. With Cauchy integral, 0.75th order derivative can be obtained as

$$_0\mathscr{D}_t^{0.75} f(t) = 3^{0.75} \sin(3t + 1 + 0.75\pi/2).$$

With Grünwald–Letnikov definition, `glfdiff9()` function can be used. Therefore, the following statements can be used to compute the two derivatives, as shown in Figure 10.5:

```
>> t=0:0.01:pi; y=sin(3*t+1); y1=3^0.75*sin(3*t+1+0.75*pi/2);
   y2=glfdiff9(y,t,0.75,4); plot(t,y1,t,y2)    % comparisons
```

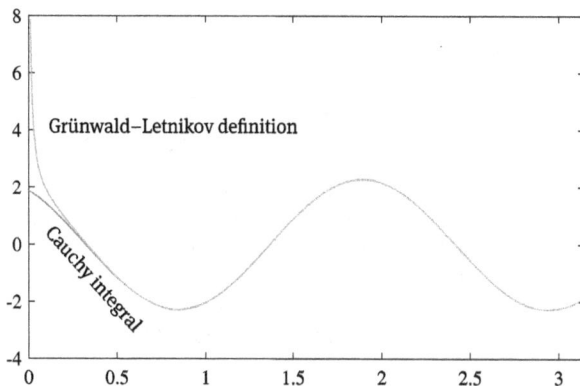

Figure 10.5: Fractional-order derivatives using different definitions.

It can be found through the comparison, that Cauchy integral formula does not have a jump at the initial instance. The two definitions are almost identical in the area other than in the region of $t = 0$. For $t \leqslant 0$, Grünwald–Letnikov definition assumes $y = 0$. It is obvious that at the initial value of $t = 0^+$, the function y is jumping from 0 to sin 1, therefore the derivative at this time should be ∞. The influence of the jump may last for a certain period of time. In Cauchy integral formula, assume that for $t \leqslant 0$, the function also satisfies $y(t) = \sin(3t + 1)$, therefore there is no sudden jump at $t = 0^+$.

10.3.3 Quantitative comparisons of different algorithms

It has been pointed out that the high precision algorithm proposed earlier may reach the accuracy of $o(h^p)$. Now through a known example, we validate the accuracy of the algorithm.

Example 10.5. Compute the fractional-order derivative $y_1(t) = {}_0^{RL}\mathscr{D}_t^{0.6}e^{-t}$ in the interval $[0,5]$, and assess the accuracy.

Solutions. The analytical solution of the fractional-order derivative can be written as $y_1(t) = t^{-0.6}E_{1,0.4}(-t)$, where $E(\cdot)$ is a Mittag-Leffler function with two parameters. Selecting a relatively large step-size $h = 0.01$, different orders p can be tried with the function glfdiff9(). The results can be compared with the theoretical solution $y = t^{-0.6}E_{1,0.4}(-t)$, and the errors are measured as shown in Table 10.1.

```
>> t10=0.5:0.5:5; t=0:0.01:5;
   y10=t10.^-0.6.*ml_func([1,0.4],-t10,0,eps);
   f=@(t)exp(-t); ii=[51:50:length(t)]; T=t10';
   for p=1:6, y1=glfdiff9(f,t,0.6,p); T=[T [y1(ii)-y10]']; end
```

Table 10.1: Computation error for different orders p.

t	$p = 1$	$p = 2$	$p = 3$	$p = 4$	$p = 5$	$p = 6$
0.5	−0.00180	1.194×10^{-5}	-8.893×10^{-8}	7.066×10^{-10}	-5.85×10^{-12}	4.8×10^{-14}
1	−0.00172	1.148×10^{-5}	-8.586×10^{-8}	6.848×10^{-10}	-5.69×10^{-12}	4.8×10^{-14}
1.5	−0.00151	1.005×10^{-5}	-7.522×10^{-8}	6.005×10^{-10}	-4.99×10^{-12}	4.4×10^{-14}
2	−0.00129	8.613×10^{-6}	-6.446×10^{-8}	5.147×10^{-10}	-4.28×10^{-12}	4.7×10^{-14}
2.5	−0.00111	7.385×10^{-6}	-5.527×10^{-8}	4.413×10^{-10}	-3.7×10^{-12}	3.0×10^{-15}
3	−0.00096	6.395×10^{-6}	-4.785×10^{-8}	3.820×10^{-10}	-3.23×10^{-12}	2.1×10^{-14}
3.5	−0.00084	5.611×10^{-6}	-4.197×10^{-8}	3.350×10^{-10}	-2.93×10^{-12}	7×10^{-14}
4	−0.00075	4.993×10^{-6}	-3.734×10^{-8}	2.980×10^{-10}	-2.83×10^{-12}	1.4×10^{-13}
4.5	−0.00068	4.503×10^{-6}	-3.366×10^{-8}	2.688×10^{-10}	-3.03×10^{-12}	4.7×10^{-13}
5	−0.00062	4.110×10^{-6}	-3.072×10^{-8}	2.453×10^{-10}	-3.46×10^{-12}	4.0×10^{-13}

It can be seen that when $p = 1$, the result is the same as that from the function glfdiff(). For other orders p, the accuracy is much higher than from function glfdiff(), and time taken is similar. Although $h = 0.01$ is already rather large, the high precision algorithm may lead to very accurate solutions. For instance, when $p = 6$ is selected, the maximum error may reach the level of 10^{-13}.

Now, if an even larger step-size $h = 0.1$ is tried, the errors for different p are measured as shown in Table 10.2.

```
>> t10=0.5:0.5:5; y10=t10.^-0.6.*ml_func([1,0.4],-t10,0,eps);
   t=0:0.1:5; y=exp(-t); ii=[6:5:51]; T=t10';
   for p=1:6, y1=glfdiff9(y,t,0.6,p); T=[T [y1(ii)-y10]']; end
```

It can be seen that although such a large step-size is selected, extremely high-precision results can be obtained, which is not possible with any other existing algorithms.

Table 10.2: Error comparison for a larger step-size of $h = 0.1$.

t	$p = 1$	$p = 2$	$p = 3$	$p = 4$	$p = 5$	$p = 6$
0.5	−0.017078	0.0010339	-6.9464×10^{-5}	4.5263×10^{-6}	-1.9794×10^{-7}	-3.0685×10^{-9}
1	−0.016764	0.0010808	-7.8225×10^{-5}	5.9813×10^{-6}	-4.7312×10^{-7}	3.7396×10^{-8}
1.5	−0.014743	0.00096247	-7.0585×10^{-5}	5.5076×10^{-6}	-4.4747×10^{-7}	3.7261×10^{-8}
2	−0.012641	0.0008275	-6.0864×10^{-5}	4.7719×10^{-6}	-3.9004×10^{-7}	3.279×10^{-8}
2.5	−0.010832	0.0007085	-5.208×10^{-5}	4.0837×10^{-6}	-3.3396×10^{-7}	2.8098×10^{-8}
3	−0.0093663	0.00061127	-4.4842×10^{-5}	3.51×10^{-6}	-2.8658×10^{-7}	2.407×10^{-8}
3.5	−0.0082052	0.00053401	-3.9073×10^{-5}	3.0507×10^{-6}	-2.4846×10^{-7}	2.0816×10^{-8}
4	−0.0072896	0.00047309	-3.4522×10^{-5}	2.6881×10^{-6}	-2.1832×10^{-7}	1.8241×10^{-8}
4.5	−0.0065645	0.00042489	-3.0926×10^{-5}	2.4019×10^{-6}	-1.9455×10^{-7}	1.6211×10^{-8}
5	−0.0059845	0.00038644	-2.8063×10^{-5}	2.1744×10^{-6}	-1.7569×10^{-7}	1.4603×10^{-8}

Further increasing the precision p, the error can be measured as shown in Table 10.3. Due to the limitations of double precision framework, it does not immediately imply that the higher the p, the better. For this example, $p = 8$ yields the best result.

```
>> t10=0.5:0.5:5; y10=t10.^-0.6.*ml_func([1,0.4],-t10,0,eps);
   t=0:0.1:5; y=exp(-t); ii=[6:5:51]; T=t10';
   for p=7:11, y1=glfdiff9(y,t,0.6,p); T=[T [y1(ii)-y10]']; end
```

Table 10.3: Error comparisons for higher orders p.

t	$p = 7$	$p = 8$	$p = 9$	$p = 10$	$p = 11$
0.5	-8.1653×10^{-11}	-2.9697×10^{-12}	-1.3595×10^{-13}	-1.3572×10^{-14}	9.714×10^{-17}
1	-2.9148×10^{-9}	2.4836×10^{-10}	-2.0282×10^{-11}	7.7977×10^{-13}	4.8572×10^{-15}
1.5	-3.12×10^{-9}	2.4794×10^{-10}	-2.0988×10^{-11}	2.4357×10^{-12}	-2.0356×10^{-13}
2	-2.847×10^{-9}	2.5309×10^{-10}	-1.6299×10^{-11}	-4.0665×10^{-13}	-1.5415×10^{-13}
2.5	-2.4454×10^{-9}	2.501×10^{-10}	-4.5112×10^{-11}	6.7932×10^{-12}	5.3905×10^{-13}
3	-2.0587×10^{-9}	1.5191×10^{-10}	-6.3581×10^{-12}	3.1815×10^{-11}	-1.0038×10^{-11}
3.5	-1.7566×10^{-9}	4.4916×10^{-11}	2.3519×10^{-10}	-3.3692×10^{-10}	8.9195×10^{-11}
4	-1.5471×10^{-9}	2.2712×10^{-10}	-6.0421×10^{-10}	9.424×10^{-10}	-3.9803×10^{-10}
4.5	-1.3943×10^{-9}	4.9411×10^{-10}	-1.9055×10^{-9}	7.2915×10^{-9}	-3.4967×10^{-9}
5	-1.2581×10^{-9}	-2.1665×10^{-10}	1.1419×10^{-8}	-8.3347×10^{-8}	9.9516×10^{-8}

10.4 Numerical computation of Caputo derivatives

It is known that the integrals in the Caputo and Grünwald–Letnikov definitions are exactly the same, therefore, `glfdiff9()` function can be used to evaluate fractional-order integrals. Let us now consider the computation of Caputo derivatives. If $\alpha > 0$, it can be seen that the compensation formula in (10.2.4) can be used to compute Caputo derivatives. A new function `caputo9()` can be write high precision Caputo derivatives

```
function dy=caputo9(y,t,gam,p)
if gam<0, dy=glfdiff9(y,t,gam,p); return; end
h=t(2)-t(1); t=t(:); n=length(t); y=y(:); q=ceil(gam);
r=max(p,q); R=sym(fliplr(vander((0:(r-1))'*h)));
c=double(inv(R)*y(1:r)); u=0; du=0;
for i=1:r, u=u+c(i)*t.^(i-1); end
if q<r
    for i=(q+1):p, du=du+c(i)*t.^(i-1-gam)*gamma(i)/gamma(i-gam);
end, end
v=y(:)-u(:); dv=glfdiff9(v,t,gam,p); dy=dv(:)+du(:);
```

The syntax of the function is y_1=caputo9(y,t,a,p), where, if $\alpha > 0$, the Caputo derivatives can be evaluated, while if $\alpha \leqslant 0$, Grünwald–Letnikov and Caputo integral can be returned. The argument p can be specified, such that the error may reach $o(h^p)$.

Example 10.6. Consider again the function $f(t) = \sin(3t + 1)$ in Example 10.4. Compute its 0.3rd, 1.3rd, and 2.3rd derivatives under different definitions, and assess their precisions.

Solutions. It is known that at $t = 0$, the initial value of $f(t)$ is sin 1, therefore the difference between the Caputo and Grünwald–Letnikov definitions of 0.3rd order derivative is $d(t) = t^{-0.3} \sin 1/\Gamma(0.7)$. The 0.3rd derivatives under the two definitions can be obtained as shown in Figure 10.6. It can be seen that the difference between the two definitions is rather large in the initial period.

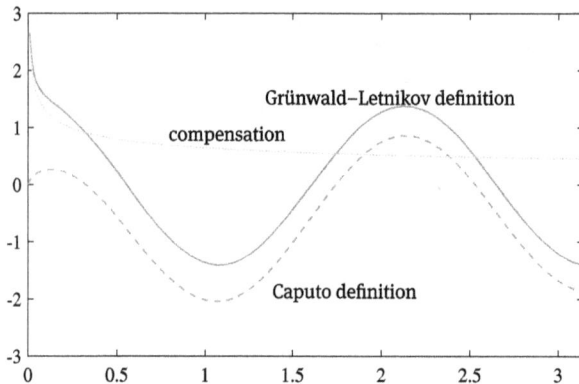

Figure 10.6: The 0.3rd order fractional-order derivatives.

```
>> t=0:0.01:pi;
   y=sin(3*t+1); d=t.^(-0.3)*sin(1)/gamma(0.7);   % compensation
   y1=glfdiff9(y,t,0.3,4); y2=caputo9(y,t,0.3,4); % two derivatives
   plot(t,y1,t,y2,'--',t,d,':') % derivatives and compensation
```

The derivatives ${}_0^C\mathscr{D}_t^{2.3}y(t)$ and ${}_0^C\mathscr{D}_t^{1.3}y(t)$ can easily be evaluated with the following statements, as shown in Figure 10.7. The advantage of caputo9() function is that there is
no need to know $y'(0)$, $y''(0)$ before the function call.

```
>> y1=caputo9(y,t,1.3,4); y2=caputo9(y,t,2.3,4);
   plotyy(t,y1,t,y2) % Caputo derivatives of different orders
```

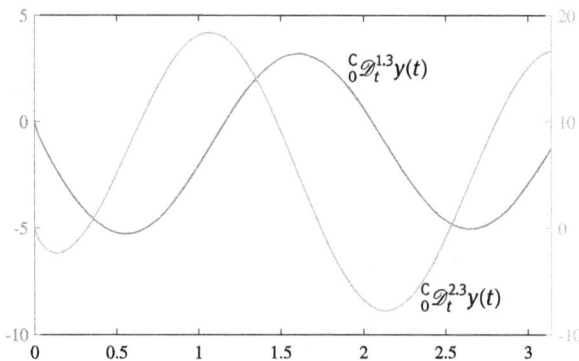

Figure 10.7: Caputo derivatives of different orders.

Example 10.7. Now let us consider the 0.6th order Caputo derivative of function $f(t) =$
e^{-t}. It is known that the analytical solution is $y_0(t) = -t^{0.4}E_{1,1.4}(-t)$. We try different
step-sizes and values p, and assess the accuracy of the results.

Solutions. Selecting again a step-size of $h = 0.01$, the Caputo derivatives for different p can be obtained. Compared with the analytical solutions given, the errors can
be obtained as shown in Table 10.4. It can be seen that the smallest error appears at
$p = 6$, with the maximum error of 10^{-13}. Further increase of the value of p is unlikely
to increase the accuracy under double precision framework. On the contrary, it is also
possible to make the results worse.

```
>> t0=0.5:0.5:5; t=0:0.01:5; y=exp(-t); ii=[51:50:501]; % samples
   y0=-t0.^0.4.*ml_func([1,1.4],-t0,0,eps); T=[];         % analytical
   for p=1:7, y1=caputo9(y,t,0.6,p); T=[T [y1(ii)-y0']]; end
```

Further increasing the step-size to $h = 0.1$, the 0.6th order Caputo derivatives can also
be found for different values of p, and the errors are given in Table 10.5. It can be seen

Table 10.4: Computation errors with $h = 0.01$.

t	$p = 1$	$p = 2$	$p = 3$	$p = 4$	$p = 5$	$p = 6$
0.5	−0.0018	1.19×10^{-5}	-8.89×10^{-8}	7.07×10^{-10}	-5.85×10^{-12}	4×10^{-14}
1	−0.00172	1.15×10^{-5}	-8.59×10^{-8}	6.85×10^{-10}	-5.69×10^{-12}	4×10^{-14}
1.5	−0.00151	1.01×10^{-5}	-7.52×10^{-8}	6.00×10^{-10}	-4.99×10^{-12}	4×10^{-14}
2	−0.00129	8.61×10^{-6}	-6.45×10^{-8}	5.15×10^{-10}	-4.28×10^{-12}	5×10^{-14}
2.5	−0.0011	7.39×10^{-6}	-5.53×10^{-8}	4.41×10^{-10}	-3.7×10^{-12}	1×10^{-14}
3	−0.00096	6.40×10^{-6}	-4.78×10^{-8}	3.82×10^{-10}	-3.25×10^{-12}	2×10^{-14}
3.5	−0.00084	5.61×10^{-6}	-4.20×10^{-8}	3.35×10^{-10}	-2.96×10^{-12}	1.4×10^{-13}
4	−0.00075	4.99×10^{-6}	-3.73×10^{-8}	2.98×10^{-10}	-2.92×10^{-12}	1×10^{-14}
4.5	−0.00068	4.50×10^{-6}	-3.37×10^{-8}	2.68×10^{-10}	-2.72×10^{-12}	4.7×10^{-13}
5	−0.00062	4.11×10^{-6}	-3.07×10^{-8}	2.45×10^{-10}	-3.41×10^{-12}	5.3×10^{-13}

Table 10.5: Computation errors when $h = 0.1$.

t	$p = 4$	$p = 5$	$p = 6$	$p = 7$	$p = 8$	$p = 9$
0.5	4.53×10^{-6}	-1.98×10^{-7}	-3.07×10^{-9}	-8.17×10^{-11}	-2.97×10^{-12}	-1.3×10^{-13}
1	5.98×10^{-6}	-4.73×10^{-7}	3.74×10^{-8}	-2.91×10^{-9}	2.48×10^{-10}	-2.03×10^{-11}
1.5	5.51×10^{-6}	-4.47×10^{-7}	3.73×10^{-8}	-3.12×10^{-9}	2.48×10^{-10}	-2.10×10^{-11}
2	4.77×10^{-6}	-3.90×10^{-7}	3.28×10^{-8}	-2.85×10^{-9}	2.53×10^{-10}	-1.63×10^{-11}
2.5	4.08×10^{-6}	-3.34×10^{-7}	2.81×10^{-8}	-2.45×10^{-9}	2.50×10^{-10}	-4.51×10^{-11}
3	3.51×10^{-6}	-2.87×10^{-7}	2.41×10^{-8}	-2.06×10^{-9}	1.52×10^{-10}	-6.34×10^{-12}
3.5	3.05×10^{-6}	-2.48×10^{-7}	2.08×10^{-8}	-1.76×10^{-9}	4.49×10^{-11}	2.35×10^{-10}
4	2.69×10^{-6}	-2.18×10^{-7}	1.82×10^{-8}	-1.55×10^{-9}	2.27×10^{-10}	-6.05×10^{-10}
4.5	2.40×10^{-6}	-1.95×10^{-7}	1.62×10^{-8}	-1.39×10^{-9}	4.94×10^{-10}	-1.91×10^{-9}
5	2.17×10^{-6}	-1.76×10^{-7}	1.46×10^{-8}	-1.26×10^{-9}	-2.17×10^{-10}	1.14×10^{-8}

that even with such a large step-size, when $p = 8$, the maximum error may still be as low as 10^{-10}.

```
>> t0=0.5:0.5:5; t=0:0.1:5; y=exp(-t); T=[];        % samples
   y0=-t0.^0.4.*ml_func([1,1.4],-t0,0,eps); ii=[6:5:51]; % exact
   for p=3:9, y1=caputo9(y,t,0.6,p); T=[T [y1(ii)-y0']]; end
```

10.5 Oustaloup filter algorithms and applications

The numerical evaluations of fractional-order derivatives studied so far were based on the assumption that $f(t)$ is a given function, while in real applications, this is not always true, since the signal may come from other parts of the system. Other methods should be introduced to compute fractional-order derivatives. For instance, filters can be designed to simulate the behaviors of fractional-order operators.

Filters can be continuous or discrete. Continuous filters can be used to approximate Laplace operator s^y. The output of the filter can be regarded as the Riemann–Liouville derivative of the input signal.

10.5.1 Oustaloup filter approximations

Many filters are presented in [29, 17]. Here Oustaloup filter is presented as it is among those with very good behavior.[16]

Theorem 10.9. *Assume that the frequency range of interest is selected as (ω_b, ω_h). Then the Nth order transfer function, known as Oustaloup filter, can be designed as*

$$G_f(s) = K \prod_{k=1}^{N} \frac{s + \omega_k'}{s + \omega_k} \qquad (10.5.1)$$

where the poles, zeros, and gain can be designed from

$$\omega_k' = \omega_b \omega_u^{(2k-1-\gamma)/N}, \quad \omega_k = \omega_b \omega_u^{(2k-1+\gamma)/N}, \quad K = \omega_h^\gamma, \qquad (10.5.2)$$

and $\omega_u = \sqrt{\omega_h/\omega_b}$.

Based on the above algorithm, a MATLAB function is written to design the continuous filter. With such a filter, if an $f(t)$ signal is fed into the filter, the output signal of the function can be regarded as $^{\text{RL}}\mathscr{D}_t^y f(t)$.

```
function G=ousta_fod(gam,N,wb,wh)
if round(gam)==gam, G=tf('s')^gam;   % if the order is an integer
else, k=1:N; wu=sqrt(wh/wb);         % compute base frequency
    wkp=wb*wu.^((2*k-1-gam)/N);
    wk=wb*wu.^((2*k-1+gam)/N);  % zeros and poles
    G=zpk(-wkp,-wk,wh^gam); G=tf(G);  % construct integer-order TF
end
```

The syntax of the function is G_1=ousta_fod(y,N,ω_b,ω_h), where y is the order of the derivative, which can be positive or negative; N is the order of the filter; ω_b and ω_h are respectively the user-selected lower and upper frequency bounds. Normally, the fitting to the fractional-order operator is satisfactory within the user selected interval, and poor outside the interval.

Example 10.8. Assuming that the bounds are selected as $\omega_b = 0.01\,\text{rad/s}$, $\omega_h = 1\,000\,\text{rad/s}$, and the order of the filter is selected as $N = 5$, design an Oustaloup filter. For $f(t) = e^{-t}\sin(3t + 1)$, compute its 0.5th order derivative, and compare the results.

Solutions. The filter can be designed easily with

```
>> G=ousta_fod(0.5,5,0.01,1000), zpk(G), bode(G) % Oustaloup filter
```

and the result is

$$G(s) = \frac{31.623(s + 177.8)(s + 17.78)(s + 1.778)(s + 0.1778)(s + 0.01778)}{(s + 562.3)(s + 56.23)(s + 5.623)(s + 0.5623)(s + 0.05623)}.$$

Bode diagram comparison of the Oustaloup filter and the exact result can also be made with the above commands, as shown in Figure 10.8. If the input to the Oustaloup filter is selected as $f(t) = e^{-t} \sin(3t+1)$, the output of the filter is the expected fractional-order derivative. Also the Grünwald–Letnikov derivative can also be evaluated, and comparison of the two curves can be obtained, as shown in Figure 10.9. It can be seen that the results from filter are satisfactory.

```
>> t=0:0.001:pi; y=exp(-t).*sin(3*t+1);                    % generate inputs
   y1=lsim(G,y,t); y2=glfdiff(y,t,0.5); plot(t,y1,t,y2) % output
```

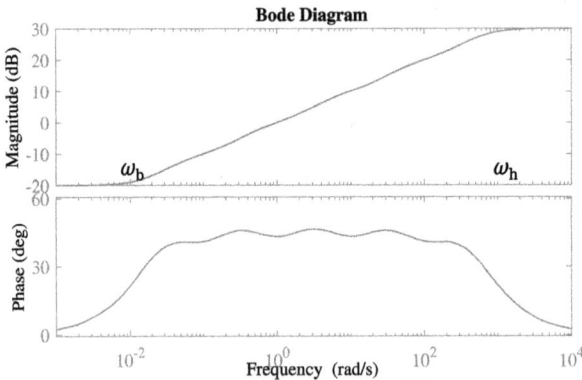

Figure 10.8: Bode diagram comparisons of Oustaloup filters.

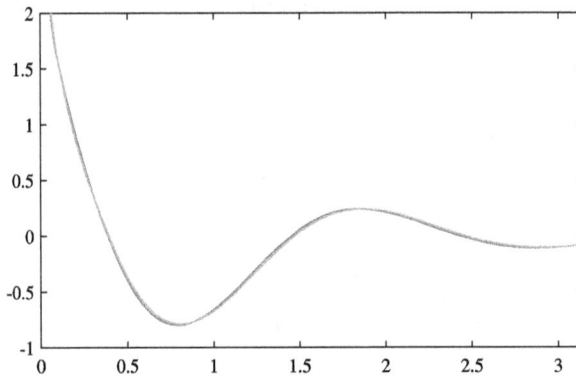

Figure 10.9: Comparison of fractional-order derivatives.

Larger frequency range and high-order of the filter can be assigned to achieve better fitting. The fitting in the frequency interval $(10^{-4}, 10^4)$ can be obtained, as shown in Figure 10.10, from which it can be seen that for the large frequency interval, $N = 5$ is not a good choice. Higher order N should be selected, such as $N = 11$.

```
>> G=ousta_fod(0.5,5,1e-4,1e4); G1=ousta_fod(0.5,7,1e-4,1e4);
   G2=ousta_fod(0.5,9,1e-4,1e4); G3=ousta_fod(0.5,11,1e-4,1e4);
   bode(G,'-',G1,'--',G2,':',G3,'-.') % frequency response fitting
```

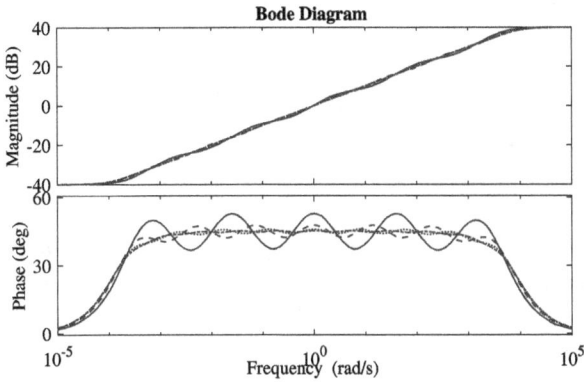

Figure 10.10: Filter approximation for different orders.

If computational load is permitted, even larger interval and order can be assigned, such as $(10^{-6}, 10^6)$ rad/s and $N = 30$. In fact, the filter designed according to the specification is very close to the expected fractional-order operator.

10.5.2 Filter for Caputo derivative fitting

The above presented Oustaloup filter can only be used to approximate Riemann–Liouville fractional-order operators, while it cannot be used to directly generate Caputo derivatives. Two important theorems will be proposed later to reconstruct Caputo derivatives with filters.

Theorem 10.10. *Caputo derivative can be obtained with Oustaloup filter as follows:*

$$ {}_{t_0}^{C}\mathscr{D}_t^{\gamma} y(t) = {}_{t_0}^{RL}\mathscr{D}_t^{-(\lceil \gamma \rceil - \gamma)}[y^{(\lceil \gamma \rceil)}(t)]. \tag{10.5.3} $$

The physical interpretation of the theorem is that, if $y^{(\lceil \gamma \rceil)}(t)$, the γth order Caputo derivative of signal $y(t)$, is expected, the integer-order derivative $y^{(\lceil \gamma \rceil)}(t)$ should be used, then the expected result can be obtained by taking $(\lceil \gamma \rceil - \gamma)$th order Riemann–Liouville integral. In other words, taking the integer-order derivative $y^{(\lceil \gamma \rceil)}(t)$ is followed by an Oustaloup filter.

Theorem 10.11. *Taking the ($\lceil y \rceil - y$)th order Riemann–Liouville derivative of (10.1.6), it can be found that*

$$\underset{t_0}{\overset{RL}{\mathscr{D}}}_t^{\lceil y \rceil - y} \left[\underset{t_0}{\overset{C}{\mathscr{D}}}_t^y y(t) \right] = y^{(\lceil y \rceil)}(t). \tag{10.5.4}$$

The physical interpretation of the theorem is that, taking the ($\lceil y \rceil - y$)th order Riemann–Liouville derivative of the yth order Caputo derivative $\underset{t_0}{\overset{C}{\mathscr{D}}}_t^y y(t)$, the integer-order derivative $y^{(\lceil y \rceil)}(t)$ can be obtained. In other words, computing Caputo derivative followed by an Oustaloup filter may generate integer-order derivatives. It might be better to implement the two theorems with Simulink environment, which will be addressed later.

Example 10.9. Compute the 1.3th order Caputo derivative with the filter method for the function defined in Example 10.6.

Solutions. If 1.3th order Caputo derivative is expected, $y''(t)$ signal should be generated first. Then the -0.7th order derivative, i. e., 0.7th order integral, should be taken to generate the final signal. The 0.7th order integral can be obtained through the Oustaloup filter, while the input is $y''(t)$. Also, `caputo9()` function can be used to evaluate the 1.3th order Caputo derivative, and it can be seen that the results with the two methods are exactly the same, indicating the filter is applicable. Note that the specifications of Oustaloup filter should not be selected too low, which validates the correctness of Theorem 10.10.

```
>> t=0:0.01:pi; y=sin(3*t+1);
   G=ousta_fod(-0.7,30,1e-6,1e6); [y2,t2]=num_diff(y,0.01,2,4);
   y13=lsim(G,y2,t2); y13a=caputo9(y,t,1.3,4);
   plot(t,y13a,t2,y13)
```

If signal $y_{13a}(t)$ is used as input, then a 0.7th order Oustaloup filter can be designed, and second-order derivative can be restored, which validates Theorem 10.11.

```
>> G=ousta_fod(0.7,30,1e-6,1e6); y21=lsim(G,y13a,t);
   plot(t2,y2,t,y21)
```

10.5.3 Simulink-based Caputo derivative computation

The fractional-order derivatives can be evaluated with the dedicated function, also a Simulink block in FOTF Toolbox can be used to evaluate fractional-order derivatives and integrals in the Simulink framework. Typing the command `fotflib` in MATLAB command window, the FOTF blockset is opened, as shown in Figure 10.11. The block **Fractional operator** is the key block to construct Riemann–Liouville derivatives directly. The block implements a typical Oustaloup filter in Theorem 10.9. The output

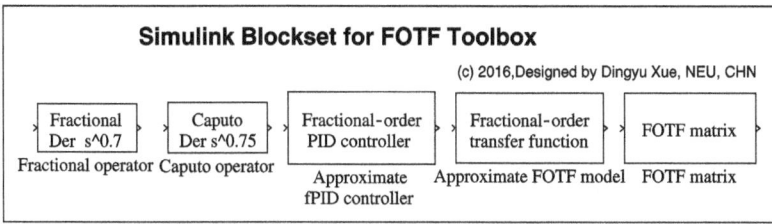

Figure 10.11: Multivariable FOTF blockset (fotflib.slx).

of the block can be regarded as the Riemann–Liouville derivative or integral. The other blocks are not presented in this book.

Example 10.10. Solve the following fractional-order Caputo equation:[31]

$$
{}^C_0\mathscr{D}^{0.7}_t y(t) = \begin{cases} t^{0.3}/\Gamma(1.3), & 0 \leqslant t \leqslant 1, \\ t^{0.3}/\Gamma(1.3) - 2(t-1)^{1.3}/\Gamma(2.3), & t > 1, \end{cases} \tag{10.5.5}
$$

where $y(0) = 0$ and $t \in (0, 2)$. The analytical solution is

$$
y(t) = \begin{cases} t, & 0 \leqslant t \leqslant 1, \\ t - (t-1)^2, & t > 1. \end{cases}
$$

Solutions. This problem is the first of five benchmark problems in [31]. This equation is, in fact, the 0.7th Caputo integral of a piecewise function, which can be computed numerically. If Simulink environment is used, the key signals $y(t)$ and $y'(t)$ can be defined with an integer-order integrator block, as shown in Figure 10.12. The signal on the right-hand side of (10.5.5) can be constructed, and through 0.3th order Riemann–Liouville derivative (Oustaloup filter), the output of the filter can be directly connected to the $y'(t)$ signal. It can be seen from Theorem 10.10 that the original Caputo equation is satisfied, where the embedded MATLAB function is

```
function y=bp1fcn(t)
y=t.^0.3/gamma(1.3)-2*(t-1).^1.3/gamma(2.3).*(t>1);
```

If the following statements are given, within 0.38 seconds simulation results can be obtained. Compared with analytical solutions, the maximum error is 6.0761×10^{-6}, and the number of points is 1 467.

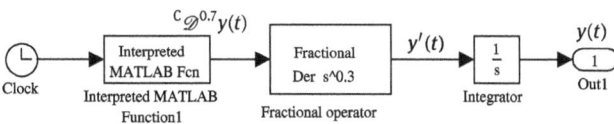

Figure 10.12: Simulink model for Caputo derivatives (bl1model.slx).

```
>> ww=[1e-5 1e5]; N=25;
   tic, [t,x,y]=sim('bp1model',2); toc, %default ode15s algorithm
   y1=t-(t-1).^2.*(t>1); max(abs(y-y1)), length(t)
```

If the algorithm is selected as Runge–Kutta–Felhberg, i.e., ode45() function algorithm, the following commands can be executed; and in about 10 seconds, the final results can be obtained. The precision is almost the same as that obtained earlier, with the number of points 78 468. Therefore, it is recommended to use ode15s algorithm to compute Caputo derivatives.

```
>> set_param('bp1model','Solver','ode45')
   tic, [t,x,y]=sim('bp1model',2); toc,
   y1=t-(t-1).^2.*(t>1); max(abs(y-y1)), length(t)
```

10.6 Numerical computation of even higher-order derivatives and integrals

The algorithms and solvers discussed so far can only be used to evaluate fractional-order derivatives of relatively low orders. If higher-order derivatives are expected, problems may be encountered. This phenomenon will be demonstrated through an example, followed by possible solution schemes. Accuracy of the schemes is also demonstrated through examples.

Example 10.11. For the given function $f(t) = e^{-t}$, generate samples, from which compute the 5.6th order Caputo derivative, and validate the results.

Solutions. It is known from Caputo derivative property that the αth order Caputo derivative of e^{-t} can be obtained. Letting $m = \lceil \alpha \rceil = 6$, $\gamma = m - \alpha = 0.4$, the analytical solution is given by

$$ {}_0^C\mathscr{D}_t^{5.6} e^{-t} = (-1)^m t^\gamma E_{1,1+\gamma}(-t) = t^{0.4} E_{1,1.4}(-t). \tag{10.6.1} $$

Similar to the cases discussed before, the derivative can be computed directly from function caputo9(). Selecting $h = 0.1$, we generate a set of samples, from which the computed results obtained are as shown in Figure 10.13. It can be seen that the result obtained diverges, and cannot be used in evaluating high-order fractional derivatives. Other feasible methods must be explored.

```
>> t0=0.5:0.5:5; t=0:0.1:5; y=exp(-t); T=[]; %generate the samples
   for p=1:9, y1=caputo9(y,t,5.6,p); plot(t,y1); hold on, end
```

How can we compute high-order fractional derivatives? It is known from Chapter 7 that a sophisticated algorithm is proposed to evaluate numerically high-order

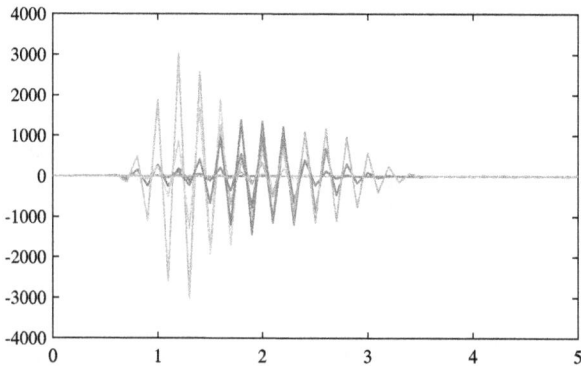

Figure 10.13: Direct computation (erroneous) results.

derivatives of given functions, with very high precision. Thus it is possible to find high precision and order Caputo derivatives from Theorem 10.10. An example is used to compute the high-order derivatives as follows.

Example 10.12. Solve the problem in Example 10.11 with the new ideas.

Solutions. Since the algorithm proposed in Chapter 7 may lose trailing samples in numerical derivatives, to better solve the problem in this example, more samples can be appended from the original function. The sixth order derivative can be evaluated under the same step-size, and based on the derivative samples, the 0.4th order Riemann–Liouville integral can be evaluated for different selections of p, such that the 5.6th order derivative can be found. Compared with analytical solution, the computational errors from the new algorithm can be obtained as shown in Table 10.6.

```
>> t0=0.5:0.5:6.5; h=0.1; t=0:h:6.5; y=exp(-t);        % generate samples
   [y01,t1]=num_diff(y,h,6,6); ii=[6:5:length(t1)];    % integer-order
   y0=t1.^0.4.*ml_func([1,1.4],-t1,eps); T=[t(ii).'];  % theoretical
   for p=1:6,   % compute the results
       y1=glfdiff9(y01,t1,-0.4,p); err=y0-y1; T=[T abs(err(ii).')];
   end
```

It can be seen that when p is chosen appropriately, e. g., when p is selected as 5 or 6, the computational error is very small. In the function num_diff() call $p = 6$ is selected, since a higher value of p may not increase the accuracy of the results. The accuracy can only be improved by setting a higher accuracy in the process of evaluating the sixth-order derivative, then selecting an appropriate p in the Caputo derivative evaluation. Interested readers may try different parameter combinations to improve the accuracy of the results.

Table 10.6: Errors in finding the 5.6th order Caputo derivative.

t	p = 1	p = 2	p = 3	p = 4	p = 5	p = 6
0.5	0.00421	0.0001825	8.6293×10^{-7}	7.0327×10^{-6}	7.064×10^{-6}	6.9727×10^{-6}
1	0.009927	0.0005446	2.6092×10^{-5}	9.0496×10^{-6}	7.5657×10^{-6}	6.4418×10^{-6}
1.5	0.015153	0.0008858	5.1906×10^{-5}	9.9643×10^{-6}	6.6693×10^{-6}	5.2587×10^{-6}
2	0.019673	0.0011831	7.4802×10^{-5}	1.0564×10^{-5}	5.6042×10^{-6}	4.0288×10^{-6}
2.5	0.023544	0.0014379	9.4323×10^{-5}	1.1203×10^{-5}	4.7954×10^{-6}	3.0834×10^{-6}
3	0.026876	0.0016568	0.00011099	1.1824×10^{-5}	4.1878×10^{-6}	2.3721×10^{-6}
3.5	0.029776	0.0018468	0.00012537	1.2412×10^{-5}	3.7151×10^{-6}	1.8062×10^{-6}
4	0.032337	0.002014	0.00013797	1.2927×10^{-5}	3.3009×10^{-6}	1.2988×10^{-6}
4.5	0.034629	0.0021631	0.00014911	1.3455×10^{-5}	3.0096×10^{-6}	9.2141×10^{-7}
5	0.036706	0.0022979	0.00015914	1.3936×10^{-5}	2.7573×10^{-6}	5.8847×10^{-7}

10.7 Exercises

10.1 For the given function $f(t) = e^{-3t} \sin(t + \pi/3) + t^2 + 3t + 2$, compute the 0.2th order derivative and the 0.7th order integral of the signal. Draw the two signals.

10.2 Design continuous filters for the fractional calculus required in Problem 10.1. Take fractional-order derivatives and integrals through the filters, compare the results with those using high precision computation, and assess the accuracy of the algorithm.

10.3 Compute the 0.5th and 1.5th order Riemann–Liouville and Caputo derivatives for the given function $f(t) = e^{-t}$. Compare the results under different step-sizes and assess the accuracy and speed. It is known that the exact expression of the αth order Riemann–Liouville derivative of $f(t)$ is $t^{-\alpha} E_{1,1-\alpha}(-t)$, while Caputo derivative is $(-1)^m t^y E_{1,1+y}(-t)$, where $m = \lceil \alpha \rceil$, $y = m - \alpha$.

10.4 The problem in Example 10.10 can also be solved with function caputo9(). Analyze the actual accuracy for different step-sizes and expected p.

10.5 Consider the problem in Example 10.12, try to select higher values of p in integer-order derivative evaluation, and by selecting an appropriate p see how accurate the result can be, when $h = 0.1$.

Bibliography

[1]   Boyer C B. The History of the Calculus and Its Conceptual Development. New York: Dover
      Publications, Inc., 1949.
[2]   Callier F M, Winkin J. Infinite dimensional system transfer functions. In Curtain R F, Bensoussan
      A, Lions J L, eds. Analysis and Optimization of Systems: State and Frequency Domain
      Approaches for Infinite-Dimensional Systems. Berlin: Springer-Verlag, 1993.
[3]   Chen C Z, Jin F L, Zhu X Y. Mathematical Analysis. Beijing: People's Education Press, 1979
      (in Chinese).
[4]   de Boor C. A Practical Guide to Splines. New York: Springer, 2001.
[5]   Demidovich B P. Problems in Mathematical Analysis. Moscow: MIR Publishers, 1970.
[6]   Ďuriš F. Infinite Series: Convergence Tests. Bachelor's thesis, Katedra Informatiky, Fakulta
      Matematiky, Fyziky a Informatiky, Univerzita Komenského, Bratislava, Slovakia, 2009.
[7]   Fornberg B. Generation of finite difference formulas on arbitrarily spaced grids. Mathematics of
      Computation, 1988, 51(184): 699–706.
[8]   Fornberg B. Calculation of weights in finite difference formulas. SIAM Review, 1998, 40(3):
      685–691.
[9]   Forsythe G E, Malcolm M A, Moler C B. Computer Methods for Mathematical Computations.
      Englewood Cliffs: Prentice-Hall, 1977.
[10] Grattan-Guinness I. Landmark Writings in Western Mathematics 1640–1940. Amsterdam:
      Elsevier, 2005.
[11] Hilfer R. Applications of Fractional Calculus in Physics. Singapore: World Scientific, 2000.
[12] Landau D P, Binder K. A Guide to Monte Carlo Simulations in Statistical Physics. Cambridge,
      MA: Cambridge University Press, 2000.
[13] Magalhaes Jr P A A, Magalhaes C A. Higher-order Newton–Cotes formulae. Journal of
      Mathematics and Statistics, 2010, 6(2): 193–204.
[14] Mathematics Handbook Group. Mathematics Handbook. Beijing: People's Education Press,
      1979 (in Chinese).
[15] Merzbachand U C, Boyer C B. A History of Mathematics. New Jersey: Wiley, 3rd edition, 2011.
[16] Oustaloup A, Levron F, Nanot F, Mathieu B. Frequency band complex non integer differentiator:
      characterization and synthesis. IEEE Transactions on Circuits and Systems I: Fundamental
      Theory and Applications, 2000, 47(1): 25–40.
[17] Petráš I, Podlubny I, O'Leary P. Analogue Realization of Fractional Order Controllers. TU Košice:
      Fakulta BERG, 2002.
[18] Podlubny I. Fractional Differential Equations. San Diego: Academic Press, 1999.
[19] Rogers R. Putting the differential back into differential calculus. In Shell-Gellasch A, Jardine D,
      eds. From Calculus to Computers. Washington DC: The Mathematical Association of America,
      2005.
[20] Song S N, Sun T, Zhang G W. Complex Functions and Integral Transforms. Beijing: Science
      Press, 2006 (in Chinese).
[21] Thomas Jr G B, Weir M D, Hass J. Thomas' Calculus. Boston: Pearson, 13th edition in SI Units,
      2016.
[22] Valsa J. Numerical Inversion of Laplace Transforms in MATLAB, MATLAB Central File ID: #32824,
      2011.
[23] Valsa J, Brančik L. Approximate formulae for numerical inversion of Laplace transforms.
      International Journal of Numerical Modelling: Electronic Networks, Devices and Fields, 1998,
      11(3): 153–166.
[24] Varberg D, Purcell E, Rigdon S. Calculus. Upper Saddle River: Prentice Hall, 9th edition,
      2006.

https://doi.org/10.1515/9783110666977-011

[25] Vinagre B M, Chen Y Q. Fractional calculus applications in automatic control and robotics. 41st IEEE CDC, Tutorial Workshop 2, Las Vegas, 2002.

[26] Wilson H, Gardner B. Numerical Integration Toolbox (NIT).

[27] Xue D. Analysis and Computer Aided Design of Nonlinear Systems with Gaussian Inputs. PhD thesis, Sussex University, UK, 1992.

[28] Xue D, Chen Y Q. Scientific Computing with MATLAB. Boca Raton: CRC Press, 2nd edition, 2016.

[29] Xue D Y. Fractional-order Control Systems – Fundamentals and Numerical Implementations. Berlin: de Gruyter, 2017.

[30] Xue D Y. FOTF Toolbox, MATLAB Central File ID: #60874, 2017.

[31] Xue D Y, Bai L. Benchmark problems for Caputo fractional-order ordinary differential equations. Fractional Calculus & Applied Analysis, 2017, 20(5): 1305–1312.

MATLAB function index

Bold page numbers indicate where to find the syntax explanation of the function. The function or model name marked by * are the ones developed by the authors. The items marked with ‡ are those down-loadable freely from Internet.

Index